# International EFES-IN2P3 Conference on Many-Body Correlations from Dilute to Dense Nuclear Systems 2011



**Journal of Physics: Conference Series Volume 321**

## Paris, France
## 15-18 February 2011

**Editors:**

**Takaharu Otsuka**
**Michael Urban**
**Taiichi Yamada**

**ISBN: 978-1-61839-291-6**
**ISSN: 1742-6588**

**Printed from e-media with permission by:**

Curran Associates, Inc.
57 Morehouse Lane
Red Hook, NY 12571

**Some format issues inherent in the e-media version may also appear in this print version.**



Printed by Curran Associates, Inc. (2011)

For permission requests, please contact the Institute of Physics
at the address below.

Institute of Physics
Dirac House, Temple Back
Bristol BS1 6BE UK

Phone: 44 1 17 929 7481
Fax: 44 1 17 920 0979

techtracking@iop.org

**Additional copies of this publication are available from:**

Curran Associates, Inc.
57 Morehouse Lane
Red Hook, NY 12571 USA
Phone: 845-758-0400
Fax: 845-758-2634
Email: curran@proceedings.com
Web: www.proceedings.com

# TABLE OF CONTENTS

## PAIRING



## AMD and FMD



## CLUSTERING



## ENERGY DENSITY FUNCTIONAL



## METHODS BEYOND MEAN-FIELD

## SHELL MODEL AND CONCLUDING REMARKS



## POSTERS

**PREFACE**

The International EFES-IN2P3 conference on "Many body correlations from dilute to dense nuclear systems" was held at the Institut Henri Poincaré (IHP), Paris, France, from 15–18 February 2011, on the occasion of the retirement of our colleague Peter Schuck.

Correlations play a decisive role in various many-body systems such as nuclear systems, condensed matter and quantum gases. Important examples include: pairing correlations (Cooper pairs) which give rise to nuclear superfluidity (analogous to superconductivity in condensed matter); particle–hole (RPA) correlations in the description of the ground state beyond mean-field theory; clusters; and α-particle correlations in certain nuclei. Also, the nucleons themselves can be viewed as clusters of three quarks. During the past few years, researchers have started to study how the character of these correlations changes with the variation of the density. For instance, the Cooper pairs in dense matter can transform into a Bose–Einstein condensate (BEC) of true bound states at low density (this is the BCS–BEC crossover studied in ultracold Fermi gases). Similar effects play a role in neutron matter at low density, e.g., in the "neutron skin" of exotic nuclei. The α-cluster correlation becomes particularly important at lower density, such as in the excited states of some nuclei (e.g., the α-condensate-like structure in the Hoyle state of $^{12}C$) or in the formation of compact stars.

In addition to nuclear physics, topics from astrophysics (neutron stars), condensed matter, and quantum gases were discussed in 48 talks and 19 posters, allowing the almost 90 participants from different communities to exchange their ideas, experiences and methods. The conference dinner took place at the Musée d'Orsay, and all the participants enjoyed the very pleasant atmosphere. One session of the conference was dedicated to the celebration of Peter's retirement. We would like to take this opportunity to wish Peter all the best and we hope that he will continue his scientific work full of creative and original ideas.

We would like to thank all those who helped to make the conference a success: Nguyen van Giai, S Fujii, J Margueron, K Hagino, and Y Kanada-En'yo for their help with the organization; the advisory committee for suggesting invited speakers; V Frois for her administrative help; L Petizon for the website; and the director of IPN Orsay, F Azaiez, for his support. We are indebted to IHP for providing the lecture hall free of charge, and we acknowledge the financial support from JSPS through its EFES core-to-core program, from CNRS (IN2P3 and INP), and from LIA France-Japon. Last but not least, we are grateful to all of the participants for making the conference exciting and successful.

Takaharu Otsuka, Michael Urban, Taiichi Yamada
**Editors of the proceedings**

# Conference Photograph (February 15-18, 2011)
# Paris, France

PEER REVIEW STATEMENT

All papers published in this volume of *Journal of Physics: Conference Series* have been peer reviewed through processes administered by the proceedings Editors. Reviews were conducted by expert referees to the professional and scientific standards expected of a proceedings journal published by IOP Publishing.

Journal of Physics: Conference Series **321** (2011) 012001 doi:10.1088/1742-6596/321/1/012001

# Pairing and alpha-like quartet condensation in N=Z nuclei

**N. Sandulescu[1], J. Dukelsky[2] and D. Negrea[1]**

[1]Institute of Physics and Nuclear Engineering, Bucharest-Magurele, Romania
[2]Instituto de Estructura de la Materia, CSIC, Serrano 123, 28006 Madrid, Spain

E-mail: sandulescu@theory.nipne.ro

**Abstract**. We discuss the treatment of isovector pairing by an alpha-like quartet condensate which conserves exactly the particle number, the spin and the isospin. The results show that the quartet condensate describes accurately the isovector pairing correlations in the ground state of systems with an equal number of protons and neutrons.

One of the most debated topic in nuclear physics is the competition between the isoscalar (T=0) and the isovector (T=1) proton-neutron pairing in nuclei with neutrons and protons in the same open shell. The most common formalism to treat both T=1 and T=0 pairing is the generalized Hartree-Fock-Bogoliubov (HFB) approach [1]. The great advantage of this approach is that it can be applied to practically all nuclei, irrespective of their masses and deformations. The drawback is that HFB does not conserve exactly the particle number, the total spin and the total isospin of the nucleus.

A simple alternative to restore the particle number, the spin and the isospin is to work not with pairs but with alpha-like 4-body clusters composed by two neutrons and two protons coupled to total isospin T=0 and total spin J=0 [2]. The existence of alpha-like structures in nuclei, usually related to low-energy alpha emission threshold, regularities in the masses of light nuclei and to specific features of low-energy excitation spectra of N=Z nuclei [3-6], is one of the oldest issue in nuclear structure (for a recent review see [3]) .

Various studies rose the question of whether a superfluid condensate of alpha-like quartets could exist in the ground state or in some excited states of self-conjugate nuclei [7-11]. Here some clarifying remarks are in order. In the ground state of nuclei the quartets are not expected to be tight clusters as the alpha particles (nuclei of $^{4}$He) but rather 4-body structures correlated in angular momentum and isospin space. Thus, in the ground state of heavy nuclei a superfluid condensate of alpha-like quartets is expected to have a small fraction of extended correlated quartets rather than to be a Bose condensate of alpha-like bound quartets. However, according to several calculations, in some specific excited states such as the Hoyle state of $^{12}$C, the quartets could become bound alpha particles and they might do form a boson condensate [12]. A boson condensate of alpha particles has been also predicted in dilute nuclear matter [13].

One of the first microscopic models of quartet condensation in nuclei, proposed by Flowers et al. [7], it is based on a BCS-type function written in terms of quartets . Recently this model was extended by including in the BCS function both Cooper pairs and quartets [11]. A theory of quartet condensation based on a BCS-type function has the advantage of simplicity but its applicability to real nuclei is hindered by the fact that it does not conserve exactly the particle number.

The scope of this paper is to show how the isovector pairing can be described by an alpha-like quartet condensate which conserves exactly the particle number, the spin and the isospin and which is simple enough for performing realistic calculations.

The isovector pairing Hamiltonian is taken in the standard form

$$H = \sum \varepsilon_i (N_i^{(\nu)} + N_i^{\pi}) - g_1 \sum_{ij,\tau} P_{i\tau}^+ P_{j\tau} , \qquad (1)$$

where the first term is the single-particle Hamiltonian and the second term is the isovector pairing interaction of constant strength. The interaction is expressed by the operators $P_{i\tau}^+ \propto [a_i^+ a_i^+]_\tau^{T=1,J=0}$ where $\tau = \{1,0,-1\} \equiv \{\nu\nu, \pi\pi, \nu\pi\}$ denotes the projections of the isospin corresponding to nn, pp and pn pairs respectively. The Hamiltonian (1) has a local SO(5) symmetry and it can be solved exactly for any distribution of single-particle levels [14]. The most common approximation is the generalized HFB approach [5], which has the drawback that it does not conserve exactly the particle number and the total isospin. An alternative to restore the total isospin is to introduce the quartet operators

$$A_{ij}^+ = [P_i^+ P_j^+]^{T=0} \propto 2P_{i,\nu\nu}^+ P_{j,\pi\pi}^+ - P_{i,\nu\pi}^+ P_{j,\nu\pi}^+ .$$

With these operators we construct a collective quartet operator

$$A^+ = \sum_{ij} x_{ij} (2P_{i,\nu\nu}^+ P_{j,\pi\pi}^+ - P_{i,\nu\pi}^+ P_{j,\nu\pi}^+) . \qquad (2)$$

Because the operator (2) is formed by two neutrons and two protons with the same quantum numbers as in an alpha particle, it will be called an alpha-like quartet. Since the calculations with the collective quartet operator (2) are difficult to perform, we assume that the mixing amplitudes are separable in the indices i and j, i.e., we take $x_{ij} = x_i x_j$. With this ansatz the quartet operator can be written as

$$A^+ = (2\Gamma_{\nu\nu}^+ \Gamma_{\pi\pi}^+ - \Gamma_{\nu\pi}^+ \Gamma_{\nu\pi}^+) , \qquad (3)$$

where $\Gamma_\tau^+ = \sum x_i P_{i,\tau}^+$ are the collective pair operators for the isovector nn, pp and pn pairs.

The quartet operator (3) is used to construct a quartet condensate

$$| \Psi >= A^{+n_q} | 0 >= \sum_{k=1}^{n_q} (-2)^{n_q-k} \begin{pmatrix} k \\ n_q \end{pmatrix} \left[ \Gamma_{\nu\nu}^+ \Gamma_{\pi\pi}^+ \right]^{n_q-k} \left[ \Gamma_{\nu\pi}^+ \right]^{2k} | 0 > , \qquad (4)$$

It can be seen that the function (4) is a particular superposition of condensates of collective nn, pp and pn pairs which combine so as to give a total isospin equal to T=0.

The quartet condensate (4) is used as a trial eigenstate for describing the isovector pairing correlations in the ground state of even-even self-conjugate (N=Z) nuclei. Its structure depends on the mixing parameters $x_i$ . They are determined by minimizing the average of the Hamiltonian (1) on the quartet condensate (4) and by imposing the normalization condition $< \Psi | \Psi >= 1$.

How accurate is the quartet condensate (4) for the description of the ground state of the Hamiltonian (1) can be seen from Table 1 below. Table 1 shows the correlation energies (the differences between the total energies and the HF energies) for a system of 8 proton-neutron pairs distributed in 16 four-fold degenerate single-particle states with the energies $\varepsilon_i = (i-1)/2$. In the third column are shown the results corresponding to the quartet condensate (4) while in the second column are given the exact results [15]. In the last column are shown the results corresponding to the PBCS approximation [15]

$$| PBCS >= (\Gamma_{\nu\nu}^+ \Gamma_{\pi\pi}^+)^{n_q} | 0 > . \qquad (5)$$

It can be seen that the quartet condensate (4) describes accurately the correlations energies. Thus, for intermediate and strong coupling the errors do not exceed 1%. These errors are comparable with the errors of the PBCS approximation for pairing between like particles [16]. On the other hand, as seen in Table I, the errors corresponding to the quartet condensate are of about one order of magnitude lower than in the case of the PBCS approximation (5) for the isovector pairing Hamiltonian.

Table 1. Correlations energies, in units of the single-particle level spacing, for N=Z=8. In brackets are given the errors relative to the exact results (second column). The third column corresponds to the condensate (4) wile the last to the PBCS approximation (5).

| g | Exact | Quartet (errors) | PBCS (errors) |
|---|---|---|---|
| 0.2 | 1.755 | 1.674 (4.6%) | 1.239 (29.4%) |
| 0.4 | 9.917 | 9.820 (0.9%) | 8.643 (12.8%) |
| 0.6 | 23.431 | 23.388 (0.12%) | 21.571 (7.9%) |
| 0.8 | 39.414 | 39.394 (0.05%) | 36.968 (6.2%) |
| 1.0 | 56.586 | 56.574 (0.02 %) | 53.546 (5.3%) |

Now we shall discuss the quartet correlations for systems with N=Z=odd. In these systems there is a pn pair which is not included in the quartet condensate. These systems can be described by

$$|\Psi>=\tilde{\Gamma}^{+}_{\nu\pi}A^{+n_q}|0>, \tag{6}$$

where $\tilde{\Gamma}^{+}_{\nu\pi}=\sum\tilde{x}_i P^{+}_{i,\nu\pi}$ describes the residual isovector pn pair. It can be noticed that the odd collective pair has different mixing amplitudes as compared with the pairs which are a part of the quartet condensate. It is expected that the structure of the odd pn pair is quite different to the structure of the pairs belonging to the quartet condensate in the weak coupling limit, and that they tend to be rather similar in the strong coupling regime. In order to examine this feature we show in Table 2 the correlations energies calculated for a system with 7 proton-neutron pairs distributed in 14 double degenerate and equidistant levels. The second column shows the exact results while in the last column are given the results corresponding to the PBCS approximation [15]

$$|PBCS>=(\Gamma^{+}_{\pi\nu})^{7}|0>, \tag{7}$$

which for N=Z=odd gives a lower energy compared to (5). In the third column are shown the results corresponding to the quartet condensate (6) assuming that the odd pair has the same structure as the pairs belonging to the quartet condensate. As seen from Table 2, this assumption is unrealistic for the weak pairing but it works very well in the intermediate and strong coupling regimes.

Table 2. Correlations energies, in units of the single-particle level spacing for N=Z=7. The third column corresponds to the quartet condensate (6); it is supposed that the odd pn pair has the same structure as the pairs in the condensate. The second column gives the exact results while the last corresponds to the results of PBCS approximation (7).

| g | Exact | Quartet (errors) | PBCS (errors) |
|---|---|---|---|
| 0.2 | 1.35 | 0.430 (68.1%) | 0.585 (56.7%) |
| 0.4 | 7.59 | 7.496 (1.2 %) | 3.749 (54.2%) |
| 0.6 | 17.87 | 17.85 (0.11%) | 10.73 (39.9%) |
| 0.8 | 30.99 | 30.99 (0.002%) | 19.71 (36.4%) |
| 1.0 | 43.06 | 43.06 (<0.002%) | 29.65 (31.1%) |

In conclusion, the present study indicates that an alpha-like quartet condensate describes accurately the isovector pairing correlations in the ground state of the Hamiltonian (1). Preliminary results show that this conclusion is also valid for N=Z nuclei calculated with realistic pairing interactions and single-particle levels [17].

**Acknowledgments** This work has been supported by the Romanian Ministry for Research and Education through the grant PN 09 37 01 02/2010.

Journal of Physics: Conference Series **321** (2011) 012001 doi:10.1088/1742-6596/321/1/012001

## References

[1] Goodman A L 1979 *Advances in Nuclear Physics* **11** 263

[2] Lane A M 1964 *Nuclear Physics* (Benjamin, New York)

[3] Oertzen W et al 2006 *Phys. Rep.* **432** 43

[4] Arima A, Gillet V, Ginocchio J 1970 *Phys. Rev. Lett.* **25** 1043

[5] Arima A and Gillet V 1971 *Ann. of Phys* **66** 117

[6] Daley H J et al 1986 *Phys. Rev Lett* **57** 1986

[7] Flowers B H and Vujicic M 1963 *Nucl. Phys.* **49** 585

[8] Eichler J and Yamamura M 1972 *Nucl. Phys.* A **182** 33

[9] Gambhir Y K, Ring P , Schuck P 1983 *Phys. Rev. Lett.* **51** 18

[10] Dobes J and Pittel S 1998 *Phys. Rev* C **57** 688

[11] Sen'kov R and Zelevinsky V 2010 *arXiv*: 1011.6250

[12] Tohsaki A, Horiuki H, Schuck P, Ropke G 2001 *Phys. Rev. Lett.* **87** 192501

[13] Roepke G, Schnell A, Schuck P, Nozieres P et al 1998 *Phys. Rev. Lett.* **80** 3177

[14] Dukelsky J, Gueorguiev V G, Isacker P, Dimitrova S, Errea B, and Lerma SH 2006 *Phys. Rev. Lett.* **96** 072503

[15] Sandulescu N, Errea B and Dukelsky J 2009 *Phys. Rev* C **80** 044335

[16] Sandulescu N and Bertsch G F 2008 *Phys. Rev.* C **78** 064318

[17] Sandulescu N et al, in preparation

Journal of Physics: Conference Series **321** (2011) 012002 doi:10.1088/1742-6596/321/1/012002

# New type of Pairing interactions in nuclear matter and finite nuclei

**H Sagawa**[1], **J Margueron**[2] **and K Hagino**[3]

[1]Center for Mathematics and Physics, University of Aizu, Aizu-Wakamatsu, 965-8580 Fukushima, Japan

[2] Institut de Physique Nucléaire, Université Paris-Sud, IN2P3-CNRS, F-91406 Orsay Cedex, France

[3]Department of Physics, Tohoku University, Sendai, 980-8578, Japan

E-mail: sagawa@u-aizu.ac.jp

**Abstract.** We propose new types of density dependent contact pairing interaction which reproduce pairing gaps in a wide range of nuclear mass table.

We discuss also the relation between the proposed paring interactions and the pairing gaps in symmetric and neutron matters obtained by a microscopic treatment based on a bare nucleon-nucleon interaction. It is shown that the isovector type pairing interaction is necessary on top of the isoscalar term to reproduce systematically nuclear empirical pairing gaps. The BCS-BEC crossover of neutrons pairs in symmetric and asymmetric nuclear matters is studied by using these contact interactions. It is shown that the bare and screened pairing interactions lead to different features of the BCS-BEC crossover in symmetric nuclear matter.

## 1. Introduction

It has been known that the pairing correlations play an important role in finite and also infinite nuclear systems. There are mainly two different approaches for a calculation of pairing correlations in finite nuclei. The first approach is based on phenomenological pairing interactions whose parameters are determined using some selected data [1], while the second approach starts from a bare nucleon-nucleon interaction and eventually includes the effect of phonon coupling [2]. The latter approach has shown that the medium polarization reduces the pairing gaps in neutron matter while the neutron pairing gaps in symmetric matter are much enlarged at low density compared to that of the bare calculation. This enhancement takes place especially for neutron Fermi momenta $k_{\rm Fn} < 0.7$ fm$^{-1}$.

In this report, we propose effective density-dependent pairing interactions which reproduce both the neutron-neutron ($nn$) scattering length at zero density and the neutron pairing gap in uniform matter. In order to simultaneously describe the density dependence of the neutron pairing gap for both symmetric and neutron matter, it is necessary to include an isospin dependence in the effective pairing interaction. Depending on whether the medium polarization effects on the pairing gap given in Ref. [3] are taken into account or not, we invent two different density dependences in the pairing interaction. Then, we apply these interactions to study BCS-BEC cross-over in infinite systems and also pairing gaps in semi-magic finite nuclei, such as Ca, Ni, Sn and Pb isotopic chains.

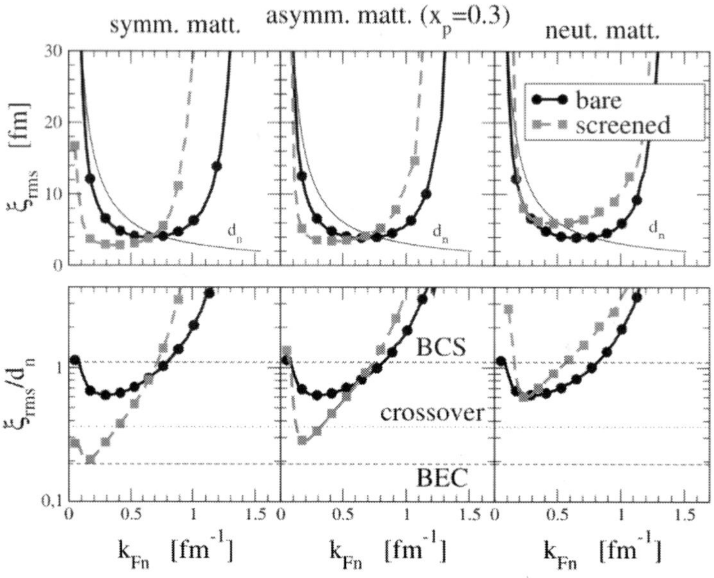


**Figure 1.** (Color online) Top panels: Comparison between the rms radius $\xi_{rms}$ of the neutron pair and the average inter-neutron distance $d_n = \rho_n^{-1/3}$ (thin line) as a function of the neutron Fermi momentum $k_{Fn}$ in symmetric (left panel), asymmetric (central panel) and neutron matters (right panel). Bottom panels: The order parameter $\xi_{rms}/d_n$ is plotted as a function of $k_{Fn}$. The boundaries of the BCS-BEC crossover are represented by the two dashed lines, while the unitary limit is shown by the dotted line. The two pairing interactions defined in Eqs. (1) and (2) are used for the calculations.

## 2. Isospin dependent pairing interaction and BCS-BEC crossover in nuclear matter

The density–dependent pairing interaction can be read as

$$v_{pair}(1,2) = \frac{1 - \mathrm{P}_\sigma}{2} v_0 \, \mathrm{g}[\rho, I] \, \delta(\vec{\mathrm{r}}_1 - \vec{\mathrm{r}}_2). \tag{1}$$

where $\rho$ is the nuclear density and $I$ is defined as $I = (\rho_n - \rho_p)/\rho \cdot \tau_z$ [4]. In Ref. [4], an isovector dependence in the density-dependent term g is separated into two parts $\mathrm{g} = \mathrm{g}^1 + \mathrm{g}^2$. The function $\mathrm{g}^1$ is determined to mimic the bare pairing gaps in nuclear matter and the function $\mathrm{g}^2$ takes care of the medium polarization effect. The functional form of $\mathrm{g}^1$ is given by

$$\mathrm{g}^1[\rho, I] = 1 - \mathrm{f_s}(I)\eta_\mathrm{s}\left(\frac{\rho}{\rho_0}\right)^{\alpha_\mathrm{s}} - \mathrm{f_n}(I)\eta_\mathrm{n}\left(\frac{\rho}{\rho_0}\right)^{\alpha_\mathrm{n}}, \tag{2}$$

where $\rho_0$=0.16 fm$^{-3}$ is the saturation density of symmetric nuclear matter and the functions $\mathrm{f_s}(I)$ and $\mathrm{f_n}(I)$ are $\mathrm{f_s}(I) = 1 - \mathrm{f_n}(I)$ and $\mathrm{f_n}(I) = I = (\rho_\mathrm{n}(r) - \rho_\mathrm{p}(r))/\rho(r) \cdot \tau_z$. The values of parameters $\eta_\mathrm{s}, \eta_\mathrm{n}$ and powers of density dependence $\alpha_\mathrm{s}, \alpha_\mathrm{n}$ are given elsewhere [4].

We study the Fermi momentum dependence (equivalently density dependence) of BCS-BEC crossover by solving the BCS gap equation [6]

$$\Delta_n = -\frac{v_0 \mathrm{g}}{2(2\pi)^3} \int d^3k \frac{\Delta_n}{E_n(k)} \theta(k, k) \,, \tag{3}$$

for a given Fermi momentum. The Fourier transform of the Cooper pair wave function $\Psi_{pair}(k) = Cu_k v_k$ is used to evaluate the rms radius $\xi_{rms}$ of the Cooper pair in nuclear matter. In Fig. 1, the rms radius $\xi_{rms}$ is shown as a function of the neutron Fermi momentum $k_{Fn}$ as well as the order parameter $\xi_{rms}/d_n$ where $d_n = \rho_n^{-1/3}$. The rms radius of the Cooper pair is less than 5 fm in the region $k_{Fn} \sim (0.4 - 0.9)$ fm$^{-1}$ ($\rho_n/\rho_0 \sim 0.01 - 0.15$) in the three panels for the bare interaction. The screened interaction gives different effects in symmetric and asymmetric matters: it increases the rms radius for the neutron matter, while the rms radius stays small around 4 fm even at very low density at $k_{Fn} \sim 0.15$ fm$^{-1}$ ($\rho_n/\rho_0 \sim 0.0007$) in symmetric matter. In the lower panels is shown the ratio of the rms radius to the average distance between neutrons $d_n$. For the bare interaction, the size of the Cooper pair becomes smaller than $d_n$ for the Fermi momentum $k_{Fn} < 0.8$ fm$^{-1}$ ($\rho_n/\rho_0 \sim 0.1$) in general. There are substantial differences for symmetric and asymmetric matters in the case of the screened interaction. The boundaries of BCS and BEC phases are indicative because the phase transition is smooth at the boundaries being of the second order. The crossover region becomes smaller for the neutron matter, while the crossover region increases in the cases of asymmetric ($x_p = 0.3$) and symmetric matters. Especially, the correlations become strong in symmetric matter and the Cooper pair reaches almost the BEC boundary at $k_{Fn} \sim 0.2$ fm$^{-1}$ ($\rho_n/\rho_0 \sim 0.002$). Even if the nuclear matter does not enter into the BEC regime, it is shown that the Cooper pair wave function is already very similar to the BEC one when it is close [4].

In ref. [7], the wave functions of di-neutrons participating to the halo nuclei $^{11}$Li and $^{6}$He have been analyzed to see the analogous behavior to the BCS-BEC crossover in terms of the correlation length. It has been shown that as the distance between the center of mass of the two-neutrons and the core increases, the wave function changes from the weak coupling BCS regime to the strongly correlated BEC regime. This is due to the fact that the pairing correlations are strongly density dependent and the distance between the two-neutrons and the core provides a measure of the pairing strength.

## 3. Isospin Dependence of Pairing Gaps in Finite Nuclei

We perform Hartree-Fock-Bogoliubov (HFB) calculations for semi-magic Calcium, Nickel, Tin and Lead isotopes using these density-dependent pairing interactions in Eq. (1) derived from a microscopic nucleon-nucleon interaction. Our calculations reproduce well the neutron number dependence of experimental data for binding energy, two-neutron separation energy, and odd-even mass staggering of these isotopes [5]. Especially the interaction IS+IV Bare without the medium polarization effect gives satisfactory results for all the isotopes as is seen in Fig. 2. It is clear in the comparison between IS bare and IS+IV bare that the isospin dependence of the pairing interaction plays an important role in the pairing gaps of neutron-rich nuclei. The isospin dependent pairing interaction is further applied to both even-even and even-odd nuclei by using EV8-odd program. The results successfully reproduce the empirical isotope and isotone dependences of the odd-even mass differences of several medium-heavy and heavy nuclei [8].

## 4. Summary

A new type of density-dependent contact pairing interaction was obtained to reproduce the microscopic pairing gaps in symmetric and neutron matter [3]. We have applied these density-dependent pairing interactions to study the BCS-BEC crossover phenomenon in symmetric and asymmetric nuclear matters. We found that the spatial di-neutron correlation is strong in general in a wide range of low matter densities, up to $k_{Fn} \sim 0.9$ fm$^{-1}$ ($\rho_n/\rho_0 \sim 0.15$). This result is independent of the pairing interactions, either bare or screened one, as well as of the asymmetry of the uniform matter.

We performed also HFB calculations for semi-magic Calcium, Nickel, Tin and Lead isotopes using these density-dependent pairing interactions. Our calculations reproduce well the neutron

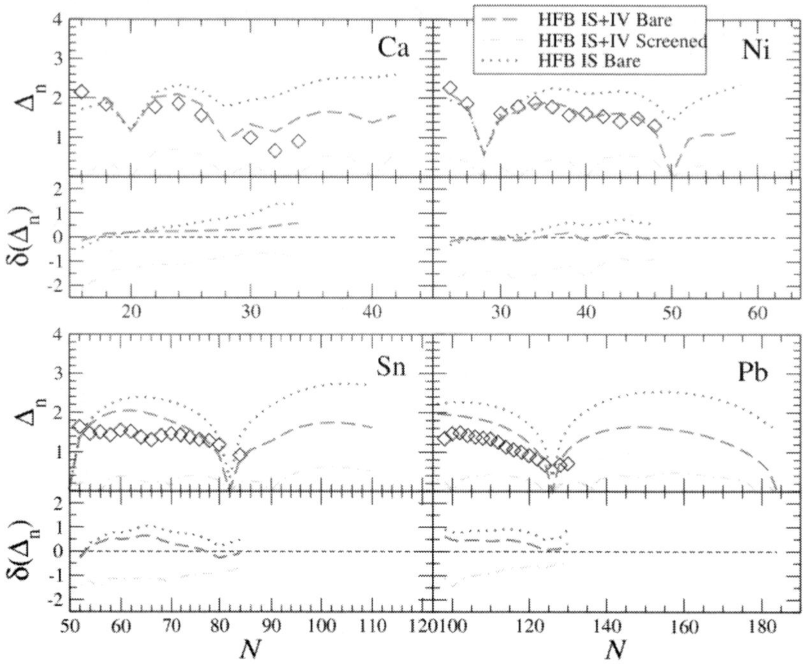


**Figure 2.** Comparison of the neutron HFB pairing gaps $\Delta_n$ with the odd-even mass staggering given by the three-point formula $\Delta^{(3)}$. The dotted line shows the results of the pairing interactions IS+IV Bare, while the short dashed, and long dashed lines are obtained with the pairing interactions IS+IV Screened and IS Bare, respectively. The difference $\delta(\Delta_n)$ is defined as $\delta(\Delta_n) = \Delta_n(\text{th.}) - \Delta_n(\text{exp.})$. All units are given in MeV.

number dependence of experimental data for binding energy, two-neutron separation energy, and odd-even mass staggering of these isotopes with the interaction IS+IV Bare. Recently, the isospin dependent interaction is extended to introduce the quadratic terms of isospin and applied to a wide region of nuclei in the mass table [9]. These results suggest that by introducing the isovector term in the pairing interaction, one can construct a global effective pairing interaction which is applicable to nuclei in a wide range of the nuclear chart.

## Acknowledgments

This work was supported by the Japanese Ministry of Education, Culture, Sports, Science and Technology by Grant-in-Aid for Scientific Research under the program numbers (C) 22540262 and 20540277.

## References

[1] Dobaczewski J, Nazarewicz W, and Reinhard P-G 2001 *Nucl. Phys.* **A693**, 361
[2] Barranco F et al. 1999 *Phys. Rev. Lett.* **83**, 2147
[3] Cao L G, Lombardo U, and Schuck P 2006 *Phys. Rev.* C **74**, 064301
[4] Margueron J, Sagawa H, and Hagino K 2007 *Phys. Rev.* C **76**, 064316
[5] Margueron J, Sagawa H, and Hagino K 2008 *Phys. Rev.* C **77**, 054309
[6] Engelbrecht J R, Randeria M, Sá de Melo C A R 1997 *Phys. Rev* **B 55**, 15153
[7] Hagino K, Sagawa H, Carbonell J and Schuck P 2007 *Phys. Rev. Lett.* **99**, 025506
[8] Bertulani C A, Lu H F and Sagawa H 2009 *Phys. Rev.* C **80**, 027303
[9] Yamagami M, Shimizu Y R and Nakatsukasa T 2009 *Phys. Rev.* C **80**, 064301

# Two-particle correlations in continuum dipole transitions in Borromean nuclei

**K Hagino**[1] **and H Sagawa**[2]

[1] Department of Physics, Tohoku University, Sendai 980-8578, Japan
[2] Center for Mathematics and Physics, University of Aizu, Aizu- Wakamatsu, Fukushima 965-8560, Japan

E-mail: `hagino@nucl.phys.tohoku.ac.jp`

**Abstract.** We use a three-body model to discuss the role of dineutron correlation in the ground state and the dipole excitations of typical weakly-bound Borromean nuclei, $^{11}$Li and $^{6}$He. We first show that the wave function for the neutron pair in the ground state is strongly concentrated on the nuclear surface for both nuclei. We then show that the energy distributions for the two emitted neutrons from the dipole excitation, on the other hand, are considerably different between the two nuclei, even though they show similar strong dineutron correlations in the ground state to each other. Our calculation indicates that this different behaviour of the energy distribution primarily reflects the interaction between the neutron and the core nucleus, rather than the interaction between the valence neutrons.

## 1. Introduction

Neutron-rich nuclei have attracted much interest during the past decades. These nuclei are characterized by a small binding energy, and many new features originating from the weakly bound property have been found. A halo and skin structures with a large spatial extension of the density distribution, and strong concentration of electric dipole ($E1$) transition at low excitation energies are well-known examples.

Among neutron-rich nuclei, two-neutron halo nuclei are particularly intriguing systems to study. Their structure has often been described as a three-body system consisting of two valence neutrons interacting with each other and with the core nucleus. Some light neutron-rich nuclei, such as $^{11}$Li and $^{6}$He, do not have a bound state in the two-body subsystem with a valence neutron and a core nucleus. These nuclei are referred to as Borromean nuclei, and their properties have been studied extensively both experimentally and theoretically.

One of the most important current open questions concerning the Borromean nuclei is to clarify the characteristic nature of *correlations* between the two valence neutrons, which do not form a bound state in the vacuum. A strong dineutron correlation, where the two neutrons take a spatially compact configuration, has been theoretically predicted [1, 2, 3, 4, 5, 6]. Although the recent experimental observation of the strong low-lying dipole strength distribution in the $^{11}$Li nucleus [7] has provided an experimental signature of the existence of dineutron correlation in this nucleus, its direct evidence has not yet been obtained.

In this contribution, we present the results of three-body model calculations with a density-dependent contact interaction between the two valence neutrons [4]. We fist discuss the dineutron correlation in the ground state of $^{11}$Li and $^{6}$He nuclei, and then consider the dipole excitation of these Borromean nuclei. We will particularly discuss the energy distribution of the two emitted neutrons.

## 2. Ground state properties

Let us first discuss the ground state properties of the $^{11}$Li and $^{6}$He nuclei. Figs. 1 and 2 show the two-particle density for these nuclei, respectively. These are plotted as a function of the neutron-core distance, $r_1 = r_2 \equiv r$ and the opening angle between the valence neutrons, $\theta_{12}$. A weight of $4\pi r^2 \cdot 2\pi r^2 \sin\theta_{12}$ has been multiplied. See Ref. [4] for the details of the calculations.

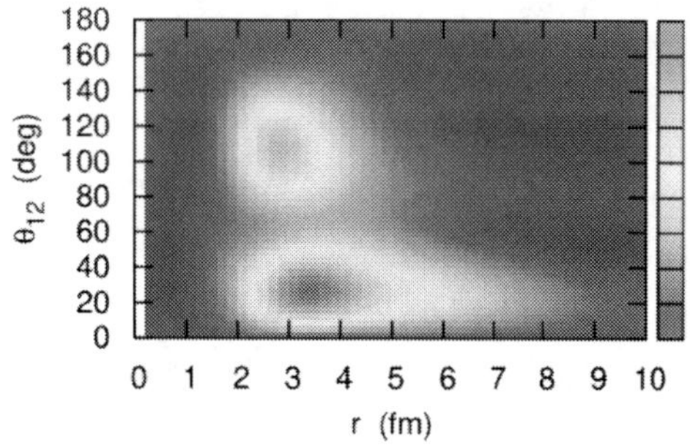


**Figure 1.** The two-particle density for the ground state of $^{11}$Li as a function of neutron-core distance, $r_1 = r_2 \equiv r$, and the opening angle between the valence neutrons, $\theta_{12}$. the density is weighted with a factor $8\pi^2 r^4 \sin\theta_{12}$.

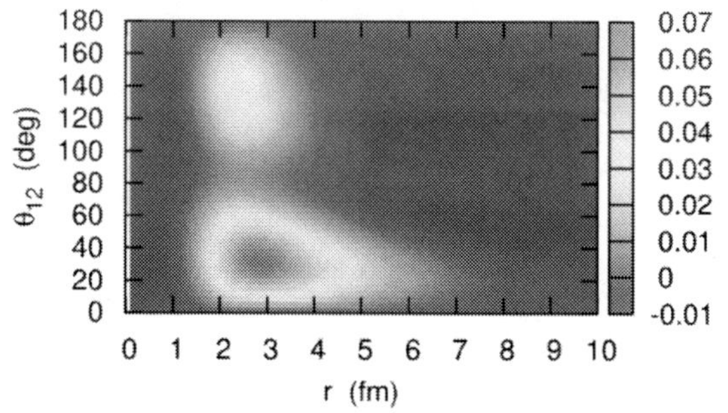


**Figure 2.** The two-particle density for the ground state of $^{6}$He as a function of neutron-core distance, $r_1 = r_2 \equiv r$, and the opening angle between the valence neutrons, $\theta_{12}$. the density is weighted with a factor $8\pi^2 r^4 \sin\theta_{12}$.

We see that a large fraction of two-particle density is concentrated on the region with small opening angle $\theta_{12}$. This is nothing but a manifestation of strong dineutron correlation. We recently pointed out that a similar correlation, that is, the diproton correlation, exists also in proton-rich nuclei, such as $^{17}$Ne [8].

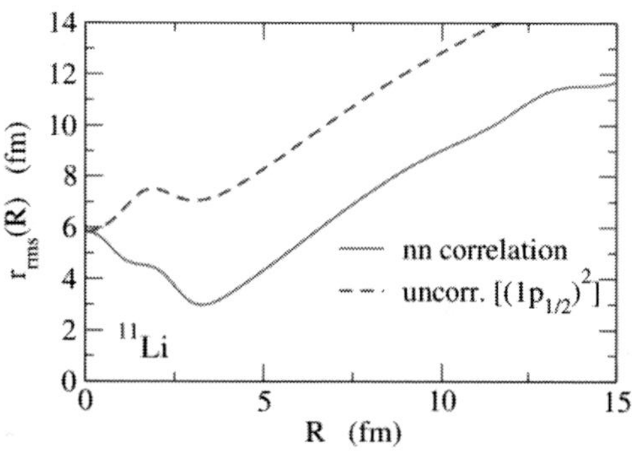


**Figure 3.** The root mean square distance $r_{\rm rms}$ for the neutron Cooper pair in $^{11}$Li as a function of the nuclear radius $R$. The solid line shows the result of three-body model calculation, while the dashed line is obtained by switching off the neutron-neutron interaction and assuming the $[(1{\rm p}_{1/2})^2]$ configuration, respectively.

The size of neutron pair takes a distinct minimum at the peak of two-particle density [9, 10]. Figure 3 shows the local coherence length of the Cooper pair in $^{11}$Li as a function of the nuclear radius $R$ obtained with and without the $nn$ interaction. For the uncorrelated calculations, we consider the $[(1{\rm p}_{1/2})^2]$ configuration and adjust the single-particle potential so that the corresponding single-particle energy is $-0.15$ MeV. One can see that, in the non-interacting case, the Cooper pair continuously expands, as it gets farther away from the center of the nucleus. In marked contrast, in the interacting case it becomes smaller going from inside to the surface before expanding again into the free space configuration.

## 3. Dipole excitation

Let us now discuss the dipole excitation of the Borromean nuclei. Recently, the strong low-lying dipole strength distribution has been observed experimentally in the $^{11}$Li nucleus [7], which strongly suggests the existence of dineutron correlation in this nucleus. The Coulomb dissociation measurement has also been carried out for the $^{6}$He nucleus [11]. Our three-body model calculations well reproduce the experimental data, as shown in Figs. 4 and 5 [12].

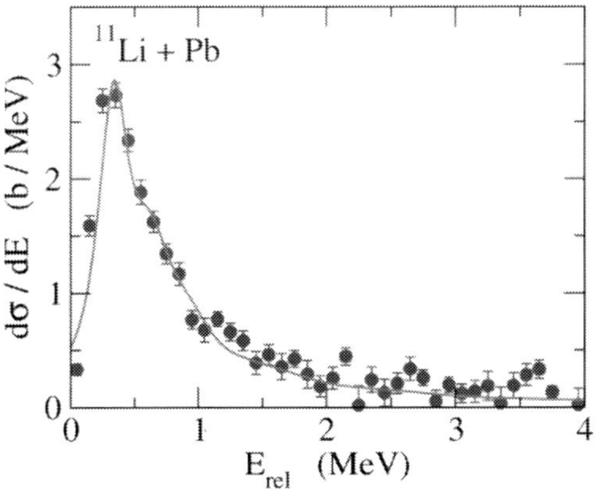


**Figure 4.** Coulomb breakup cross sections for the $^{11}$Li +Pb at 70 MeV/nucleon. The calculated results are smeared with an energy dependent width of $\Gamma = 0.25 \cdot \sqrt{E_{\rm rel}}$ MeV. The experimental data are taken from Ref. [7].

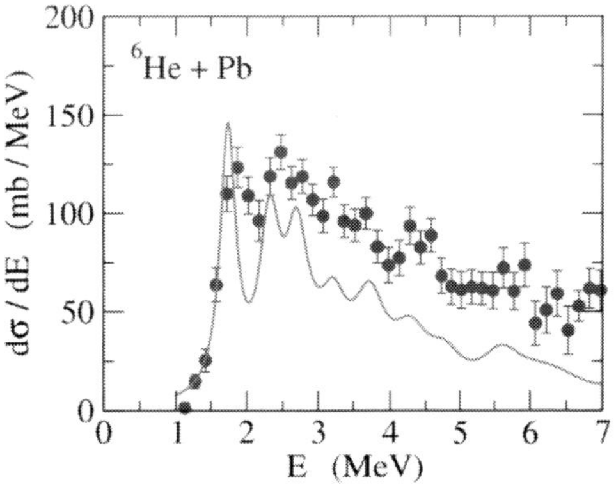


**Figure 5.** Coulomb breakup cross sections for $^{6}$He+Pb at 240 MeV /nucleon. The calculated results are smeared with an energy dependent width of $\Gamma = 0.15 \cdot \sqrt{E_{\rm rel}}$ MeV. The experimental data are taken from Ref. [11].

One may ask whether a more direct information on the dineutron correlation can be obtained by measuring energy and angular distributions of two emitted neutrons. Notice that the operator which induces the E1 excitation is proportional to the center of mass coordinate of the two valence neutrons, $\vec{R} = (\vec{r}_1 + \vec{r}_2)/2$ [2, 13]. Therefore, the relative motion of the two neutrons, $\vec{r} = \vec{r}_1 - \vec{r}_2$, is not affected by the E1 excitations. It is thus interesting to ask how the energy and angular distributions from the E1 excitation reflect the ground state properties of the Borromean nuclei, especially the correlation for the relative motion of the neutrons, that is, the dineutron correlation.

Figs. 6 and 7 show the dipole strength distribution, $d^2B(E1)/de_1de_2$, as a function of the energies of the two emitted neutrons for the $^{11}$Li and $^{6}$He nuclei, respectively[14]. Here, $e_1$ ($e_2$) is the relative energy between the first (second) neutron and the core nucleus. These distributions are obtained using the Green's function method [13]. One immediately notices that the strength distribution is considerably different between $^{11}$Li and $^{6}$He. For $^{11}$Li, a large concentration of the strength appears at about $e_1$=0.375 MeV and $e_2$=0.075 MeV (and at $e_1$=0.075 MeV and $e_2$=0.375 MeV), with a small ridge at an energy of about 0.5 MeV. On the other hand, for $^{6}$He, only a large ridge at about 0.7 MeV appears, and the strength is largely concentrated around $e_1 = e_2 = 0.7$ MeV.

In Ref. [14], we have shown that the different behaviours in the strength distribution originate primarily from the property of the neutron-core interaction. The ridge structures reflect the

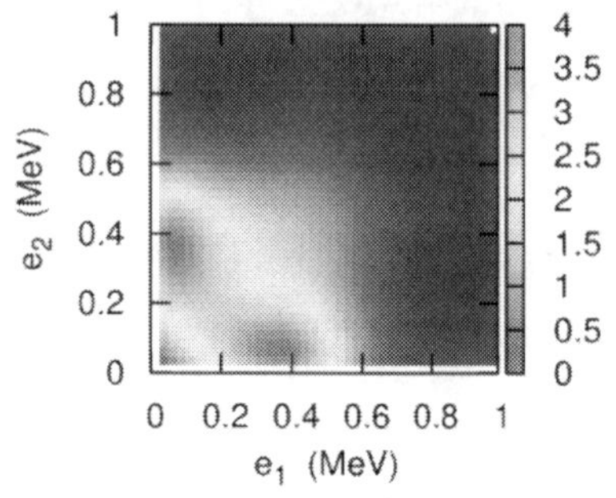


**Figure 6.** The dipole strength distribution, $d^2B(E1)/de_1de_2$, of $^{11}$Li as a function of the energies of the two emitted neutrons relative to the core nucleus. It is plotted in units of $e^2fm^2/MeV^2$.

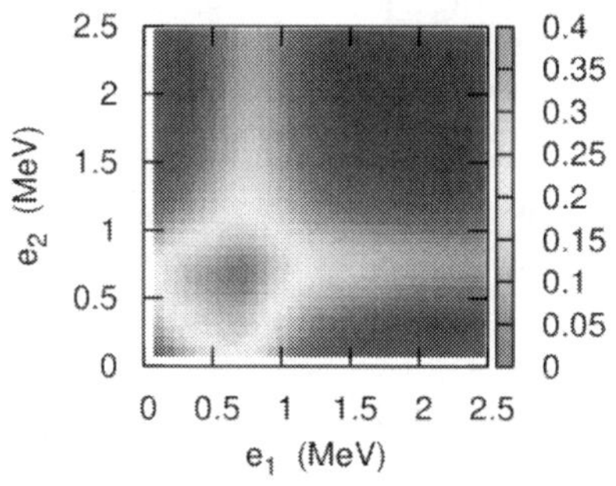


**Figure 7.** The dipole strength distribution, $d^2B(E1)/de_1de_2$, of $^6$He as a function of the energies of the two emitted neutrons relative to the core nucleus. It is plotted in units of $e^2fm^2/MeV^2$.

single-particle resonances, that is, the $p_{1/2}$ resonance around 0.54 MeV for $^{10}$Li and the $p_{3/2}$ resonance at 0.91 MeV for $^5$He. These ridges correspond to the physical process in which one of the neutrons is excited by the dipole field while the other remains near the resonance state as a spectator in the neutron-core system [2, 13]. For $^{11}$Li, in addition to the ridge, the dipole strength is concentrated in the region in which one of the neutrons has an energy close to zero. This reflects the *s*-wave virtual state in $^{11}$Li close to zero energy.

The correlation measurements for $^{11}$Li have been recently done at RIKEN[15]. We mention that the present calculation shown here is qualitatively in good agreement with the preliminary data.

## Acknowledgments

We are grateful to Peter Schuck for collaborations on the dineutron correlation in light neutron-rich nuclei. This work was supported by the Japanese Ministry of Education, Culture, Sports, Science and Technology by a Grant-in-Aid for Scientific Research under the program numbers (C) 22540262 and 20540277.

## References

[1] Hansen P G and Jonson B 1987 *Europhys. Lett.* **4** 409
[2] Bertsch G F and Esbensen H 1991 *Ann. Phys. (N.Y.)* **209** 327
[3] Zhukov M V *et al.* 1993 *Phys. Rep.* **231** 151
[4] Hagino K and Sagawa H 2005 *Phys. Rev.* **C72** 044321
[5] Matsuo M, Mizuyama K and Serizawa Y 2005 *Phys. Rev.* **C71** 064326
[6] Pillet N, Sandulescu N and Schuck P 2007 *Phys. Rev.* **C76** 024310
[7] Nakamura T *et al.* 2006 *Phys. Rev. Lett.* **96** 252502
[8] Oishi T, Hagino K, and Sagawa H 2010 *Phys. Rev.* **C82** 024315
[9] Hagino K, Sagawa H, Carbonell J, and Schuck P 2007 *Phys. Rev. Lett.* **99** 022506
[10] Hagino K, Sagawa H and Schuck P 2010 *J. Phys. G* **37** 064040
[11] Aumann T *et al.* 1999 *Phys. Rev.* **C59** 1252
[12] Hagino K and Sagawa H 2007 *Phys. Rev.* **C76** 047302
[13] Esbensen H and Bertsch G F 1992 *Nucl. Phys.* **A542** 310
[14] Hagino K, Sagawa H, Nakamura T and Shimoura S 2009 *Phys. Rev.* **C80** 031301(R)
[15] Nakamura T, private communication.

# Heavy-ion two-particle transfer reactions as a probe of pairing correlations

**Ilyas Inci** [1,2,3] **and Andrea Vitturi** [2,3]

[1] Institute of Science, Erciyes University, Kayseri, Turkey

[2] Dipartimento di Fisica, Universita' di Padova, Padova, Italy

[3] Istituto Nazionale di Fisica Nucleare, Sezione di Padova, Padova, Italy

E-mail: `vitturi@pd.infn.it`

**Abstract.**
Two-particle transfer processes induced by light and heavy ions are known to be an ideal dynamical tool for extracting information on nuclear pairing correlations. The procedure is however not unique and different reaction mechanism models (ranging from microscopic correlated successive one-particle transfer to collective macroscopic models) can be introduced to establish a link with the structural aspects of initial and final states. In the case of reactions involving weakly-bound nuclei the low-density features characterizing the vicinity of the drip line and the suggested consequent enhancement of the pairing correlations should show up as clear signature in the two-particle transfer cross sections. From a theoretical point of view in the weakly-bound case the focus is centered on the specific role of continuum states, both at the nuclear structure level, to give properly correlated wave functions, and at the reaction level, for example to treat the unbound states in the intermediate A±1 system.

If one wants to use dynamics to study pairing correlations the main road is clearly provided by two-particle transfer processes induced by light ions (reactions as (t,p), (p,t), ($^3$He,n), ($\alpha$,d)) or heavy ions, which are both exploring precisely the pair radial correlations. Unfortunately, the situation is different, for example, from low-energy one-step Coulomb excitation, where the excitation probability is directly proportional to the B(E$\lambda$) values. Here the reaction mechanism is much more complicated and the possibility of extracting spectroscopic information on the pairing field is not obvious. The situation is actually more complicated even with respect to other processes (as inelastic nuclear excitation) that may need to be treated microscopically, but where the reaction mechanism is somehow well established.

It is often assumed that the cross section for two-particle transfer will scale with the square of the matrix element of the pair creation (or removal) operator. In this perspective in order to define and measure the collectivity of pairing modes one could compare with single-particle pair matrix elements to define some pairing single-particle units and therefore pairing enhancement factors. But the two-particle transfer process in not sensitive to just the pair matrix element. We have to look at the radial dependence of the pair transition density, which is relevant for the reaction mechanism associated with pair transfer processes. As examples we show in Figs. 1 and 2 the comparison with pure single-particle configurations in the case of the addition mode around $^{16}$O (i.e. the ground state of $^{18}$O) and the removal mode around $^{208}$Pb (i.e. the ground state of $^{206}$Pb). The comparison is done in Fig. 1 directly on the local transition density ($\delta\rho_P(r,r)$). In Fig. 2 the comparison is done on the non-local transition density $\delta\rho_P(r_1,r_2)$ by fixing the position on one particle and plotting as contour plots in space the probability of finding the second one. One immediately sees that while for pure configuration one has equal probablity of finding the second particle on the same or on the opposite side with respect to the core, the pairing correlation induces a correlation in space, favouring the clustering of the pair.

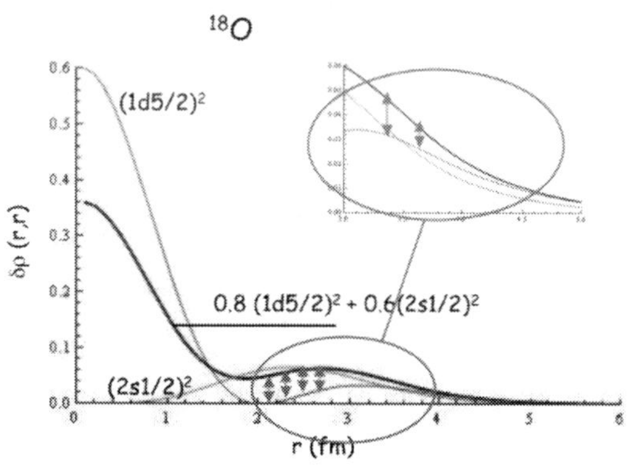


**Figure 1.** Pair transition density in $^{18}$O. The cases of pure $(1d5/2)^2$ and $(2s1/2)^2$ are compared with the case of correlated wave function.

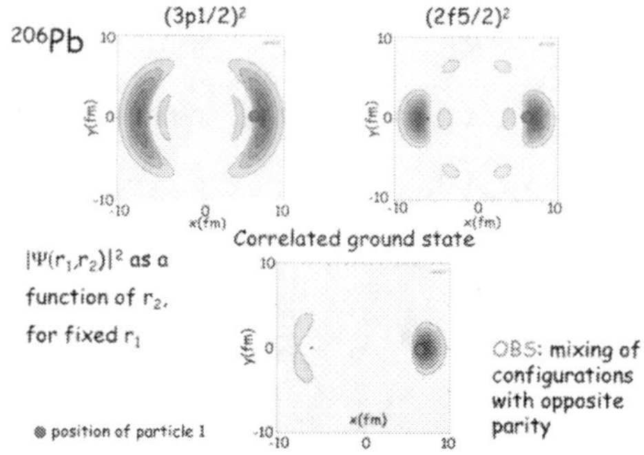


**Figure 2.** Squared wave functions for the ground state of $^{206}$Pb, plotted fixing the position of one particle as a function of the position of the second one. Pure configurations are compared with the case of pairing correlated wave function

The difficult point is to clarify how the structure properties enter into the reaction mechanism. To this end one needs reliable models for the description of the reaction mechanism. There are actually in the literature a large number of different approaches, all trying to reduce the actual complexity of the problem, which is a genuine four-body scattering (the two cores plus the two transferred particles). Note also that different approaches are used in the case of heavy-ion induced reactions (where one may need detailed information on the wave functions in both target and projectile, but on the same time the reaction is mainly surface-peaked and semiclassical descriptions may apply) and in the case of light ions (as (p,t) reactions).

As an example, in a semi-microscopic approach the reaction mechanism is described as a one-step di-neutron (cluster) transfer. In this case the microscopy amounts to construct a formfactor obtained by double-folding the microscopic pair transition densities of initial and final states with some nucleon-nucleon interaction or by simple folding of microscopic pair density in the target with the one-body mean field of the projectile. Such an approach only involves the local transition densities and does not exploit the full non-local features of the pair correlation.

A fully microscopic approach is instead the one in which the two-particle transfer reaction is described by sequential two-step process (at each step one transfers one particle)[1, 2]. In this case the pairing enhancement comes from the coherent interference of the different paths through the different intermediate states in (a-1) and (A+1) nuclei, due to the correlations in initial and final wave functions. Basis ingredients of the calculation are the individual formfactors for one-particle transfer and the microscopic pairing correlated wave functions. Since such fully quantal calculations are rather complex (taking into account full recoil), semiclassical approximation scheme are commonly used (although with approximate treatment of recoil).

As an example, we consider the two-particle transfer reaction $^{208}$Pb($^{16}$O, $^{18}$O$_{gs}$)$^{206}$Pb$_{gs}$, assuming that the two particles are transferred from the 3p1/2 level in $^{208}$Pb into a correlated ground state of $^{18}$O, linear combination of $(1d5/2)^2$ and $(2s1/2)^2$ (cf. Fig. 3). The effect of the

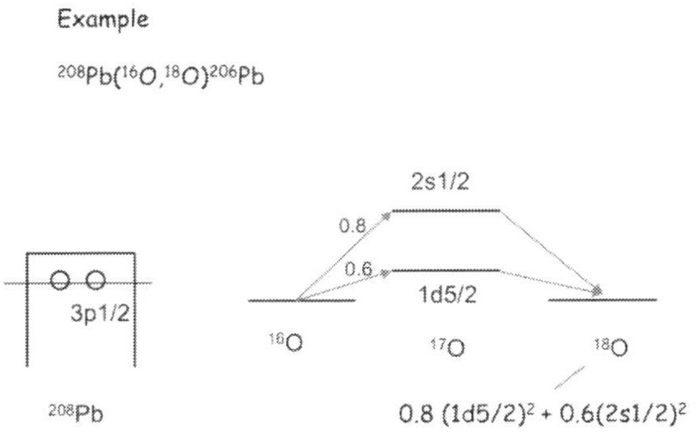


**Figure 3.** Scheme of two-particle transfer etc in $^{18}$O.

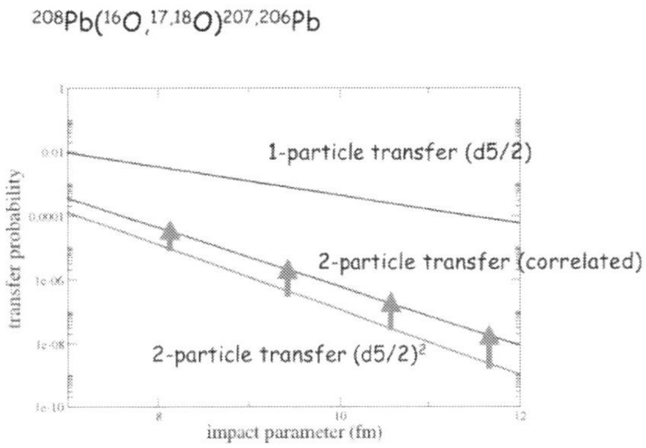


**Figure 4.** Probabilities for one-, two- (uncorrelated) and two-particle (correlated) transfer as a function of the impact parameter

correlation on the two-particle transfer probability is evident from Fig. 4, where the result for the correlated case is compared with the corresponding value for the uncorrelated pure $(1d5/2)^2$ case.

As we move closer or even beyond the drip lines the previous picture is inevitably modified. As schematically shown in Figs. 5-7, we move from a situation in which all relevant intermediate states are bound (Fig. 5), to situations in which only part of these states are bound (Fig. 6), finally reaching at the drip line situations in which no bound intermediate states are present (Fig. 7) and only due to the pairing interaction the (A+2) system becomes bound. Typical examples are given by the so-called Borromean nuclei such as $^{11}$Li or $^{6}$Li whose grond states are bound in spite of no available bound states in $^{10}$Li or $^{5}$He.

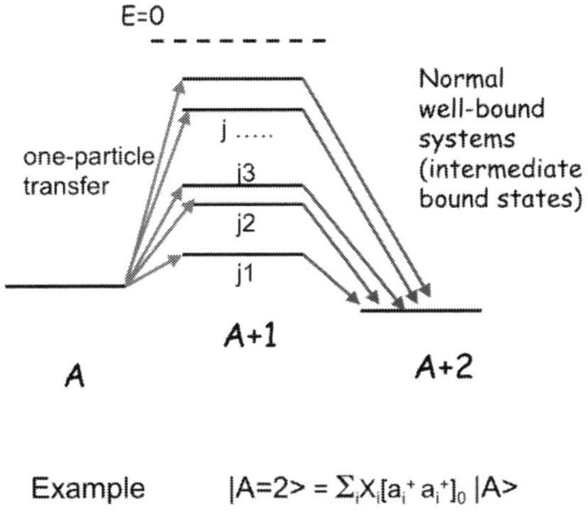


**Figure 5.** Scheme of two-particle transfer for normal systems

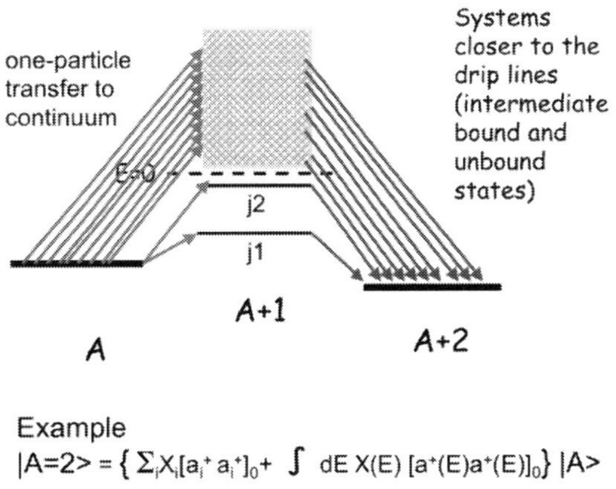


**Figure 6.** Scheme of two-particle transfer for systems towards the drip line

For weakly-bound systems at the drip lines it is mandatory to include in the models the positive energy part of the spectrum. If one wants to still use the same machinary used with bound states, the most popular approach is the so-called discretization of the continuum (cf.

Fig. 8). But the discretization must be used in a consistent way both in the structure and in reaction parts. All discretization procedures are equivalent as long as a full complete basis is used. In practice all procedudes contain a number of parameters and criteria, that make them not all equally applicable in practical calculations. Computational constraints may in fact become a severe problem. As examples we can consider the case of HO wave functions, diagonalization in a box, the case of discretized wave functions with scattering boundary conditions (CDCC) and Gamow states (complex energies).

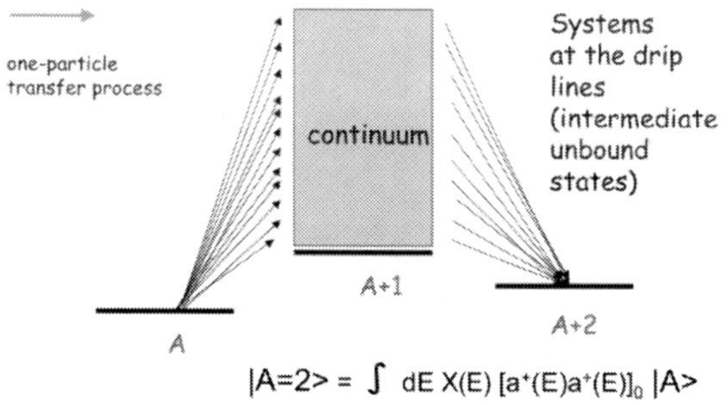


**Figure 7.** Scheme of two-particle transfer for systems at the drip line

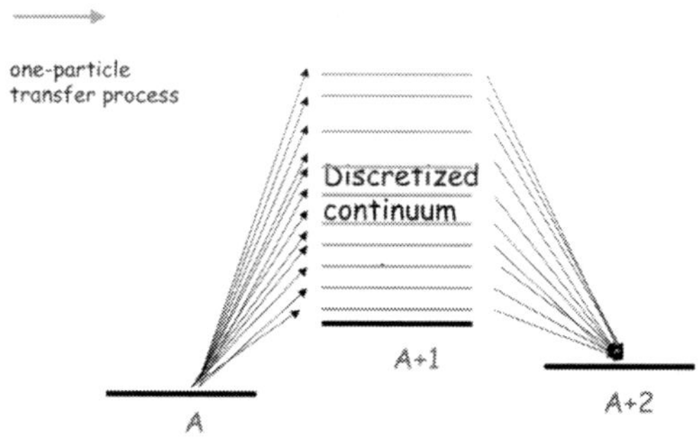


**Figure 8.** Scheme of two-particle transfer with the discretization procedure

We have tested [3, 4] in simple structure and reaction models the validity and the convergence of different discretization procedures. The comparison with the exact calculations shows that descriptions based on discretization of the continuum can be accurate only when a proper choice is made of the number of discrete states, of the energy mesh and of the energy cutoff. This may imply, even in simplified cases, the use of a rather large (and unpracticable) number of channels. The use of a more restricted number of channels may lead to rather misleading results.

In particular, while energies and other observables may converge rather rapidly to the correct value, the discretization procedures normally lead to single-particle and pair wave functions characterized by with incorrect behaviours on the tail. This produces one-particle transfer formfactors which are rather different. So even if the properties of the pair wave functions seem to be well described by all methods with relatively small basis, the predicted probabilities for two-particle transfer turn out to converge only if an extremely large basis (often impossible to treat) is used. Examples will be given in a forthcoming paper [5].

In conclusion correlated sequential one-particle transfer seems to be the proper reaction mechanism present in two-particle transfer processes and to offer a formalism able to directly test the microscopic features of the pair wave functions. This implies a knowledge of the pairing states expressed in a single particle basis, since the same single-particle states enter as intermediate states in the sequential reaction process. For systems at the drip lines this means that it is mandatory a proper description of the continuum states. The discretization procedures have to be consistently used in the structure and reaction sectors, and the behaviour outside of the nuclear radii has to carefully checked, being fundamental for the description of the reaction process.

## References

[1] R.A. Broglia and A. Winther 1991 *Heavy Ion reactions, Lecture Notes*, Addison-Wesley
[2] W. von Oertzen and A. Vitturi 2001 *Rep. Prog. Phys.* **64** 1247
[3] C.H. Dasso and A. Vitturi 2009 *Phys. Rev.* C **79** 064620
[4] A. Vitturi and F. Perez-Bernal 2010 *Nucl.Phys.* A**834** 428c
[5] I. Inci and A. Vitturi, to be published

# Multiparticle-multihole Gogny energy density functional for low-lying spectroscopy applications

**N. Pillet[1], V. Zelevinsky[2,3], M. Dupuis[1], J.F. Berger[1], J.M. Daugas[1]**

[1] CEA,DAM,DIF, F-91297 Arpajon, France

[2] Department of Physics and Astronomy, Michigan State University, East Lansing, Michigan 48824-2320, USA

[3] National Superconducting Cyclotron Laboratory, Michigan State University, East Lansing, Michigan 48824-1321, USA

E-mail: nathalie.pillet@cea.fr

**Abstract.** In this work, we present a few spectroscopic results deduced from a multiparticle-multihole Gogny energy density functional. Both pairing correlations in the ground state and low-lying spectroscopy of a few tin isotopes are discussed. All the results have been obtained using the same D1S Gogny interaction in the one-body mean-field and the residual part associated with a two-body and density-dependent Hamiltonian.

## 1. Introduction

Mean field approaches are ones of the most powerful methods to tackle an approximate solution of the nuclear many-body problem. Nowadays, a lot of efforts is applied to go beyond the mean field approximation to account for missing correlations. Special attention is paid to the restoration of broken symmetries in mean field approaches [1]-[7], for example using projection technics. An alternate way is to develop a theory in which the trial wave functions preserve certain symmetries. In particular, this is achieved by multiconfiguration methods widely used in various domains of physics, including atomic physics, molecular physics or condensed matter physics. When the interaction is known, this kind of approach provides a very accurate description of a system. In a previous work, we have proposed a variational and self-consistent derivation of multiconfiguration methods in the nuclear physics context with two-body density-dependent and finite range interaction [8]. In section 2, we recall briefly the basics of the method. Results concerning pairing correlations in the ground states and the effect of particle number conservation in tin isotopes are presented in section 3. In section 4, preliminary spectroscopic results for tin isotopes are discussed. All the results are obtained with the D1S Gogny interaction [9, 10]. Conclusions and perspective are given in section 5.

## 2. Multiparticle-multihole configuration mixing formalism

Multiparticle-multihole (MPMH) configuration mixing method aims to unify nuclear long range correlations in a symmetry preserving context. The trial wave function is built as a superposition of Slater determinants, not restricted to a single shell:

$$|\Psi\rangle = \sum_{\alpha_\nu,\alpha_\pi} A_{\alpha_\nu,\alpha_\pi} \quad [\ |\Phi_{\alpha_\nu}\rangle \otimes |\Phi_{\alpha_\pi}\rangle\ ] \tag{1}$$

where $\{A_{\alpha_\nu,\alpha_\pi}\}$ represents the set of mixing coefficients and $\{|\Phi_{\alpha_\pi}\rangle, |\Phi_{\alpha_\nu}\rangle\}$ are proton and neutron Slater determinants, respectively, built on single-particle functions $\varphi_i$. In this formulation, particle numbers are conserved and Pauli principle is fully respected. Moreover, as the wave function contains a lot of correlations through multiple particle-hole configurations, the total angular momentum may be chosen as a good quantum number.

The mixing coefficients and the single-particle orbitals are the variational parameters. If one defines the energy functional $\mathcal{F}(\rho)$ as

$$\mathcal{F}(\rho) = \langle\Psi|\hat{H}(\rho)|\Psi\rangle - \lambda\langle\Psi|\Psi\rangle, \quad (2)$$

the variational parameters are determined applying a variational principle:

$$\frac{\partial\mathcal{F}(\rho)}{\partial A^*_{\alpha_\nu\alpha_\pi}} = 0 \quad \text{and} \quad \frac{\partial\mathcal{F}(\rho)}{\partial\varphi^*_i} = 0. \quad (3)$$

In Eq. (2), the Hamiltonian $\hat{H}(\rho) = \hat{T} + \hat{V}(\rho)$, where $\hat{V}(\rho)$ represents the two-body Gogny effective interaction plus the Coulomb term. In addition, $\hat{T}$ and $\hat{V}(\rho)$ contain 1 and 2-body center-of-mass corrections, respectively.

The set of equations in Eq. (3) are equivalent to a set of non-linear secular equations and inhomogeneous Hartree-Fock (HF) equations:

$$\sum_{\alpha'_\pi\alpha'_\nu} \mathcal{H}_{\alpha_\pi\alpha_\nu,\alpha'_\pi\alpha'_\nu}\, A_{\alpha'_\pi\alpha'_\nu} = \lambda A_{\alpha_\pi\alpha_\nu} \quad (4)$$

$$[\, h(\rho,\sigma)\ ,\ \rho\,] = G(\sigma). \quad (5)$$

In Eq. (4), $\mathcal{H}$ contains contributions from $\hat{H}(\rho)$ and rearrangement terms coming from the density-dependent part of the Gogny interaction. In Eq. (5), $h$ represents the effective one-body Hamiltonian deduced from $\hat{H}(\rho)$, and $\sigma$ is the two-body correlation matrix. The fully self-consistent solution is obtained by the simultaneous solution of systems (4) and (5). Practically, an iterative procedure is applied.

## 3. Beyond BCS pairing correlations in tin ground states

In this part, we discuss pairing correlations. We have investigated pairing properties of the ground states of tin isotopes using the Hartree-Fock-Bogoliubov approximation and the MPMH configuration mixing approach. In both cases, the D1S Gogny interaction has been used. Before discussing the results, it is important to emphasize that the MPMH configuration mixing approach can be equivalent to a particle number projection before variation. For example, starting from a BCS wave function, it is easy to establish that the BCS wave function projected on good particle number is equivalent to a MPMH wave function restricted to particular configurations [8]. Specifically, one has to introduce only configurations corresponding to excitations of pairs of nucleons, where a pair is understood as two nucleons in time-reversed states. The proton-neutron pairing is not included in the present study.

We discuss mean field properties of selected tin isotopes. Figs. 1-2 display the evolution of the total Hartree-Fock-Bogoliubov energy (arbitrary scale) and the pairing energy, respectively, according to the dimensionless deformation parameter $\beta$ for $^{100}$Sn, $^{106}$Sn and $^{116}$Sn (energies are expressed in MeV). One concludes that the minimum of total energy is found at $\beta$=0 for the three isotopes. The potential energy curve is less rigid for $^{116}$Sn, where one can also see the appearance of the second minimum at $\beta > 0$, than for $^{100}$Sn and $^{106}$Sn. Concerning pairing energies, at $\beta$=0, no proton pairing is found. However, concerning neutron pairing, one can

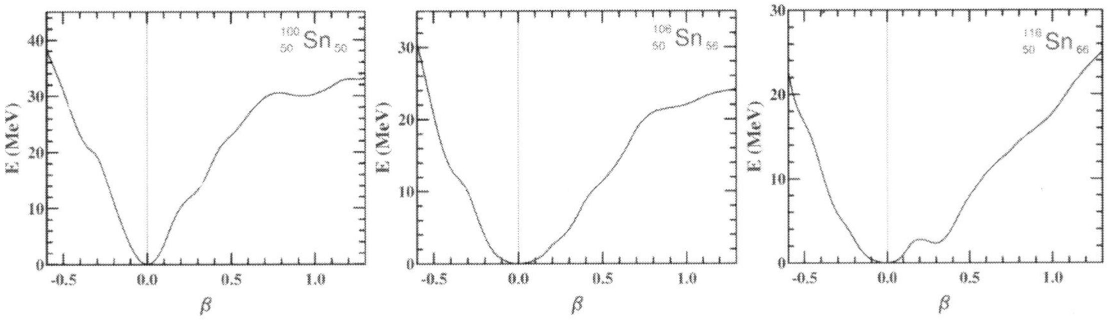


**Figure 1.** Axial potential energy curves for $^{100}$Sn, $^{106}$Sn and $^{116}$Sn in an arbitrary scale (in MeV).

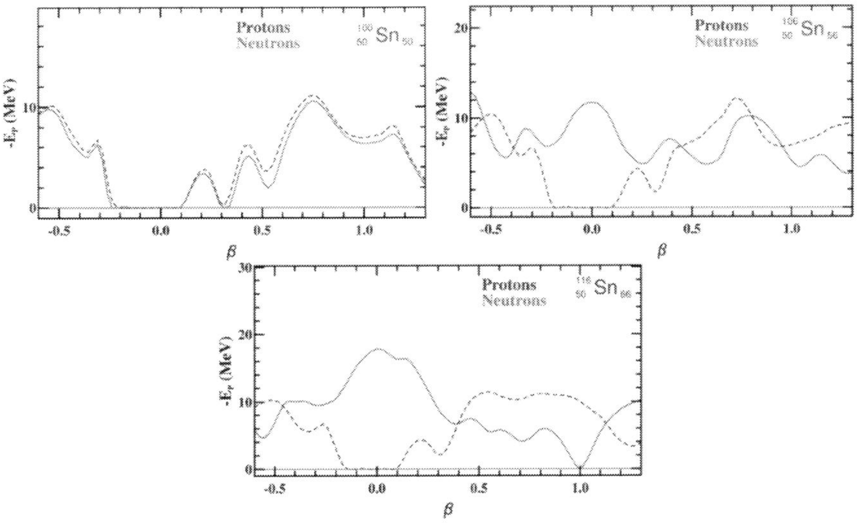


**Figure 2.** Axial pairing energy curves for $^{100}$Sn, $^{106}$Sn and $^{116}$Sn (in MeV).

say that these three Sn isotopes simulate three pairing regimes: weak, medium and strong, respectively, for $^{100}$Sn (0 MeV), $^{106}$Sn (~10 MeV) and $^{116}$Sn (~20 MeV).

One quantity we have been interested in is the correlation energy $E_{corr}$. It is taken as the difference between the total energy $\mathcal{E}(\Psi) = \langle\Psi|\hat{H}|\Psi\rangle$ of the correlated system and the energy of the simple HF method $\mathcal{E}_0^{HF} = \langle\Phi_0|\hat{H}|\Phi_0\rangle$:

$$E_{corr} = \mathcal{E}(\Psi) - \mathcal{E}_0^{HF} \quad (6)$$

Table 1 summarizes the results obtained solving Eq. (4) of the MPMH configuration mixing method (no renormalization of the single-particle orbitals). The total correlation energy (expressed in MeV) found with the MPMH wave function, as well as the neutron contribution, are indicated. The BCS correlation energy is also shown. One sees that with the MPMH configuration mixing approach, the correlation energy is never vanishes, even for the doubly-magic nucleus $^{100}$Sn for which it is equal to 3.67 MeV. The contribution due to protons is found similar for the tin isotopes. It is interesting to notice that the correlation energy found with the MPMH wave function is systematically stronger than the one found in the BCS approximation.

The difference between both approaches is much larger when the pairing regime is weak. In other words, from an energy point of view, the BCS approximation can serve as a good approximation in the strong pairing regime. It is no longer the case in medium and weak pairing regimes. This is a well known result in exactly solvable models of the pairing Hamiltonian [11, 12, 13].

**Table 1.** Total and neutron correlation energies (in MeV) obtained from the MPMH configuration mixing for $^{100}$Sn, $^{106}$Sn and $^{116}$Sn. The last column indicates the correlation energy deduced from a BCS approximation.

| Nucleus | $\lvert E_{corr}^{total}\rvert$ | $\lvert E_{corr}^{neutron}\rvert$ | $\lvert E_{corr}^{BCS}\rvert$ |
|---|---|---|---|
| $^{100}$Sn | 3.67 | 1.90 | 0.00 |
| $^{106}$Sn | 4.62 | 2.88 | 1.37 |
| $^{116}$Sn | 5.44 | 3.74 | 3.25 |

It is also instructing to discuss the content of the wave function. In order to have a measure of the amount of correlations in the MPMH wave function $|\Psi\rangle$, we define the quantity $T(i,j)$:

$$T(i,j) = \sum_{\alpha_\pi \alpha_\nu}^{ij} |A_{\alpha_\pi \alpha_\nu}|^2 \tag{7}$$

The first and second arguments of $T$ stand for the number of excited proton and neutron pairs included in $|\Psi\rangle$, respectively. Results are presented in Table 2. No renormalization of single-particle orbitals have been taken into account in those results. In the case of $^{100}$Sn, the HF approximation is good in the sense that 90% of the wave function correspond to the HF component. Configurations corresponding to the excitation of one pair of nucleons constitute the rest of the wave function. Among those configurations, none are dominant.The dissymmetry between proton and neutron configurations comes from the Coulomb term. $^{106}$Sn and $^{116}$Sn nuclei are much more correlated. The HF component is now contributing to only ~66% of the wave function. In the case of $^{106}$Sn, there is a dominant configuration corresponding to the diffusion of one neutron pair from the $2d_{5/2}$ shell toward the $1g_{7/2}$ shell. In the case of $^{116}$Sn, there are two dominant configurations corresponding to the diffusion of the pair contained in $3s_{1/2}$ toward the $2d_{3/2}$ and $1h_{11/2}$ shells. For $^{106}$Sn and $^{116}$Sn, the actual wave function is far from the HF approximation.

**Table 2.** Components of ground state wave functions for $^{100}$Sn, $^{106}$Sn and $^{116}$Sn without single particle renormalization.

| Nucleus | $T$(0,0) | $T$(0,1) | $T$(1,0) | $T$(0,2) | $T$(1,1) | $T$(2,0) |
|---|---|---|---|---|---|---|
| $^{100}$Sn | 90.85 | 5.02 | 3.70 | 0.16 | 0.18 | 0.09 |
| $^{106}$Sn | 67.44 | 25.29 | 3.63 | 2.54 | 0.99 | 0.11 |
| $^{116}$Sn | 65.38 | 26.04 | 4.50 | 2.68 | 1.23 | 0.17 |

Finally, we discuss possible effects coming from the renormalization of single particle orbitals. In order to partly account for it, we solved Eq. (5) in an approximate way, by neglecting terms associated with the two-body correlation matrix $\sigma$. Doing so, a non-negligible gain for the correlation energy has been obtained, as expected. What is more interesting is the effect on correlated wave functions. The results are summarized in Table 3. One sees that for $^{100}$Sn and $^{106}$Sn, no striking effect is obtained. However, this is completely different for $^{116}$Sn. In particular, the main contribution to the correlated wave function comes from configurations corresponding to the excitation of one neutron pair. The HF component is now provides to 42% of the wave function. This huge change comes from a ~400keV reduction of the neutron gap between the $1d_{3/2}$ and $1h_{11/2}$ shells. The renormalization by correlations of single-particle orbitals is a very important effect, in particular for the prediction of the ground state spins in odd-$A$ nuclei. For example, it is known that a single-particle spectrum obtained with the D1S Gogny interaction in the context of Hartree-Fock-Bogoliubov calculation using self-consistent blocking is unable to provide the right spin for $^{121}$Sn, $^{129}$Sn and $^{131}$Sn.

**Table 3.** Components of ground state wave functions for $^{100}$Sn, $^{106}$Sn and $^{116}$Sn with single particle renormalization.

| Nucleus | $T$(0,0) | $T$(0,1) | $T$(1,0) | $T$(0,2) | $T$(1,1) | $T$(2,0) |
|---|---|---|---|---|---|---|
| $^{100}$Sn | 88.19 | 6.36 | 4.74 | 0.27 | 0.29 | 0.15 |
| $^{106}$Sn | 62.90 | 28.65 | 3.54 | 3.62 | 1.17 | 0.11 |
| $^{116}$Sn | 42.09 | 44.28 | 3.00 | 8.43 | 2.09 | 0.11 |

## 4. Low-lying spectroscopy of a few silicon isotopes

In this section, we report very preliminary results that are analyzed and completed in a forthcoming publication [14]. They concern the low-lying spectroscopy of a few silicon isotopes using the MPMH configuration mixing approach combined with the D1S Gogny interaction. From the mean-field and beyond mean-field analysis, silicon isotopes display strong deformation effects [7, 15]. In Table 4, we show the energies of a few excited states in $^{26}$Si and $^{28}$Si. We compare experimental energies with those calculated from the MPMH configuration mixing and from a 5 dimensional collective Hamiltonian approach (5DCH) [7]. For MPMH configuration mixing results, the mixing has been restricted to the $sd$-shell. All the configurations characterized by a given total spin and a given parity are included in the correlated wave function. The renormalization of single-particle orbitals is not taken into account and is not discussed in this contribution. The equations are solved using an harmonic oscillator basis with eleven major shells. With this size, the excitation energies are converged.

One sees that the excitation energies calculated with the MPMH configuration mixing are in a good agreement with experimental values. For those two isotopes, results are better within the MPMH configuration mixing than within the 5DCH approach. This indicates clearly that correlations more general than the quadrupole ones are important to explain the structure of these nuclei. Moreover, the coupling to intrinsic degrees of freedom that exist in the MPMH configuration mixing is important. The deficiency of the 5DCH approach already noticed in [7] is also exhibited here if one considers the $0_2^+$ states.

Journal of Physics: Conference Series **321** (2011) 012005 doi:10.1088/1742-6596/321/1/012005

**Table 4.** Excitation energies of the lowest states in $^{26}$Si and $^{28}$Si. Experimental energies as well as those deduced from the MPMH configuration mixing (MPMH) and collective Hamiltonian (5DCH) are indicated.

| Nucleus | States | Experiment | MPMH | 5DCH |
|---|---|---|---|---|
| $^{26}$Si | | | | |
| | $2_1^+$ | 1.795 | 1.502 | 2.426 |
| | $2_2^+$ | 2.783 | 2.567 | 5.124 |
| | $0_2^+$ | 3.332 | 3.740 | 8.146 |
| $^{28}$Si | | | | |
| | $2_1^+$ | 1.779 | 1.993 | 2.469 |
| | $4_1^+$ | 4.618 | 5.372 | 6.446 |
| | $0_2^+$ | 4.980 | 4.409 | 10.591 |

## 5. Conclusion and perspectives

In conclusion, we have firstly discussed a realistic description of pairing correlations in various pairing regimes in the context of multiparticle-multihole configuration mixing approach. In this study, the ground states of a few even-even tin isotopes have been investigated. We have demonstrated that correlations beyond the BCS approximation are important. The renormalization of single-particle orbitals by those correlations can produce strong variations in correlated wave functions and hence in calculated observables. A natural extension of this study would be for odd-$A$ isotopes in order to provide the odd-even mass differences that is an observable directly connected with pairing properties in nuclei. Secondly, we have presented a few results concerning the low-lying spectroscopy of silicon isotopes. The encouraging results indicate the need to perform a precise analysis of the structure of correlated wave functions as well as the $T = 0$ residual channel of the Gogny interaction.

## References

[1] P. Ring and P. Schuck, *The Nuclear Many-Body Problem* (Springer-Verlag, New-York, 1980).
[2] M. Bender *et al*, Phys. Rev. C **73**, 034322 (2006).
[3] M. Bender and P.-H. Heenen, Phys. Rev. C **78**, 024309 (2008).
[4] T.R. Rodriguez, J.L. Egido, L.M. Robledo, and T. Rodriguez-Guzman, Phys. Rev. C **71**, 044313 (2005).
[5] T. Rodriguez-Guzman, L. Egido, and L.M. Robledo,, Nucl. Phys. **A709**, 201 (2002).
[6] T.R. Rodriguez and J.L. Egido, Phys. Rev. C **81**, 064323 (2010).
[7] J.-P. Delaroche, M. Girod, J. Libert, H. Goutte, S. Hilaire, S. Péru, N. Pillet, and G.F. Bertsch, Phys. Rev. C **81**, 014303 (2010).
[8] N.Pillet, J.-F. Berger, and E. Caurier, Phys. Rev. C **78**, 024305 (2008).
[9] J. Dechargé and D. Gogny, Phys. Rev. C **21**, 1568 (1980).
[10] J.-F. Berger, M. Girod, and D. Gogny, Comp. Phys. Commun. **63**, 365 (1991).
[11] R.W. Richardson and N. Sherman, Nucl.Phys. **A52**, 221, 253 (1964); R.W. Richardson, Phys. Rev. **141**, 949 (1966).
[12] A. Volya, B.A. Brown and V. Zelevinsky, Phys. Lett. B **509**, 37 (2001).
[13] N. Pillet, N. Sandulescu, Nguyen Van Giai, and J.-F. Berger, Phys. Rev. C **71** (2005) 044306.
[14] N.Pillet, V. Zelevinsky, M. Dupuis, J.-F. Berger, and J.-M. Daugas, to be submitted.
[15] http://wwww-phynu.cea.fr/HFB-5DCH-table.htm

Many-body correlations from dilute to dense nuclear systems (MBC 2011) IOP Publishing
Journal of Physics: Conference Series **321** (2011) 012006 doi:10.1088/1742-6596/321/1/012006

# Brueckner-AMD study of light nuclei

**Kiyoshi Katō[1], Yuhei Yamamoto[2] and Tomoaki Togashi[3]**

[1] Department of Physics, Faculty of Science, Hokkaido University, Sapporo 060-0810, Japan
[2] Division of Physics, Graduate School of Science, Hokkaido University, Sapporo 060-0810, Japan
[3] Graduate School of Advanced Life Science, Hokkaido University, Sapporo, 060-0812, Japan

E-mail: kato@nucl.sci.hokudai.ac.jp

**Abstract.** In many states of light nuclei, the cluster structure is observed in addition to the shell structure. To understand the mechanism of clustering, we apply the Brueckner theory to the Antisymmetrized Molecular Dynamics (AMD) based on realistic nuclear interactions. The Bethe-Goldstone equation in the Brueckner theory is solved for every nucleon pair described by wave packets of AMD, and the $G$-matrix is calculated with single-particle orbits in AMD self-consistently. We show applicability of this method not only to self-conjugate nuclei but also to $N \neq Z$ nuclei with $A \leq 12$. It is confirmed that these results present reasonable description of cluster structures and energy-level schemes in comparison with the experimental ones in light nuclei. For $^8$Be having a typical $\alpha + \alpha$ cluster structure, the structure dependence of the $G$-matrix is investigated and the role of tensor force is shown to be important in understanding the clustering mechanism.

## 1. Introduction

Cluster structure of nuclei has been extensively studied. In light nuclei, many cluster states are observed in low excitation energies. These observations mean that much energy is not needed to divide a nucleus to clusters of smaller nuclear systems. Therefore the clustering mechanism has been studied based on the overall saturation property of the binding energy.

About three decades ago, to study the mechanism of $\alpha$ clustering in $^8$Be and $^{12}$C, Bandō *et al.* [1] applied the Brueckner theory to the $2\alpha$ and $3\alpha$ models, respectively, using the molecular orbits as single-particle states and discussed the $G$-matrix based on the realistic nuclear forces in the cluster model. Their pioneer works to study the cluster structures based on the realistic nuclear interactions have also provided us with the promising method of studying the clustering mechanism. In their study, they adopted the Brink wave function [2] to $^8$Be and $^{12}$C where the $\alpha$-cluster of four nucleons forming the $(0s)^4$ configuration was assumed at a localized center. They also discussed the important contributions of the tensor force through analyses of the triplet-even term of the $G$-matrices in the Brueckner theory.

In the last decade, the antisymmetrized molecular dynamics (AMD) [3, 4] calculations have resulted in understanding and predicting the properties of both stable and unstable nuclei while the phenomenological effective interactions have been used. A great advantage of AMD is that AMD can be used to microscopically describe both shell and cluster structures without any assumption of configurations as results of the energy variation. Recently, we have developed AMD into a framework starting from realistic nuclear forces combining the Brueckner theory with the AMD+Hartree-Fock method [5, 6] where the single-particle orbits in AMD can be defined. We call this framework "Brueckner-AMD".[7, 8, 9, 10, 11, 12]

In section 2 of this presentation, we briefly explain the framework of the Brueckner-AMD, and show its applicability to nuclear structure of several light nuclei based on realistic nuclear interactions in section 3. Using the method, we elucidate the effects of the tensor force on the $\alpha$-$\alpha$ clustering of $^8$Be in section 4.

## 2. Framework of Brueckner-AMD

### 2.1. *Antisymmetrized Molecular Dynamics (AMD)*

For a mass number $A$ nucleons, the Antisymmetrized Molecular Dynamics (AMD) wave function is expressed as follows [3, 4]:

$$|\Phi\rangle = \mathcal{A}\left\{\prod_j^A |\phi_j\rangle\right\} = \mathcal{A}\left\{\prod_j^A |\mathbf{Z}_j\rangle \cdot |\chi_j\rangle\right\}, \tag{1}$$

$$\langle \mathbf{r}|\mathbf{Z}_j\rangle = \left(\frac{2\nu}{\pi}\right)^{3/4} \exp\left[-\nu\left(\mathbf{r} - \frac{\mathbf{Z}_j}{\sqrt{\nu}}\right)^2 + \frac{\mathbf{Z}_j^2}{2}\right]. \tag{2}$$

Here $\mathcal{A}$ is the antisymmetrization operator, and $|\phi_j\rangle = |\mathbf{Z}_j\rangle \cdot |\chi_j\rangle$ is the single-particle wave function consisting of a spatial part $|\mathbf{Z}_j\rangle$ and a spin-isospin part $|\chi_j\rangle$. The spatial part $|\mathbf{Z}_j\rangle$ is defined by a Gaussian form as shown in Eq. (2).

In the Brueckner theory [13], the single-particle orbits and energies are needed to determine the $G$-matrix. Using the AMD+Hartree-Fock method[5, 6], we obtain the single-particle orbits $|\tilde{f}_p >= \sum_j \tilde{g}_{jp}|\phi_j\rangle$ and energies $\epsilon_p$.

### 2.2. *The Single-Particle Energies in the Brueckner-AMD*

We cannot calculate straightforward the expectation value of the realistic interaction with the AMD wave functions because of its strong repulsive core at a short-range distance. We construct the $G$-matrix originated from the realistic nuclear interaction by applying the Brueckner theory. We perform this procedure using the method proposed by Bandō *et al.* [1]. In Brueckner-AMD, we construct the $G$-matrix, in the usual sense, solving the Bethe-Goldstone equation [14, 15]:

$$\hat{G} = \hat{v}_2 + \hat{v}_2 \frac{Q}{\epsilon_1 + \epsilon_2 - (\hat{t}_1 + \hat{t}_2)} \hat{G}. \tag{3}$$

Details of the solving method of Eq.(3) were discussed in our papers [7, 8, 9, 10, 11]. Using the $G$-matrix, the single-particle energy $\epsilon_p$ is given as

$$\epsilon_p = \langle \tilde{f}_p|\hat{t}|\tilde{f}_p\rangle + \sum_q \langle \tilde{f}_p\tilde{f}_q|\hat{G}|\mathcal{A}\{\tilde{f}_p\tilde{f}_q\}\rangle. \tag{4}$$

We calculate the $G$-matrix using this single-particle energy $\epsilon_p$ and orbits $|\tilde{f}_p >$, and again solve the single-particle orbits and energies applying this $G$-matrix to the AMD+Hartree-Fock equation as an effective interaction. This procedure is repeated until the convergence is obtained.

The $G$-matrix can be represented with the correlation function $\hat{F}_{ij}$[8] as

$$\psi_{ij} = \phi_{ij} + \frac{Q}{\epsilon_i + \epsilon_j - (\hat{t}_i + \hat{t}_j)} \hat{G} \cdot \phi_{ij} \equiv \hat{F}_{ij} \cdot \phi_{ij}, \tag{5}$$

and then,

$$\langle \phi_{ij}|\hat{G}|\phi_{ij}\rangle = \langle \phi_{ij}|\hat{v}_2|\psi_{ij}\rangle = \langle \phi_{ij}|\hat{v}_2 \cdot \hat{F}_{ij}|\phi_{ij}\rangle, \tag{6}$$

where $\psi_{ij}$ is the solution of the Bethe-Goldstone equation and $\phi_{ij}$ is the relative part between two wave packets $|\phi_i\rangle$ and $|\phi_j\rangle$ in AMD.

The single-particle potential energy $\varepsilon_p^{\rm pot}$ is presented as

$$\begin{aligned}\varepsilon_p^{\rm pot} &= \sum_q \langle \tilde{f}_p \tilde{f}_q | \hat{G} | \mathcal{A}\{\tilde{f}_p \tilde{f}_q\}\rangle \\ &= \sum_q \sum_{ijkl} \tilde{g}^*_{ip}\, \tilde{g}^*_{jq}\, \tilde{g}_{kp}\, \tilde{g}_{lq}\, \langle \phi_i\phi_j | \hat{v}_2 \cdot \hat{F}^{pq}_{kl} | \mathcal{A}\{\phi_k\phi_l\}\rangle. \end{aligned} \tag{7}$$

When the realistic nuclear interaction $\hat{v}_2$ has the central $\hat{v}_2^C$, the tensor $\hat{v}_2^T$, and the spin-orbital $\hat{v}_2^{LS}$ forces, we decompose the single-particle potential energy into the contributions $\varepsilon_p^C$, $\varepsilon_p^T$, and $\varepsilon_p^{LS}$ from the central, the tensor, and the spin-orbital forces, respectively;

$$\varepsilon_p^{\rm pot} = \varepsilon_p^C + \varepsilon_p^T + \varepsilon_p^{LS}, \tag{8}$$

where each term is expressed as follows:

$$\varepsilon_p^X = \sum_q \sum_{ijkl} \tilde{g}^*_{ip}\, \tilde{g}^*_{jq}\, \tilde{g}_{kp}\, \tilde{g}_{lq}\, \langle \phi_i\phi_j | \hat{v}_2^X \cdot \hat{F}^{pq}_{kl} | \mathcal{A}\{\phi_k\phi_l\}\rangle \quad (X = C, T, LS). \tag{9}$$

*2.3. The Cooling Method with a Constraint*

We determine a configuration of a nucleus solving the following equation of the constraint cooling method [16] which is an energy variation method:

$$\frac{d\mathbf{Z}_i}{dt} = -\frac{\partial H}{\partial \mathbf{Z}_i^*} - \frac{\partial C}{\partial \mathbf{Z}_i^*}, \qquad C = C_0(\langle \mathcal{W}\rangle - W)^2, \tag{10}$$

where $H = \langle\Phi|\hat{H}|\Phi\rangle/\langle\Phi|\Phi\rangle$ is an expectation value of the Hamiltonian $\hat{H} = \sum_i t_i + \sum_{i>j} \hat{v}_2 \cdot \hat{F}_{ij}$, and $C_0$ is the positive constant and the term $W$ is a value of the constraint quantity $\mathcal{W}$. In the case of $^8$Be, we adopt the distance between quasi-clusters as the constraint quantity. It is defined as

$$\langle d^2\rangle = |\mathbf{R}_n - \mathbf{R}_m|^2, \tag{11}$$

where the $\mathbf{R}_n(\mathbf{R}_m) = \frac{1}{A_{n(m)}} \sum_{i\in C_{n(m)}} \frac{ReZ_i}{\sqrt{\nu}}$ represent the center of mass for the quasi-cluster $C_n(C_m)$ consisting of $A_n(A_m)$ nucleons.

In Eq. (10), the derivative $\partial H/\partial Z_i$ cannot be calculated analytically, and then we calculate this derivative as the difference in $\mathbf{Z}_i$. We use a common Gaussian width value for every wave packet; $\nu = 0.195$ fm$^{-2}$ in Eq.(2). We perform the parity projection alone before variation, that is, $H$ in Eq. (10) is replaced by $H = \langle \hat{P}^\pm\Phi|\hat{H}|\hat{P}^\pm\Phi\rangle/\langle\hat{P}^\pm\Phi|\hat{P}^\pm\Phi\rangle$, where $\hat{P}^\pm$ is a parity projection operator. When we calculate a fixed spin state, the spin projection is further performed in addition to the parity projection.

## 3. Binding Energies and Low-Lying Levels of Light Nuclei

We apply the Brueckner-AMD method to various light nuclei with $A \leq 12$. In this report, we present the results of $^4$He, $^7$Li, $^8$Be, $^9$Be, $^{11}$B and $^{12}$C. In these calculations, we use the Argonne $v8'$ (AV8') [17, 18] as the realistic nuclear force and switch off Coulomb interactions.

The binding energies 24.6 MeV and 44.0 MeV are calculated for $^4$He and $^8$Be. This result indicates that the experimental values 28.3 MeV and 56.5 MeV for $^4$He and $^8$Be, respectively, are still large and suggests the necessity of a three-nucleon force. On the other hand, the $^8$Be binding energy from two $\alpha$ threshold is -5.2 MeV. A preliminary calculation for $^8$Be suggests

Journal of Physics: Conference Series **321** (2011) 012006 doi:10.1088/1742-6596/321/1/012006

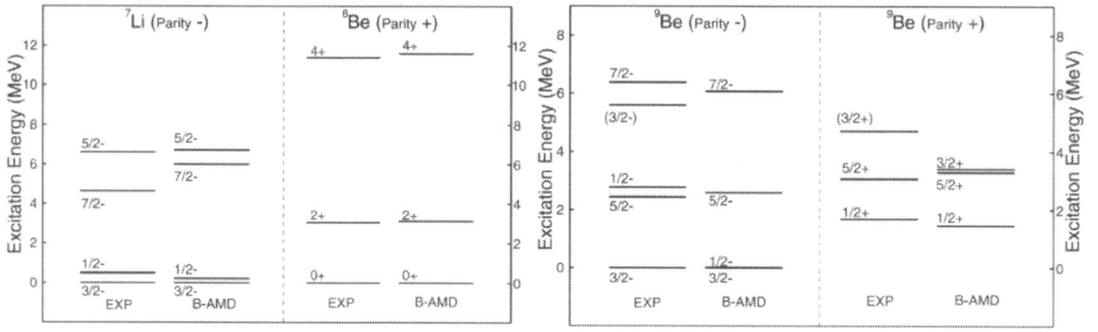


**Figure 1.** Low-lying energy levels of $^{7}$Li, $^{8}$Be and $^{9}$Be. "EXP" and "B-AMD" show the experiments and the Brueckner-AMD calculations, respectively.

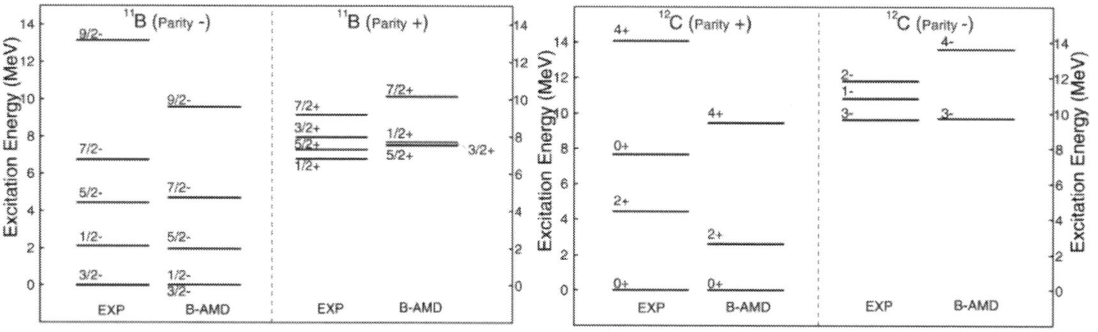


**Figure 2.** Low-lying energy levels of $^{11}$Be and $^{12}$C. "EXP" and "B-AMD" show the experiments and the Brueckner-AMD calculations, respectively.

that about 5 MeV binding energy is added by solving the zero point oscillation of the relative motion between two $\alpha$-clusters. Thus the present calculation reproduces the reasonable $^{8}$Be binding energy measured from the two-$\alpha$ threshold.

The ground and excited energy levels in $p$-shell nuclei for each $J^{\pm}$ state are shown in Figs. $1-2$, where the energies of the lowest-energy states corresponding to the experimental ground states are normalized to zero energy.

From Figs. 1 and 2 of the energy-level schemes, we see that the results of the Brueckner-AMD with the spin and parity projection (B-AMD) well agree with the experimental data (EXP) on the whole. Here, note that the excitation energies of the excited parity states in these calculations are consistent with the experimental data, while their excitation energies in the shell model calculations [19, 20] generally become much higher than the experimental ones. In our results, the $3/2^{-}$ and $1/2^{-}$ states in $^{9}$Be and $^{11}$B, and the $5/2^{+}$ and $3/2^{+}$ states in $^{9}$Be are almost degenerated. This means that the effect of the spin-orbit interaction is not taken into account appropriately in the present calculation.

## 4. Tensor Force Effects on $\alpha$-$\alpha$ Clustering of $^{8}$Be

The calculated $^{8}$Be binding energy of the parity-projected intrinsic state is shown in the left panel of Fig. 3 as a function of the distance between two quasi-$\alpha$ clusters. The energy minimum appears at the large distance $\sim 3.0$ fm, which indicates the realization of the $\alpha$-$\alpha$ structure.

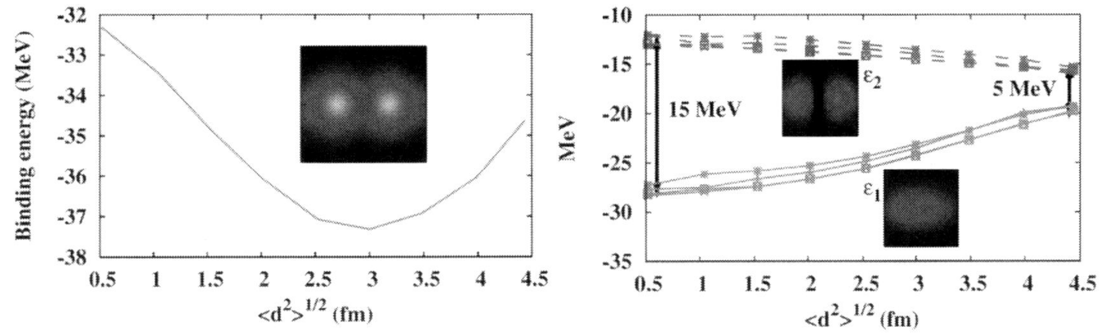


**Figure 3.** Left panel: The energy surface as a function of the distance of quasi-clusters. Inside figure presents the density distribution at $d = 3$ fm. Right panel: Single-particle energies of the eight-nucleon orbits in $^8$Be. Inside figures present single particle densities of $\epsilon_1$ and $\epsilon_2$, respectively, at $d = 3$ fm.

The single-particle energies degenerate into almost two sets of energies $\varepsilon_{1,2}$ as seen in the right panel of Figure 3. These $\varepsilon_{1,2}$ are understood to correspond to $0s$- and $0p$-orbits of a one-center potential at a small distance but to $0s$-orbits of a two-center potential at a large distance.

We investigate the mechanism why the energy minimum appears at the large distance $d \sim 3.0$ fm from the view of the nuclear force.[12] For this purpose, we decompose the single-particle potential energies $\varepsilon_{1,2}^{pot}$ in Eq.(7) into the channel components for such as $^1E$, $^3E$, $^1O$ and $^3O$ channels. Although this $d$ dependence is caused by wave function and $G$-matrix, our interest exists in the $d$ dependence of the effective interaction $G$. To see the $d$ dependence of the $G$-matrix on the single-particle energies, we compare the $d$ dependence of the single-particle energies, $\varepsilon_{1,2}^{3E,T}[G(d)]$, with $\varepsilon_{1,2}^{3E,T}[G(d = 4.5)]$ obtained using the no state-dependent $G$-matrix fixed at $d = 4.5$ fm. In the left panel of Figure 4, we show the single-particle energies of the $^3E$ tensor channel which has the largest $d$ dependence among all channel components included in

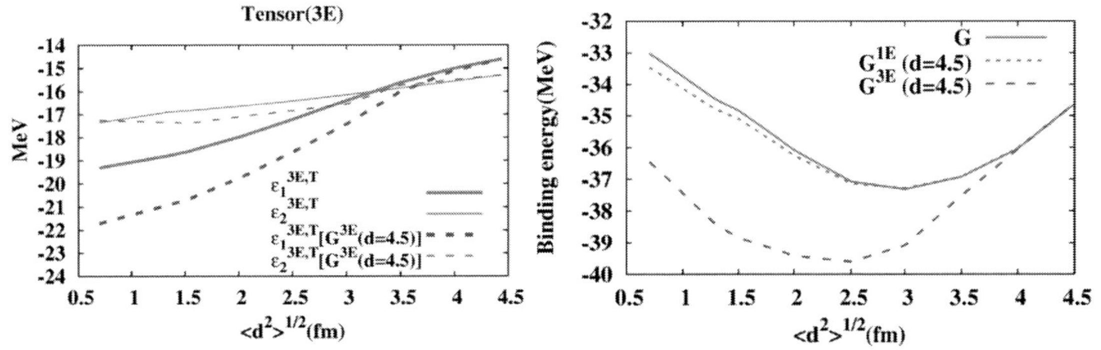


**Figure 4.** Left panel: single-particle energy components $\varepsilon_{1,2}^{3E,T}(G(d))$ of the tensor force in the $^3E$ channel is compared with $\varepsilon_{1,2}^{3E,T}(G(d = 4.5))$ which are calculated using the fixed $G$-matrix for the $^3E$ channel at $d = 4.5$fm. Right panel: The solid line, $G$, is the total energy surfaces same as Figure 3, and $G^{1E}[d = 4.5]$ and $G^{3E}[d = 4.5]$ are the total energy surfaces calculated using the fixed $G$-matrices at $d = 4.5$(fm) for $^1E$ and $^3E$ channel, respectively.

the AV8' force. In the right panel, the total energy surface is also shown in comparison with calculations using the fixed $G$-matrices at $d = 4.5$ fm for $^1E$ or $^3E$ channel.

Comparing $\varepsilon_{1,2}^{3E,T}$ and $\varepsilon_{1,2}^{3E,T}[G(d = 4.5)]$ in the left panel of Fig. 4, $\varepsilon_2$ shows a small difference but $\varepsilon_1$ indicates a large difference at a small distance $d$. This result means that the tensor force causes the strong state-dependence in the $^3E$ channel, and makes the binding of the $\varepsilon_1$ orbit weak. As a result of the weak binding of this single-particle orbit, the total energy surface becomes shallow at the small distance ($d$) region in comparison with the fixed-$G$-matrix energy surface described by $G^{3E}(d = 4.5)$ in the right panel in Figure 4. This is understood as the result of the state-dependent $G$-matrix dominantly coming from the weakened tensor-force contribution. We call this effect of the state-dependence in the $G$-matrix as a "tensor suppression".

## 5. Summary

We showed the framework of the Brueckner-AMD and the procedure of the spin-parity projection. The $G$-matrix elements in a single Slater determinant are calculated for single-particle orbits in AMD. In the projection procedure, the $G$-matrix between the different Slater determinants is defined by the two-body correlation functions derived from the solutions of the Bethe-Goldstone equation. We applied this framework to some light nuclei and confirmed that it can reproduce low excited levels based on the realistic nuclear force AV8'. Furthermore, we investigated the reason why an $\alpha$-$\alpha$ cluster structure is formed in $^8$Be. It is shown that the single-particle energies $\varepsilon_1^{3E,T}$ of the $^3E$ channel become shallow at the small distance of quasi-clusters. This feature of $^3E$ is caused by the tensor force, and such a property of the tensor force is understood as the suppression of the tensor force role.[1]

The Brueckner-AMD can be extended easily to the multi-configuration calculation calculating the $G$-matrix with the correlation functions. We attempt to describe, for example, the second or higher $0^+$ states in $^{12}$C.

## References

[1] Bandō H, Nagata S and Yamamoto Y 1970 *Prog. Theor. Phys.* **44** 646; 1971 *Prog. Theor. Phys.* **45** 1515
[2] Brink D, 1965 *Proc. Int. School of Physics "Enrico Fermi "* Course 36 ed. C. Bloch (Academic Press, New York and London, 1966), p. 247
[3] Kanada-En'yo Y, Horiuchi H and Ono A 1995 *Phys. Rev.* C **52** 628
[4] Kanada-En'yo Y and Horiuchi H 2001 *Prog. Theor. Phys.* Suppl. No. **142** 205
[5] Doté A, Kanada-En'yo Y and Horiuchi H 1997 *Phys. Rev.* C **56** 1844
[6] Doté A and Horiuchi H 2000 *Prog. Theor. Phys.* **103** 91
[7] Togashi T and Katō K 2007 *Prog. Theor. Phys.* **117** 189
[8] Togashi T, Murakami T and Katō K 2009 *Prog. Theor. Phys.* **121** 299
[9] Togashi T, Murakami T and Katō K 2008 *Int. J. Mod. Phys.* E **17** 2081; 2008 J. Phys. Conference Series **111** 012026
[10] Togashi T, Katō K and Murakami T 2009 *Int. J. Mod. Phys.* A **24** 2142
[11] Murakami T, Togashi T and Katō K 2009 *Mod. Phys. Lett.* A **24** 1013
[12] Yamamoto Y, Togashi T and Katō K 2010 *Prog. Theor. Phys.* **124** 315
[13] K. A. Brueckner K A, Levinson C A and Mahmoud H M 1954 *Phy. Rev.* **95** 217
[14] Bethe H A 1956 *Phy. Rev.* **103** 1353
[15] Goldstone J 1957 *Proc. R. Soc.* A **239** 267
[16] Taniguchi Y, Kimura M and Horiuchi H 2004 *Prog. Theor. Phys.* **112** 475
[17] Pudliner B S, Pandharipande V R, Carlson J, Pieper S C and Wiringa R B 1997 *Phys. Rev.* C **56** 1720
[18] Wiringa R B and Pieper S C 2002 *Phys. Rev. Lett.* **89** 182501
[19] Navrátil P, Vary J P and Barrett B R 2000 *Phys. Rev. Lett.* **84** 5728
[20] Forssén C, Navrátil P, Ormand W E and Caurier E 2005 *Phys. Rev.* C **71** 044312

# Microscopic calculation of the $^3$He($\alpha$,$\gamma$)$^7$Be reaction rate using realistic interactions

**Thomas Neff, Hans Feldmeier, and Karlheinz Langanke**
GSI Helmholtzzentrum für Schwerionenforschung GmbH, Planckstraße 1, 64291 Darmstadt, Germany

E-mail: t.neff@gsi.de

**Abstract.** The radiative capture cross sections for the $^3$He($\alpha$,$\gamma$)$^7$Be and the $^3$H($\alpha$,$\gamma$)$^7$Li reactions are calculated in the fully microscopic fermionic molecular dynamics approach. A realistic effective interaction derived in the unitary correlation operator method is employed. The model space consists of frozen cluster configurations at large distances and polarized configurations, obtained by variation of the many-body wave functions, in the interaction region. These polarized configurations are essential for the description of the $^7$Be and $^7$Li bound states and for the $S$- and $D$-wave scattering phase shifts. The calculated capture cross section agrees with both the normalization and the energy dependence of the modern $^3$He($\alpha$,$\gamma$)$^7$Be data. In case of the $^3$H($\alpha$,$\gamma$)$^7$Li reaction the energy dependence of the most recent data is described well but the normalization is larger by about 15% compared to experiment.

## 1. Introduction

The $^3$He($\alpha$,$\gamma$)$^7$Be reaction is one of the key reactions in the solar proton-proton chains [1, 2] and determines the production of the solar $^7$Be and $^8$B neutrinos. In recent years the capture cross section has been remeasured by a number of groups [3, 4, 5, 6, 7]. The new data improve significantly on older results but are still not able to reach the low energies relevant for solar burning.

From the theoretical side this reaction has been studied over the years using simple potential models where $^3$He and $^4$He are treated as point-like particles interacting via an effective nucleus-nucleus potential, e.g., [8] or microscopic cluster models, e.g., [9, 10] where the $^7$Be bound and scattering states are built from microscopic $^3$He and $^4$He clusters interacting via an effective nucleon-nucleon interaction. *Ab-initio* calculations using variational Monte Carlo [11] and no-core shell model wave functions [12] were used to calculate asymptotic normalization coefficients but relied on potential models for the scattering states and the calculation of the capture cross sections. In [13] we presented the first consistent calculation of the capture cross section using a realistic effective interaction.

## 2. Effective Interaction

Starting from the realistic Argonne V18 interaction [14] we derive an effective low-momentum interaction using the unitary correlation operator method (UCOM). The basic idea of the UCOM approach is to explicitly include short-range central and tensor correlations into the many-body state by means of a unitary operator [15, 16, 17]. The correlation operators can also be mapped on the Hamiltonian to define a correlated interaction which not only contains two-body but also higher order contributions. In our applications we only consider the two-body part of the correlated Hamiltonian. This effective two-body interaction gives the same phase shifts as the original interaction but behaves differently in many-body systems. No-core shell model calculations show that the two-body UCOM interaction gives a good description of $s$- and light $p$-shell nuclei [17].

## 3. Many-Body Approach

We employ fermionic molecular dynamics (FMD) as our many-body approach. FMD uses Slater determinants

$$|Q\rangle = \mathcal{A}\left\{|q_1\rangle \otimes \ldots \otimes |q_A\rangle\right\} \tag{1}$$

as intrinsic many-body basis states. The single-particle states are Gaussian wave-packets

$$\langle \vec{x}|q_k\rangle = \exp\left\{-\frac{(\vec{x}-\vec{b}_k)^2}{2a_k}\right\} \otimes |\chi_k^\uparrow, \chi_k^\downarrow\rangle \otimes |\xi_k\rangle\,, \tag{2}$$

where the complex parameters $\vec{b}_k$ encode the mean positions and momenta of the wave-packets and $a_k$ their widths. The spins can assume any direction, isospin is $\pm 1$ denoting a proton or a neutron. The wave-packet basis is very flexible – it contains harmonic oscillator shell model and Brink-type cluster states as special limiting cases. To restore the symmetries of the Hamiltonian the intrinsic basis states are projected on parity, angular momentum and total linear momentum. The total linear momentum projection is necessary to decouple the internal motion of the clusters from their relative motion and the total center-of-mass motion.

For the $^3$He($\alpha$,$\gamma$)$^7$Be reaction we divide the model space into two regions. In the external region bound and scattering states are described by $^3$He and $^4$He clusters in their FMD ground states. These microscopic wave functions can also be expressed with resonating group (RGM) wave functions. This allows us include the boundary conditions for bound and scattering states by matching to Whittaker and Coulomb functions at the channel radius ($a$=12 fm) employing the microscopic $R$-matrix method developed by the Brussels group [18]. In the interaction region we include additional FMD many-body configurations that are obtained by variation after parity and angular momentum projection on spin-parity $1/2^+, 3/2^+, 5/2^+$ and $3/2^-, 1/2^-, 7/2^-, 5/2^-$. With this procedure we generate polarized cluster configurations. Furthermore a constraint on the radius of the intrinsic states is used to vary the distance between the (polarized) clusters. In Fig. 1 density distributions for some selected frozen and polarized configurations are shown.

## 4. Results

Up to energies of about 2.5 MeV only the capture from the $S$- and $D$-wave scattering states into the $3/2^-$ and $1/2^-$ bound states has to be considered. Using only the frozen configurations $^7$Be

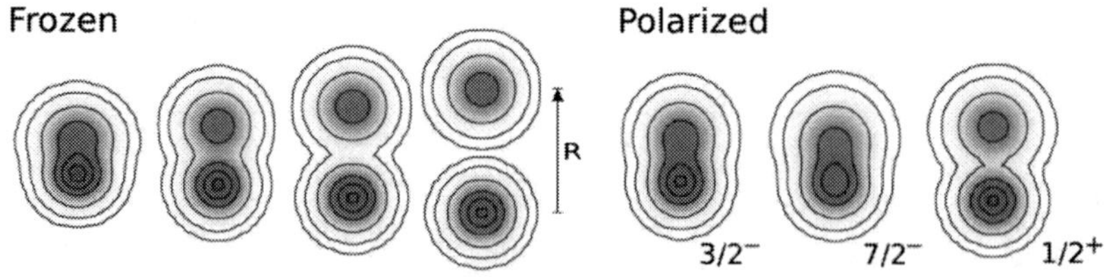


**Figure 1.** Left: Frozen configurations with $^3$He and $^4$He clusters in their ground states. Right: Polarized configurations obtained by variation after parity and angular momentum projection.

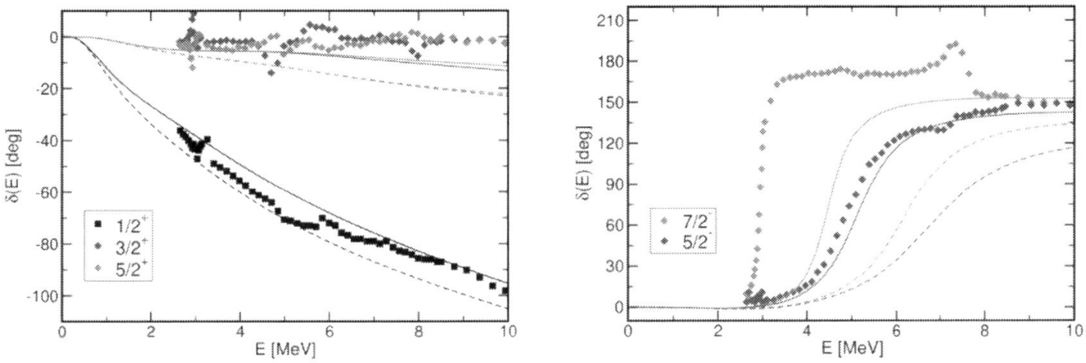


**Figure 2.** $^4$He-$^3$He scattering phase shifts. Dashed lines show results using only frozen configurations, solid lines show results with full FMD model space. Left: $S$- and $D$-wave phase shifts. Right: $F$-wave phase shifts. Experimental results are from [20] and [21].

is only bound by 200 keV. Including the polarized configurations the $3/2^-$ state is bound by 1.49 MeV and the $1/2^-$ state by 1.31 MeV with respect to the cluster threshold. The splitting between the two states is too small compared to the experimental value of 430 keV. This is related to a deficiency of our two-body interaction – additional spin-orbit strength is assumed to be coming from three-body forces. Fortunately the centroid energy agrees nicely with the experimental value. It turns out that the capture cross section depends strongly on the centroid energy but only very weakly on the splitting. Of great importance for the dipole matrix element and therefore for the capture cross section is an accurate description of the tail of the bound state wave functions. The charge radius is an observable that is sensitive to this tail and the calculated value of 2.67 fm agrees well with the experimental value of 2.647(17) fm [19].

In Fig. 2 we compare the calculated phase shifts for the $S$-, $D$- and $F$-waves with the phase shifts extracted from the scattering data [20, 21]. There is a noticeably change in the calculations going from the model space containing only the frozen configurations to the full FMD model space. In case of the $S$- and $D$-wave phase shifts we find a good agreement with the experimental data. For the $F$-waves the full calculation agrees reasonable well for the $5/2^-$ resonance but the $7/2^-$ resonance is about 1.5 MeV too high in energy. This again reflects a missing spin-orbit strength of the effective interaction. For the capture cross section this is only relevant at energies above 3 MeV where $F$-wave capture will contribute.

The total cross section for $^3$He($\alpha$,$\gamma$)$^7$Be capture in form of the astrophysical $S$-factor is shown in the left part of Fig. 3. It agrees very well with the recent experimental data both in absolute normalization and in the energy dependence. Comparing our results for example with the microscopic cluster model calculations by Kajino [10] we find that the biggest differences are found in the $S$-wave contribution to the capture cross section. Our results also deviate from the empirical correlation between radius or quadrupole moment of the $^7$Be ground state with the $S$-factor at zero energy that was found in the microscopic cluster model using different phenomenological interactions.

The $A$-body bound and scattering state wave functions can be analyzed in terms of overlap functions where the full microscopic wave function is projected on the subspace of cluster configurations. If one takes antisymmetrization between the clusters into account properly (by folding with the square-root of the RGM norm kernel) these overlap functions can be interpreted as the relative wave functions of point-like $^3$He and $^4$He clusters. It is interesting that the overlap functions deviate from the Whittaker and Coulomb functions, that describe the asymptotic behavior of the cluster motion for bound and scattering states, up to distances

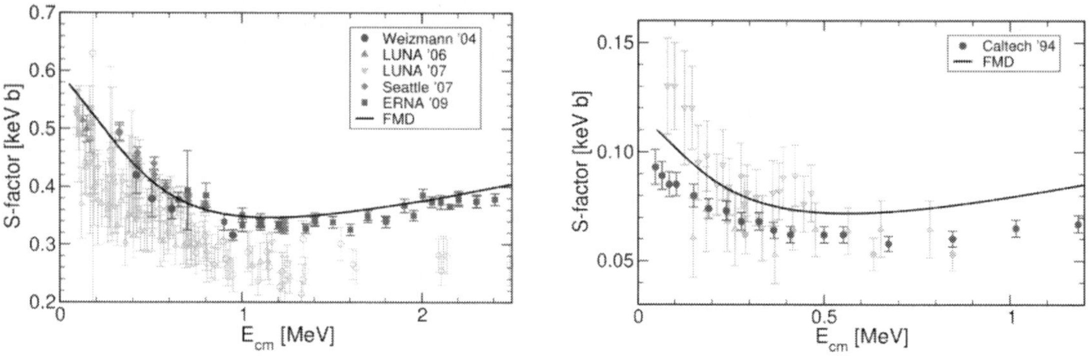


**Figure 3.** Left: $S$-factor for the $^3$He$(\alpha,\gamma)^7$Be reaction. Recent experimental data [3, 4, 5, 6, 7] are shown as colored symbols, older data as gray symbols. Right: $S$-factor for the $^3$H$(\alpha,\gamma)^7$Li reaction. Most recent data [22] is shown as colored symbols, older data as gray symbols.

of about 9 fm. These deviations from the asymptotic behavior are even more pronounced in the dipole matrix elements. The dipole matrix elements calculated from the overlap functions agree with the matrix elements from the microscopic wave functions within 2%. They have sizable contributions already for distances as small as 3 fm and deviate significantly from matrix elements calculated from the asymptotic Whittaker and Coulomb functions up to 10 fm. It is therefore not really justified to treat the reaction as a pure external capture picture.

As our model successfully describes the $^3$He$(\alpha,\gamma)^7$Be reaction we expect that it should also work well for the isospin-mirror reaction $^3$H$(\alpha,\gamma)^7$Li. Indeed we find a good agreement for the ground state properties of $^7$Li and the phase shifts for $^3$H-$^4$He scattering. For the capture cross section shown in the right part of Fig. 3 we observe that the calculated cross section has the same energy dependence as the as the data by Brune *et al.* but the absolute normalization is about 15% larger.

## References

[1] Adelberger E G *et al.* 1998 *Rev. Mod. Phys.* **70** 1265
[2] Adelberger E G *et al.* 2011 *Rev. Mod. Phys.* **83** 195
[3] Nara Singh B S, Hass M, Nir-El Y and Haquin G 2004 *Phys. Rev. Lett.* **93** 262503
[4] Bemmerer D *et al.* 2006 *Phys. Rev. Lett.* **97** 122502
[5] Confortola F *et al.* 2007 *Phys. Rev. C* **75** 065803
[6] Brown T A D *et al.* 2007 *Phys. Rev. C* **76** 055801
[7] Di Leva A *et al.* 2009 *Phys. Rev. Lett.* **102** 232502
[8] Kim B T, Izumoto T and Nagatani K 1981 *Phys. Rev. C* **23** 33
[9] Langanke K 1986 *Nucl. Phys.* **A457** 351
[10] Kajino T 1986 *Nucl. Phys.* **A460** 559
[11] Nollett K M 2001 *Phys. Rev.* **C63** 054002
[12] Navrátil P, Gueorguiev V G, Vary J P, Ormand W E and Nogga A 2007 *Phys. Rev. Lett.* **99** 042501
[13] Neff T 2011 *Phys. Rev. Lett.* **106** 042502
[14] Wiringa R B, Stoks V G J and Schiavilla R 1995 *Phys. Rev. C* **51** 38
[15] Feldmeier H, Neff T, Roth R and Schnack J 1998 *Nucl. Phys.* **A632** 61
[16] Neff T and Feldmeier H 2003 *Nucl. Phys.* **A713** 311
[17] Roth R, Neff T and Feldmeier H 2010 *Prog. Part. Nucl. Phys.* **65** 50
[18] Descouvemont P and Baye D 2010 *Rep. Prog. Phys.* **73** 036301
[19] Nörtershäuser W *et al.* 2009 *Phys. Rev. Lett.* **102** 062503
[20] Boykin W R, Baker S D and Hardy D M 1972 *Nucl. Phys.* **A195** 241
[21] Spiger R J and Tombrello T A 1967 *Phys. Rev.* **163** 964
[22] Brune C R, Kavanagh R W and Rolfs C 1994 *Phys. Rev. C* **50** 2205

# Landscape of the island of inversion studied by AMD

**M. Kimura**

Creative Research Institution Sousei (CRIS), Hokkaido University, N21W10, Sapporo 001-0021, Japan.

E-mail: masaaki@nucl.sci.hokudai.ac.jp

**Abstract.** Deformation and many particle and many hole configurations in the island of inversion are investigated with Antisymmetrized Molecular Dynamics (AMD). The result suggests the coexistence of many particle and many hole configurations with different deformation at small excitation energy. In the case of $^{33}$Mg, we suggest the negative-parity assignment for the ground state. Other particle hole configurations are predicted as the excited states.

## 1. Introduction

The breaking of neutron magic number $N = 20$ that was firstly pointed out from the observation of the anomalous binding energy and spin of $^{31}$Na [1, 2] has been one of the great interests in nuclear physics. Based on the theoretical calculations, neighboring nuclei around $^{31}$Na was named "island of inversion" [3, 4] because they are dominated by the intruder configuration in which neutrons are promoted into $pf$-shell across $N = 20$ shell gap due to strong deformation [5, 6]. Since then, numerous experimental and theoretical studies have been devoted to the island. One of the recent experimental finding is that there are coexistence of "spherical and deformed shape" or "normal and intruder configurations" in many cases [7, 8, 9, 10]. It is expected that the exploration of the coexistence will bring us further understanding of the shell structure in neutron rich $N \sim 20$ region.

The onset of deformation in the island of inversion implies that there should be the coexistence of spherical and deformed shapes and many-particle and many-hole ($mpmh$) configurations associated with these deformations. Therefore the systematic investigation of the spectrum in the island of inversion is of importance to understand the shape coexistence phenomena.

The spectroscopy of odd-neutron nuclei is of importance and interest, because their spin and parity are related to the last neutron particle's or hole's orbit. For example, we have predicted the coexistence of many-particle and many-hole configurations at small excitation energy in $^{31}$Mg ($N = 19$) [11], and some of these excited states are identified by the the proton knock-out reaction from $^{32}$Al [12]. The spin-parity of the excited states are associated with the neutron orbits $[N, n_z, l_z, j_z] = [2, 0, 2, 3/2]$, [3,3,0,1/2], [2,0,0,1/2] and [3,2,1,3/2] in terms of the Nilsson model.

In this talk, we report our study of island of inversion based on AMD. We first demonstrate the coexistence of the $mpmh$ configurations in the island and discuss the spectrum of $N = 21$ system ($^{33}$Mg).

## 2. Framework of AMD

For the detail of AMD framework, readers are directed to the references [13, 14]. The Hamiltnian used in this study is given as,

$$H = \sum_{i}^{A} t_i + \sum_{i<j}^{A} v_N(ij) + \sum_{i<j\in p} v_C(ij) - T_{c.m.}. \tag{1}$$

As an effective nuclear force $v_N$, we have used Gogny D1S force, and the Coulomb force is approximated by the sum of seven Gaussians. $T_{c.m.}$ takes care of the center-of-mass energy. In the AMD framework, the wave function of the center-of-mass motion is separable from the internal wave funciton and the calculation is free from the spurious center-of-mass energy.

We employ the parity projected wave function as the variational wave function,

$$\Phi^{\pm} = \hat{P}^{\pm}\Phi_{\text{int}} = \frac{(1 \pm \hat{P}_x)}{2}\Phi_{\text{int}}, \tag{2}$$

and the intrinsic wave function is represented by a Slater determinant,

$$\Phi_{\text{int}} = \mathcal{A}\{\varphi_1, \varphi_2, ..., \varphi_A\}, \quad \varphi_i(\mathbf{r}) = \phi_i(\mathbf{r})\chi_i\xi_i, \tag{3}$$

where $\varphi_i$ is the $i$th single particle wave packet consisting of the spatial $\phi_i$, spin $\chi_i$ and isospin $\xi_i$ parts. $\phi_i$ is described by a triaxially deformed Gaussian,

$$\phi_i(\mathbf{r}) = \exp\Big\{-\sum_{\sigma=x,y,z} \nu_\sigma \Big(r_\sigma - \frac{Z_{i\sigma}}{\sqrt{\nu_\sigma}}\Big)^2\Big\}, \quad \chi_i = \alpha_i\chi_\uparrow + \beta_i\chi_\downarrow, \quad \xi_i = \text{p or n}. \tag{4}$$

The centroids of wave packets ($\boldsymbol{Z}_i$) and direction of spins ($\alpha_i, \beta_i$) are optimized to minimize the total energy under the constraint on the matter quadrupole deformation $\beta$.

After the variation, we project out the eigenstate of the total angular momentum, $\Phi^{J\pm}_{MK}(\beta) = \hat{P}^J_{MK}\Phi^{\pm}(\beta)$. Finally, the wave functions $\Phi^{J\pm}_{MK}(\beta)$ which have the same parity and angular momentum but have different $K$ and deformation $\beta$ are superposed (GCM). Then the wave function of the system is written as

$$\Phi^{J\pm}_n = c_n\Phi^{J\pm}_{MK}(\beta) + c'_n\Phi^{J\pm}_{MK'}(\beta') + \cdots. \tag{5}$$

The coefficients $c_n$, $c'_n$,... are determined by the Hill-Wheeler equation.

## 3. Onset of the deformation in the island of inversion

Fig. 1 shows the energy curves of Ne, Mg and Al isotopes with $N = 18$, 20 and 22 calculated with AMD. Here Ne and Mg isotopes with $N = 20$ and 22 are in the island and Al isotopes are out of or on the border. In the case of $^{32}$Mg, there are three minima in the positive-parity energy curve and two in the negative-parity. These minima have different single-particle configurations. It is confirmed by the neutron single particle energies as function of quadropole deformation shown in Fig. 2. In the spherical region, the last two neutrons occupy the orbital originates in *sd*-shell, and the system has a $0p0h$ configuration (Hereafter, we denote $mpmh$ configurations with respect to $N = 20$ shell closure). Around $\beta = 0.3$ two neutrons occupy an orbital intruding from *pf*-shell and the system has a $2p2h$ configuration. Further increase of deformation causes the intrusion of another orbital from *pf*-shell ($4p4h$ configuration). Thus, the minima on the positive-parity energy curve are understood to have $0p0h$, $2p2h$ and $4p4h$ configurations in ascending order of deformation. In the same way, the minima on the negative-parity energy curve have $1p1h$ and $3p3h$ configurations.

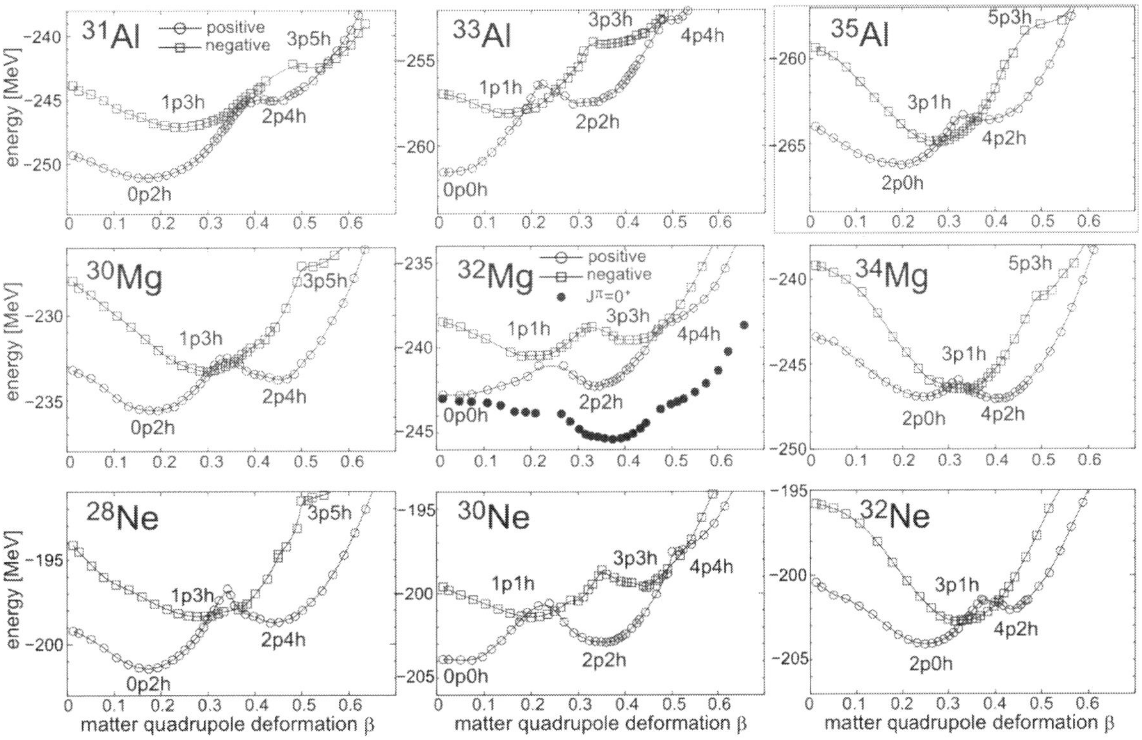


**Figure 1.** Energy curves of Ne, Mg and Al isotopes as function of quadrupole deformation. Open circles (open boxes) show the positive-parity (negative-parity) states before the angular momentum projection for $^{32}$Mg.

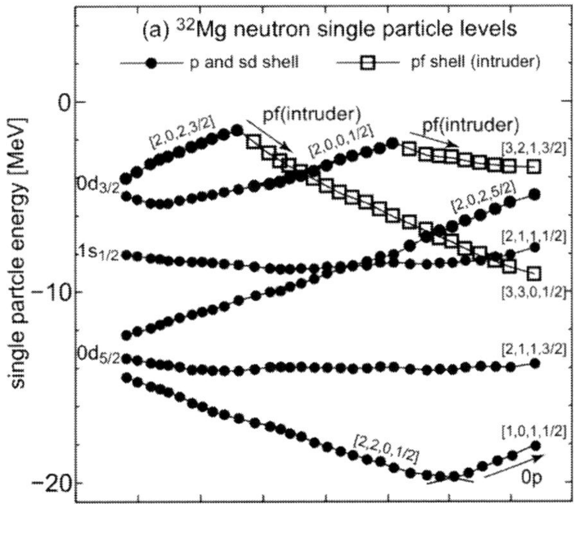


**Figure 2.** Single-particle levels for last 12 neutrons in $^{32}$Mg. Filled circles (open boxes) show the orbitals originates in the $p$- or $sd$-shell (the intruder orbital from $pf$-shell).

These are several striking features in the energy curves of Ne, Mg and Al isotopes, (i) there are several minima on the surface, (ii) number of minima depends on neutron number but not

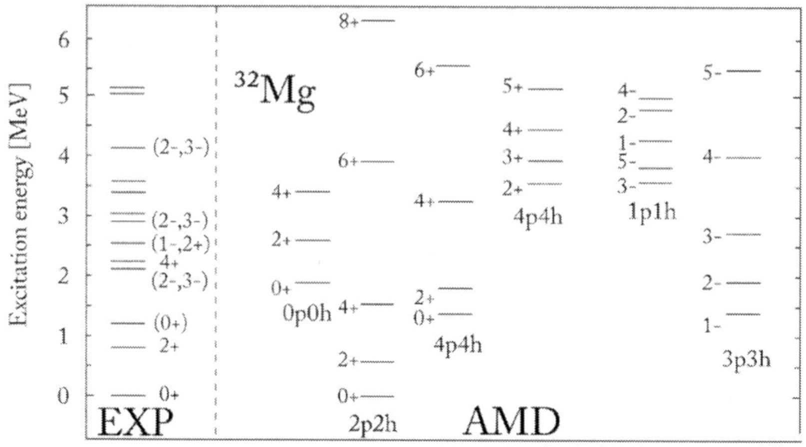

**Figure 3.** Energy spectrum of $^{32}$Mg obtained by GCM. Experimental data is taken from [7]

on proton number, (iii) relative energy between minima strongly depends on proton number. The particle-hole configuration of each minimum shown in Fig.1 tells that up to 4 particle or 4 hole configurations (from $np4h$ to $4pnh$) appear at small excitation energy. It is consistent with the behavior of neutron single-particle orbital in which two orbitals originate in the $pf$-shell and two from $sd$-shell cross as deformation becomes larger. Since the states on the energy curve are dominated by the neutron particle-hole configurations, (i) minima with different neutron configurations appear, and (ii) the single-particle configuration of each minimum depends on the neutron number. As proton number decreases, the binding energy of the spherical (or small deformed) states become smaller. It reduces the relative energy between different particle hole configurations and, in the island of inverision like $^{32}$Mg, strongly deformed configuration becomes the ground state.

Fig. 3 shows the spectrum of $^{32}$Mg after the GCM calculation compared with the observation. Note that AMD calculation predicts the coexistence of many bands with different partilcle-hole configurations as well as the strongly deformed ground band.

## 4. Low-lying level structure of $^{33}$Mg

The system with odd number neutrons may be the best example to show the coexistence of shapes and $mpmh$ configurations. Fig. 4 shows the calculated spectrum and the magnetic moment of $^{33}$Mg, and compares them with observations and the shell model calculation. There are conflicting assignments for the ground state spin-parity based on several experiments. From the observation of $\beta$ decay from $^{33}$Na [15] and to $^{33}$Al [16] the positive-parity assignment is suggested, while the negative-parity is suggested from the measurement of the magnetic moment [17] and $1n$ removal cross section [18].

Looking at the AMD result, we can see that $1p0h$, $2p1h$, $3p2h$ and $4p3h$ configurations coexist within very small excitation energy. The spin parity of these band head states are basically consistent with the last neutron single particle orbitals. It's behavior is quite similar to that of $^{32}$Mg shown in Fig.2. Namely, at small deformation, the last neutron occupies [3,3,0,1/2] orbital ($1p0h$) with $J^{\pi} = 7/2^{-}$. As deformation becomes larger, the last neutron occupies [2,0,2,3/2] ($2p1h$), [3,2,1,3/2] ($3p2h$) and [2,0,0,1/2] ($4p3h$) orbitals that respectively generate $K^{\pi} = 3/2^{+}, 3/2^{-}$ and $1/2^{+}$ bands. Among these bands, only the $J^{\pi}$=3/2$^{-}$ state has the negative-sign of the magnetic moment and consistent with the observation of the magnetic moment. Note that AMD calculation predicts the $mpmh$ configurations with different

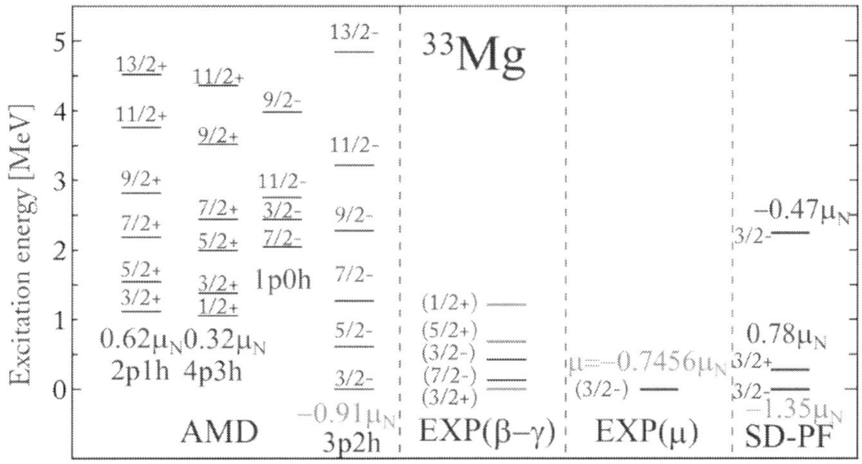


**Figure 4.** Calculated and observed low-lying spectrum and the magnetic moment of $^{33}$Mg. From left to right, it shows AMD, the observation of $\beta$-decay of $^{33}$Na [15], the measurement of magnetic moment [17] and the SD-PF shell model calculation [17].

deformation at small excitation energy and further experimental studies are needed.

## 5. Summary

We have investigated deformation and *mpmh* configurations in the island of inversion based on AMD. The energy curves show the coexistence of *mpmh* configurations with different deformation. In the case of $^{33}$Mg, these *mpmh* configurations coexist at very small excitation energy. AMD calculation supports the negative-parity assignment for the ground state suggested by the observation of the magnetic moment. The calculation also predicts different *mpmh* configurations and further experimental investigations are requested.

## References

[1] C. Thibault *et al.*, Phys. Rev. **C 12**, 644 (1975).
[2] G. Huber *et al.*, Phys. Rev. C **18**, 2342 (1978).
[3] A. Poves and J. Retamosa, PHys. Lett. B **184**, 311 (1986).
[4] E. K. Warburton, J. A. Becker and B. A. Brown, Phys. Rev. C **41**, 1147 (1990).
[5] N. Fukunishi, T. Otsuka and T. Sebe, Phys. Lett. B **296**, 279 (1992).
[6] T. Motobayashi *et al.*, Phys. Lett. B **346**, 9 (1995).
[7] S. Takeuchi *et al.*, Phys. Rev. C **79**, 054319 (2009).
[8] W. Schwerdtfeger *et al.*, Phys. Rev. Lett. **103**, 012501 (2009).
[9] Z. M. Wang *et al.*, Phys. Rev. C **81**, 064301 (2010).
[10] K. Wimmer *et al.*, Phys. Rev. Lett. **105**, 252501 (2010).
[11] M. Kimura, Phys. Rev. C **75**, 041302 (2007).
[12] D. Miller *et al.*, Phys. Rev. C **79**, 054306 (2009).
[13] Y. Kanada-En'yo, H. Horiuchi and A. Ono, PHys. Rev. **C52**, 628 (1995).
[14] M .Kimura, Phys. Rev. **C69**, 044319 (2004).
[15] S. Nummela, *et. al.*, Phys. Rev. **C64**, 054313 (2001).
[16] V. Tripathi, *et. al.*, Phys. Rev. Lett. *101*, 142504 (2008).
[17] D. T. Yordarov, *et. al.*, Phys. Rev. Lett. **99**, 212501 (2007).
[18] R. Kanungo, *et. al.*, Phys. Lett. **B685**, 253 (2010).

Many-body correlations from dilute to dense nuclear systems (MBC 2011) IOP Publishing
Journal of Physics: Conference Series **321** (2011) 012009 doi:10.1088/1742-6596/321/1/012009

# Cluster states in $^{11}$B, $^{11}$C and $^{8}$He and their similarity to $^{12}$C

**Yoshiko Kanada-En'yo, Tadahiro Suhara, and Fumihiko Kobayashi**
Department of Physics, Kyoto University, Kyoto 606-8502, Japan
E-mail: yenyo@ruby.scphys.kyoto-u.ac.jp

**Abstract.** We discuss cluster structures of $^{11}$B($^{11}$C) and $^{8}$He. The $^{11}$B($3/2^-_3$) is found to have a well developed $2\alpha + t$-cluster structure whose features are similar to those of the $^{12}$C($0^+_2$). The ground and excited states of $^{8}$He were also investigated. Dineutron correlations at the surface of $^{8}$He are analyzed. The mechanism of dineutron correlations at a nuclear surface is also discussed.

## 1. Introduction

In light stable nuclei, cluster structures are known for a long time as discussed in $^{12}$C and $^{20}$Ne. Recently, various new types of cluster structures were suggested in stable nuclei and also in unstable nuclei[1].

In C isotopes, three-center cluster structures have been attracting great interests as in $^{12}$C, in which developed $3\alpha$-cluster structures are found in excited states[2, 3, 4, 5, 6]. In 2001, Tohsaki *et al.* proposed a new interpretation of of the $^{12}$C($0^+_2$). According to their idea, this state is regarded as a cluster gas state, where three $\alpha$ particles are weakly interacting[7, 8]. Interestingly, similar dilute gas-like states of clusters were suggested in excited states of $^{11}$B($^{11}$C) and $^{16}$O [7, 9, 10, 11].

The $\alpha$ gas states in $^{12}$C and $^{16}$O are often discussed in relation with the alpha condensation in dilute nuclear matter[12] because of bosonic behavior of $\alpha$ particles in the dilute density. From the similar point of view, dineutron correlations are one of the key issues in physics of unstable nuclei. The nuclear force in the $^1S_0$ channel shows a rather strong attraction though it is not strong enough to form a two-neutron bound state. Therefore, spin-zero neutron-neutron correlations may be found in nuclear many-body systems. We here call the spin-zero ($S = 0$) $nn$ pair with strong spatial correlations "dineutron". In theoretical investigations, it has been suggested that the dineutron correlations may enhance in dilute neutron matter and also at the surface of neutron-rich nuclei[13, 14, 15, 16]. These facts may provide us a possible interpretation of dineutron correlations as virtual dineutrons formed at the nuclear surface.

In this paper, we report the cluster structure of the $3/2^-_3$ state of $^{11}$B($^{11}$C) and discuss its similarity to the $0^+_2$ of $^{12}$C based on theoretical calculations with antisymmetrized molecular dynamics(AMD)[17, 18]. The ground and excited states of $^{8}$He were also investigated with the AMD method[15]. To analyze dineutron correlations, we investigate the dineutron cluster components in the He isotopes. The mechanism of dineutron formation at nuclear surfaces is also discussed by introducing an ideal dineutron wave function for one dineutron moving around an $\alpha$ particle.

## 2. Formulation

We here briefly explain the formulation of the present calculations by means of the AMD method. The details of the AMD method for nuclear structure study are described in Refs. [17, 18].

An AMD wave function is given by a Slater determinant of Gaussian wave packets;

$$\Phi_{\rm AMD}(\mathbf{Z}) = \frac{1}{\sqrt{A!}}\mathcal{A}\{\varphi_1, \varphi_2, ..., \varphi_A\}, \tag{1}$$

where the $i$th single-particle wave function is written by a product of spatial($\phi$), intrinsic spin($\chi$) and isospin($\tau$) wave functions as,

$$\begin{aligned} \varphi_i &= \phi_{\mathbf{X}_i}\chi_i\tau_i, & (2)\\ \phi_{\mathbf{X}_i}(\mathbf{r}_j) &\propto \exp\{-\nu(\mathbf{r}_j - \frac{\mathbf{X}_i}{\sqrt{\nu}})^2\}, & (3)\\ \chi_i &= (\frac{1}{2}+\xi_i)\chi_\uparrow + (\frac{1}{2}-\xi_i)\chi_\downarrow. & (4) \end{aligned}$$

$\phi_{\mathbf{X}_i}$ and $\chi_i$ are spatial and spin functions, and $\tau_i$ is isospin function which is fixed to be a proton or a neutron. Accordingly, an AMD wave function is expressed by a set of variational parameters, $\mathbf{Z} \equiv \{\mathbf{X}_1, \mathbf{X}_2, \cdots, \mathbf{X}_A, \xi_1, \xi_2, \cdots, \xi_A\}$.

In calculations of $^{11}$B($^{11}$C) and $^{12}$C, we performed energy variation after spin parity projection(VAP) within the AMD model space in the same way as done in Ref. [10]. For $^8$He we applied the AMD+GCM, in which we performed superposition of a number of AMD wave functions obtained by energy variation with constraints.

For analysis of dineutron correlations, we introduce dineutron condensate wave functions as follows,

$$\Psi_{\rm cond}(B, b_n) \equiv n_0 \int \prod_{i=1}^{k}\left\{d^3\mathbf{S}_i \exp\left(-\frac{(\mathbf{S}_i - \mathbf{S}_C)^2}{B^2}\right)\right\}\Phi_{\rm Brink}(\mathbf{S}_C, \mathbf{S}_1, \cdots \mathbf{S}_k), \tag{5}$$

where $n_0$ is the normalization factor and $\Phi_{\rm Brink}(\mathbf{S}_C, \mathbf{S}_1, \mathbf{S}_2, \cdots \mathbf{S}_k)$ is the Brink wave function for the $C + k(^2n)$-cluster system consisting the core($C$) and $k$ dineutrons($^2n$) as,

$$\Phi_{\rm Brink}(\mathbf{S}_C, \mathbf{S}_1, \cdots \mathbf{S}_k) \equiv \mathcal{A}\left\{\phi^C(\mathbf{S}_C)\phi^{2n}_{b_n}(\mathbf{S}_1)\cdots\phi^{2n}_{b_n}(\mathbf{S}_k)\right\}. \tag{6}$$

Here, the wave function of the $i$th $^2n$, $\phi^{2n}(\mathbf{S}_i)$, is given by the $(0s)^2$ state with the size $b_n$ localized around $\mathbf{S}_i$.

## 3. The $3/2^-_3$ state of $^{11}$B($^{11}$C)

Details of the AMD calculations for $^{11}$B($^{11}$C) are described in Ref. [10]. We adopt the effective nuclear interactions consisting of the central force, the spin-orbit force and Coulomb force.

The calculated energy levels of negative-parity states in $^{11}$B are in reasonable agreement with the experimental data. The $3/2^-_3$ state at 8.5 MeV is a candidate of the cluster gas-like state. This state has characteristic features different from ordinary states, namely, it has the weak $M1$ and GT transitions and the strong monopole transitions to the lower states. These features of the $3/2^-_3$ state are reproduced well the present calculations.

By analyzing the wave functions of the excited states, we find that $2\alpha + t$ is a three-center cluster state with loosely bound $2\alpha + t$. In $^{11}$B, the root-mean-square radius of the $3/2^-_3$ state is 3.1 fm which is remarkably large compared with that of the ground state(2.5 fm), though it is not as large as that(3.3 fm) of the $^{12}$C($0^+_2$). Moreover, the wave function of the $3/2^-_3$ state can not be expressed by a dominant single Slater determinant but it is described by a linear combination of various spatial configurations of three-cluster positions. It means that the $3/2^-_3$ state is the well-developed $2\alpha + t$ cluster state, where three clusters have no geometric configuration but they are rather freely moving. The cluster features of the $3/2^-_3$ state are quite similar to those of $^{12}$C($0^+_2$, 7.65 MeV), which is known to be a dilute gas-like $3\alpha$ state. Therefore, the $3/2^-_3$ state is considered to be a weakly interacting cluster state in a dilute density, and it is likely to be a cluster gas. Recently, Yamada *et al.* suggested that two $\alpha$ clusters in the $3/2^-_3$ state do not condensate in the $S$ orbit[19]. Further analysis of the present wave functions are required to get a more definite conclusion. We will report more details in a future paper.

## 4. Dineutron correlation in $^8$He

Details of the AMD calculations for $^8$He as well as $^6$He are explained in Ref. [15].

In the calculated results, the $2^+_1$ state is the lowest excited state, and the $0^+_2$ state appears as well as the $1^-_1$ and $3^-_1$ states. An interesting result is that the $0^+_2$ state has a remarkably large neutron radius ($r_{\rm r.m.s.r.}$=3.1 fm) compared with the ground state ($r_{\rm r.m.s.r}$=2.6 fm). The reason is a well developed $^4$He + $2n$ + $2n$ structure. In the wave function of the $0^+_2$ state large amplitudes are found for AMD basis states with various spatial configurations of $^4$He + $2n$ + $2n$. This may indicate a gas-like cluster feature where dineutrons are rather freely moving around the $^4$He core.

In order to see the dineutron correlations in $0^+$ states in more detail, we assume a $^2n$ condensate wave function defined Eq.5 for the $^4\text{He} +^2 n +^2 n$ system and calculated the overlap with the obtained $^8$He($0^+$) wave functions. Here the dineutron size $b_n$ is chosen to be $b = 1/\sqrt{2\nu(^8\text{He})}$. The $^8$He($0^+_2$) has the maximum overlap of about 0.5 with the dineutron condensate wave function $\Psi_{\rm cond}(B)$ at $B = 4 \sim 5$ fm. This means that the dineutron gas-like component is rather large in the calculated $^8$He($0^+_2$).

In contrast to the excited $0^+$ state, the ground state of $^8$He shows the dominant $p_{3/2}$-shell closed neutron structure with mixing of the dineutron correlations. Here, the dineutron correlations in the $^8$He($0^+_1$) do not mean the spatially developed dineutron structure but the dineutron correlations at the nuclear surface which are described by four neutrons in the $p$ shell forming spin-zero $nn$ pairs.

To clarify the mechanism of dineutron correlations at the nuclear surface, we consider a spin-zero pair of two neutrons in $S$-wave channel around a $^4$He core. Let us calculate energy expectation values of the condensate wave function $\Psi_{\rm cond}(B, b_n)$ defined in Eq. 5 for $k = 1$ with a $^4$He core (a toy model for a $^6$He system) as a function of $b_n$ and $\beta = \sqrt{B^2 + b_n^2}$. $b_n$ indicates the dineutron size and $\beta$ means the width for the motion of the dineutron mass center. The dineutron is moving in the width-$\beta$ Gaussian distribution which is equivalent to a size-$b_n$ dineutron in the $S$ orbit around the $^4$He core. A wave function $\Psi_{\rm cond}$ with a fixed $\beta$ shows features of two neutrons virtually confined in a finite region around the core. In the energy curve for $\Psi_{\rm cond}$ with the fixed value $\beta = 3$ fm, we find an energy pocket at $b_n \sim 1.5$ fm as seen in Fig. 1. The energy pocket originates in the energy loss of the kinetic term in $b_n \sim \beta$ where two neutrons become the uncorrelated limit occupying the $S$ orbit around the $^4$He core and Pauli repulsion against the core neutrons increases. In other words, a $nn$ pair favors a compact size to avoid the Pauli blocking effect from the core. This result indicates that, when two neutrons are confined in a core potential, they favor to form a compact-size dineutron at the surface of the core.

## 5. Summary

We discussed cluster structures of $^{11}$B($^{11}$C) and $^8$He. The $^{11}$B($3/2^-_3$) was found to have the well developed $2\alpha + t$-cluster structure which shows cluster features similar to those of the $^{12}$C($0^+_2$). The ground and excited states of $^8$He were also investigated. Dineutron correlations at the

Journal of Physics: Conference Series **321** (2011) 012009 doi:10.1088/1742-6596/321/1/012009

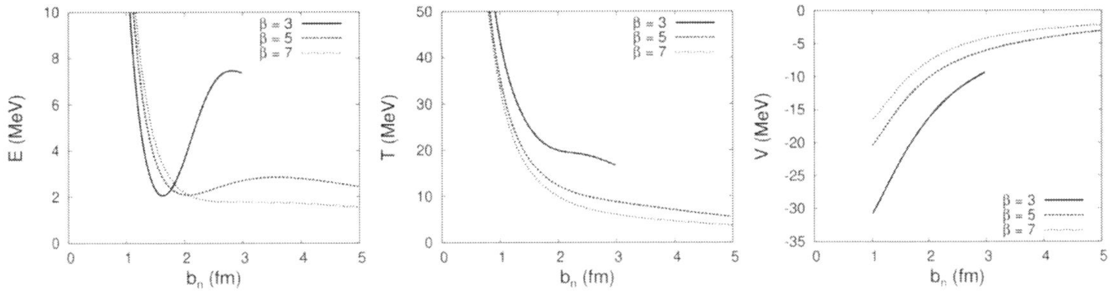


**Figure 1.** Energy expectation values of the condensate wave function $\Psi_{\rm cond}(B, b_n)$ for one dineutron($k = 1$) with a $^4$He core. The total energy(left), the kinetic energy(middle), and the potential energy(right) are plotted as a function of $b_n$ with fixed $\beta$ values, $\beta$=3, 5, and 7 fm. The energies are measured from the $\alpha + n + n$ energy.

surface of $^8$He are analyzed. The mechanism of dineutron correlation at nuclear surfaces is discussed.

## Acknowledgments

The computational calculations of this work were performed by using the supercomputers at YITP and KEK. This work was supported by Grant-in-Aid for Scientific Research from JSPS. It was also supported by the Grant-in-Aid for the Global COE Program "The Next Generation of Physics, Spun from Universality and Emergence" from MEXT of Japan.

## References

[1] von Oertzen W, Freer M and Kanada-En'yo, Y 2006 Phys. Rep. **424** 175
[2] Uegaki E, Okabe S, Abe Y and Tanaka H 1977 Prog. Theor. Phys. **57**, 1262
[3] Itagaki N, Okabe S, Ikeda K and Tanihata I 2001 Phys. Rev. C **64**, 014301
[4] Neff T and Feldmeier H 2004 Nucl. Phys. **A738**, 357
[5] Kanada-En'yo Y 2007 Prog. Theor. Phys. **117**, 655
[6] Suhara T and Kanada-En'yo Y 2010 Phys. Rev. C **82**, 044301
[7] Tohsaki A, Horiuchi H, Schuck P and Röpke G 2001 Phys. Rev. Lett. **87** 192501
[8] Funaki Y, Tohsaki A, Horiuchi H, Schuck P and Röpke G 2003 Phys. Rev. C **67** 051306(R)
[9] Kawabata T *et al.* 2007 Phys. Lett. B **646** 6
[10] Kanada-En'yo Y 2007 Phys. Rev. C **75** 024302
[11] Funaki T, Yamada T, Tohsaki A, Horiuchi H, Röpke G and Schuck P 2010 Phys. Rev. C **82** 024312
[12] Röpke G, Schnell A, Schuck P and Nozieres P 1998 Phys. Rev. Lett. **80**, 3177
[13] Matsuo M, Mizuyama K and Serizawa 2005 Phys. Rev. C **71**, 064326
[14] Pillet N, Sandulescu N and Schuck P 2007 Phys. Rev. C **76**, 024310
[15] Kanada-En'yo Y 2007 Phys. Rev. C **76** 044323
[16] Hagino K, Takahashi N and Sagawa H 2008 Phys. Rev. C **77**, 054317
[17] Kanada-En'yo Y, Horiuchi H and Ono A 1995 *Phys. Rev. C* **52**, 628; Kanada-En'yo Y and Horiuchi H 1995 *Phys. Rev. C* **52**, 647
[18] Kanada-En'yo Y and Horiuchi H 2001 Prog. Theor. Phys. Suppl.**142** 205; Kanada-En'yo Y, Kimura M and Horiuchi H 2003 C. R. Physique **4** 497
[19] Yamada T and Funaki Y 2010 Phys. Rev. C **82**, 064315

Journal of Physics: Conference Series **321** (2011) 012010 doi:10.1088/1742-6596/321/1/012010

# Predictive Power of Nuclear Cluster-Model Study

**M. Kamimura**

Department of Physics, Kyushu University, Fukuoka 812-8581, Japan and
RIKEN Nishina Center, RIKEN, Wako 351-0198, Japan

E-mail: `mkamimura@riken.jp`

**Abstract.** Predictive power of theory needs good models and accurate calculation methods to solve the Shrödinger equations of the systems concerned. In this talk, I present some examples of successful predictions based on the nuclear cluster models of light nuclei and hypernuclei and on the calculation methods (CDCC and GEM) that have been developed by Kyushu group.

## 1. Introduction

Sometimes, theoreticians are requested to make suggestive, helpfull predictions for the sake of experiments that are in a quickly developing stage, or in a preliminary stage. Sometimes, experimentalists enjoy to challenge theoreticians to make prediction before measurement which is to be done soon. Predictive power of theory needs good models and accurate calculation methods to solve the Schrödinger equations of the systems concerned.

In this talk, I shall present some examples of successful predictions by our nuclear cluster-model study of light nuclei and hypernuclei with the use of the calculation methods, CDCC (Continuum-Discretized Coupled Channels[1]) and GEM (Gaussian Expansion Method[2, 3] for few-body bound and scattering states) that have been developed by our Kyushu group.

The method of CDCC and its application to various breakup reactions up to 1986 were extensively summarized in a review volume of Ref. [1]. Nowadays, CDCC is used as one of standard methods to investigate breakup reactions with stable- and unstable-nucleus beams. Theoretical foundation of CDCC was presented in Ref. [4] in relation to the Faddeev method. In numerical calculations, the CDCC results for $^{12}\mathrm{C}(d,d)^{12}\mathrm{C}$ and $^{12}\mathrm{C}(d,pn)^{12}\mathrm{C}$ were precisely compared with that of the Faddeev calculation with a good agreement between them[5].

The nuclear cluster model has been extensively employed to perform a comprehensive study of the structure and reactions of light nuclei. The study in Japan in 1960's and 70's was performed as a long-term research project on "Comprehensive Study of Structure of Light Nuclei—Based on the Viewpoint of the Alpha-Cluster Correlations and Molecular Structure—" that was supported by Yukawa Institute for Fundamental Physics (Kyoto). The study was summarized in three volumes of the Supplements of Theoretical Physics [6, 7, 8].

To the present author's opinion, the most successful example of cluster-model wave function that has been used in many reaction calculations is our $3\alpha$-RGM (Resonating Group Method) wave function of $^{12}\mathrm{C}$ [9, 10]. The $(e,e')$ study of several states in $^{12}\mathrm{C}$, especially of the Hoyle state (the second $0^+$ state at 7.65 MeV) will be discussed in §2. The calculated form factor (predicted in 1978) of the inelastic electron scattering from the Hoyle state was perfectly reproduced by an experiment [11] in 2006.

Using the wave functions of $^{12}$C, I shall present in §2, that the Kyushu group's predictions on the differential cross sections (in *absolute* value) of $^6$Li scattering from $^{12}C(0_1^+, 2_1^+, 3_1^-, 0_2^+)$ were afterwards verified by experiments; and similarly for scattering of the unstable-nucleus $^7$Be from $^{12}$C with the use of CDCC for the projectile breakup channels.

Details of GEM are reviewed in Ref. [3] together with its applications to bound and scattering states of various few-body systems. An example of the most accurate 3-body GEM calculation is the determination of mass of antiproton as seen in the Particle Listings 2000 by the PDG; this is presented in §3. The cluster model has extensively been applied to the 4- and 5-body study of light hypernuclei using GEM. Some of successful examples of theoretical predictions on the structure of light hypernuclei will be shown in §4.

## 2. $3\alpha$ RGM wave functions of $^{12}$C and application to reactions

The $3\alpha$ RGM wave function is given by

$$\Psi_{JM}(^{12}\mathrm{C}) = \mathcal{A}\Big[\Phi^{(\alpha_1)}\Phi^{(\alpha_2)}\Phi^{(\alpha_3)}\psi_{JM}(\mathbf{r},\mathbf{R})\Big]. \tag{1}$$

Here $\mathcal{A}$ is the total antisymmetrization operator, $\Phi^{(\alpha)}$ is the intrinsic wave function of the $\alpha$ cluster with the $(0s)^4$ configuration and $\psi_{JM}(\mathbf{r},\mathbf{R})$ describes the relative motion among three $\alpha$'s. How to determine $\psi_{JM}(\mathbf{r},\mathbf{R})$ is described in Refs.[9, 10]. The wave functions should first be severely examined by the $(e,e')$ experiments. The calculated form factors of the elastic scattering and transitions to the excited states with $J^\pi = 2_1^+, 4_1^+, 1_1^-, 3_1^-$ and $0_2^+$ are compared with the observed ones in Fig. 1. In all the cases, agreement is good between our calculation (solid lines) and the experimental data. Here, it is to be emphasized that, in the case of the second $0^+$ state (the Hoyle state), our calculation was a prediction (1978) [9] before the measurement (1979). We see some differences in the second peak region. But, in 2006, the experimental data were much improved as shown in Fig. 2. I am glad to see that the latest experiment reproduced perfectly our prediction some 30 years ago (Recently, Funaki *et al.*[12] gave almost the same result as ours taking the alpha-condasation idea.)

We performed $3\alpha$ RGM calculation for scattering states ($J^\pi = 0^+, 2^+$)[13] between $\alpha$ and $^8$Be allowing the vertual excitation of $^8$Be that is essentially important in this scattering process. We reproduced the position and width of the Hoyle-state resonance and predicted a broad $2^+$ resonance at $E_\mathrm{x} \sim 9.1$ MeV with $\Gamma \sim 1$ MeV. The prediction was supported by a recent experiment[14] giving $E_\mathrm{x}(2^+) = 9.9 \pm 0.3$ MeV with $\Gamma = 1.0 \pm 0.3$ MeV.

Since the $3\alpha$ RGM wave functions of $^{12}$C are so good, it happened that, in 1983, K. Katori *et al.* at RCNP (Osaka) challenged our Kyushu group as follows: They were going to measure the differential cross sections $\sigma(\theta)$ of the elastic and inelastic scattering $^{12}\mathrm{C}(^6\mathrm{Li},^6\mathrm{Li})^{12}\mathrm{C}(0_1^+, 2_1^+, 0_2^+, 3_1^-)$ at $E(^6\mathrm{Li}) = 170$ MeV. They asked us to predict absolute values of $\sigma(\theta)$ before measurement which was planned to perform on 22 September, 1983. Our colleague Sakuragi accepted their challenge. On 16 September, he handed the experimentalists his prediction [15] of $\sigma(\theta)$ (solid lines in Fig. 2), in absolute values, which were calculated with the 4-state $(0_1^+, 2_1^+, 0_2^+, 3_1^-)$ CC method. All the CC potentials were calculated by folding the M3Y NN force into the wave functions of $^{12}$C and $^6$Li and then multiplying them by a complex factor $(N_\mathrm{R} + iN_\mathrm{I})$ with $N_\mathrm{R} = 0.6$ and $N_\mathrm{I} = 0.4$ (see Chapter VI of Ref. [1] for $N_\mathrm{R}$ and $N_\mathrm{I}$).

The observed differential cross sections [16] are shown in Fig. 2. At forward angles ($\lesssim 20°$), the experiment reproduced very well the theoretical prediction except for the $3_1^-$- excitation. It is to be emphasized that, the excitation of the second $0^+$ state, which was considered to be difficult in calculation, is successfully predicted by the $3\alpha$ cluster model. An extensive analysis of the reaction mechanism was performed in Ref.[17] with the use of CDCC which further includes the $\alpha + d$ $d$-state resonant breakup channel of $^6$Li and $3\alpha$ breakup channel of $^{12}$C.

Journal of Physics: Conference Series **321** (2011) 012010 doi:10.1088/1742-6596/321/1/012010

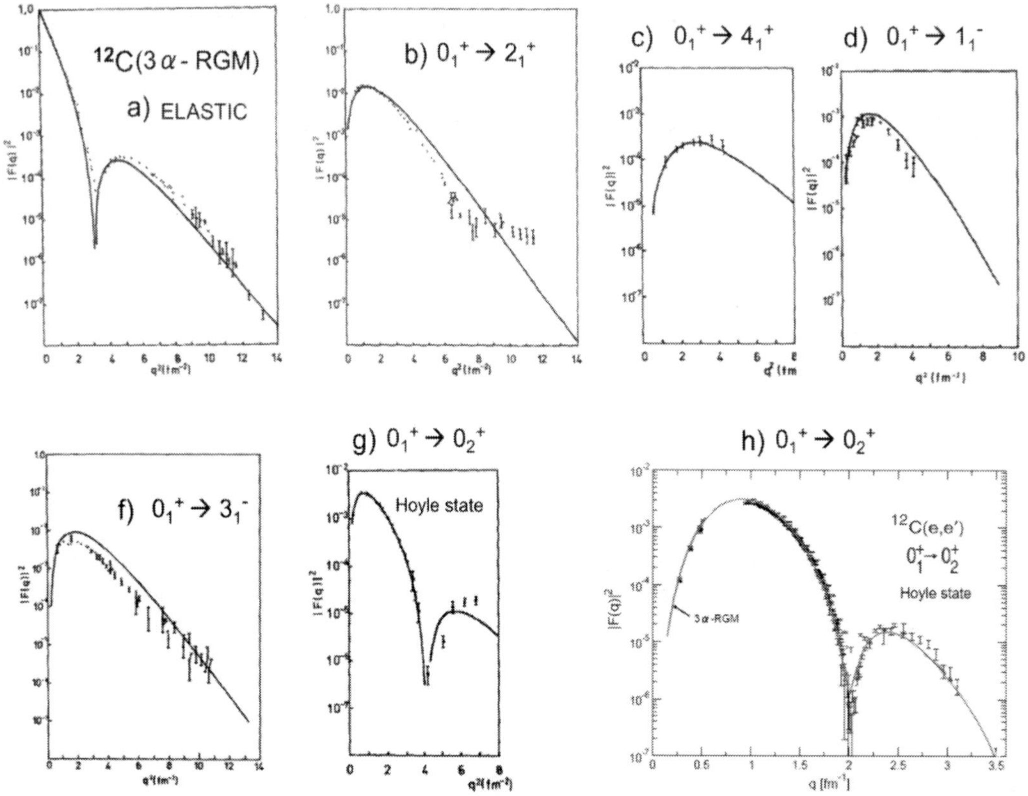


**Figure 1.** a)-g) Calculated and observed form factors of the elastic and inelastic electron scattering $^{12}\mathrm{C}(e, e')^{12}\mathrm{C}(0_1^+, 2_1^+, 4_1^+, 1_1^-, 3_1^-, 0_2^+)$. The figures are taken from [10]. h) The latest $(e, e')$ experimental data [11] for the second $0^+$ state (the Hoyle state) reproduced perfectly our $3\alpha$-RGM prediction [9, 10] some 30 years ago.

Another example of the predictive power of the cluster-model study is as follows: In 1988 at RCNP, T. Yamagata *et al.* performed the first experiment using unstable-nucleus beams in Japan, namely, the scattering $^{12}\mathrm{C}(^7\mathrm{Be},^7\mathrm{Be})^{12}\mathrm{C}$ at 140 MeV. The secondary beams of $^7$Be were produced by the reaction $^1\mathrm{H}(^7\mathrm{Li},\ ^7\mathrm{Be})n$. Since they had no experience in scattering of unstable nucleus, they requested Sakuragi to predict absolute values of $\sigma(\theta)$ which would be a good guide to set up the detectors.

The calculated differential cross section predicted before measurement is shown in Fig. 3 by the solid line [18, 19] which is a sum of the elastic $(3/2_1^-)$ and inelastic ($1/2_1^-$ at 0.48 MeV) contributions since these two were not separated in the experiment. The CDCC calculation takes the $\alpha$-$^3$He cluster-model wave functions of the bound states ($3/2_1^-$ and $1/2_1^-$) and the resonant states ($7/2^-$ and $5/2^-$) together with the non-resonant discretized continuum states. The experimental data [19] reproduced well the theoretical prediction.

Interestingly, during 1983-1988, Sakuragi accepted such types of challenge or request from experimentalists totally six times. His predictions were all successful, no failure. In that period, projectile-breakup experiments using light- and heavy-ion beams were in a quickly developing stage. I am proud of that Sakuragi was providing suggestive, helpful predictions to those experiments in such an important stage.

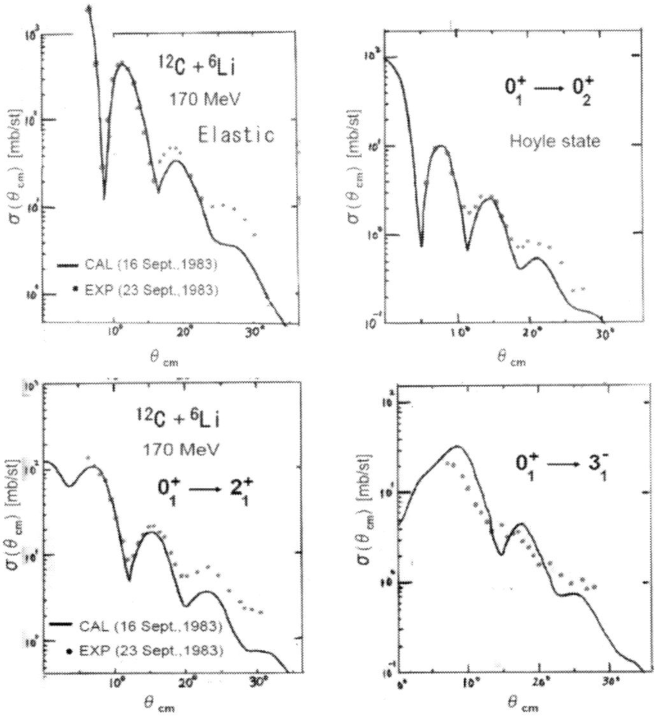


**Figure 2.** Calculated differential cross sections [15] predicted before measurement of $^{12}$C($^{6}$Li, $^{6}$Li)$^{12}$C($0_1^+, 2_1^+, 0_2^+, 3_1^-$) at 170 MeV. The final results of the theoretical and experimental analyses were published in Refs. [17] and [16], respectively.

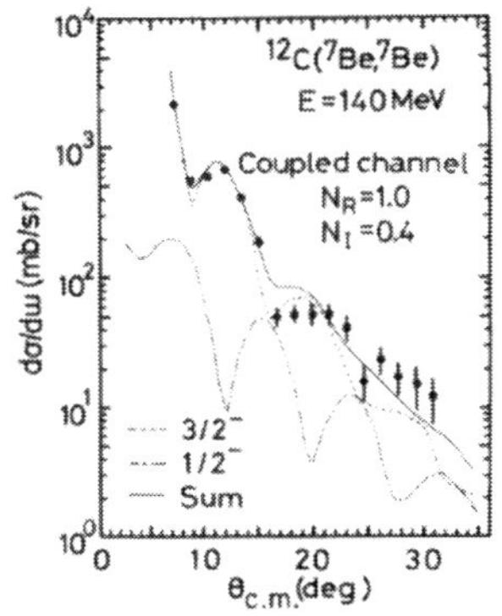


**Figure 3.** Calculated differential cross section (solid line) [18] predicted before measurement of $^{12}$C($^{7}$Be, $^{7}$Be) $^{12}$C at 140 MeV. See text for the calculation. The experimental data [19] reproduced well the theoretical prediction. This figure is taken from Ref.[19].

## 3. Determination of mass of antiproton

In order to demonstrate the accuracy of GEM, which precisely was reviewed in Ref.[3], I here present the best example, namely 3-body calculation to determine mass of anti-proton, $m_{\bar{p}}$.

Particle Listings 2000 by PDG gave the first recommended information on the antiproton mass that $|m_{\bar{p}} - m_p|/m_p < \times 10^{-7}$. This value was derived by spectroscopic study of highly-excited metastable states in antiprotonic helium atom composed of $\alpha + \bar{p} + e^-$. The sub-ppm laser spectroscopic data at CERN were analyzed by our colleague Kino making a 3-body calculation with the accuaracy of 10 significant figures. The laser experimemt for the transition between the metastable states with $(J, v) = (34, 2)$ and $(J, v) = (33, 2)$ gave the wave length $\lambda_{\rm EXP} = 470.7220(6)$ nm, which was precisely reproduced by Kino's calculation $\lambda_{\rm CAL} = 470.7220$ nm, assuming $m_{\bar{p}} = m_p$. Then Kino showed that the upper and lower bounds of $\lambda_{\rm EXP}$ were reproduced by assuming $m_{\bar{p}} = (1 \mp \varepsilon)m_p$ with $\varepsilon = 5 \times 10^{-7}$. He then considered that even if $m_{\bar{p}}$ is changed, calculated wavelength should be within the experimental error. Then, we obtain $(1-\varepsilon)m_p < m_{\bar{p}} < (1+\varepsilon)m_p$, namely, $|m_{\bar{p}} - m_p|/m_p < \varepsilon = 5 \times 10^{-7}$, which was cited in Particle Listings 2000. Very recently, this bound becomes much smaller, $2 \times 10^{-9}$, in Particle Listings 2010. This can be a test of $CPT$ invariance. I am happy to see that GEM is so accurate to determine, for the first time, mass of the elementary particle, antiproton.

A lot of transitions between excited states of antiprotonic helium atoms have been observed at the CERN. But, cost of the precise sub-ppm laser-scan search of the transition energy $\Delta E$ is very expensive. Kino was then requested to predict $\Delta E$ before measurement. A typical example[33] of the transition frequency by the Kino's prediction ($\nu_{\rm CAL}$) and the observed value

($\nu_{\rm EXP}$) is that, $\nu_{\rm CAL} = 1012445.559$ MHz versus $\nu_{\rm EXP} = 1012445.52(15)$ MHz for the transition $(32, 0) \to (31, 0)$, and $\nu_{\rm CAL} = 804633.127$ MHz versus $\nu_{\rm EXP} = 804633.11(10)$ MHz for the transition $(33, 1) \to (32, 1)$. We see how accurate is the theoretical prediction.

They say the laser spectroscopy of metastable antiprotonic helium atoms is pioneering work toward anti-matter science. I am proud of that our colleague Kino was providing suggestive, helpful predictions for anti-matter science in a preliminary stage.

## 4. Application of cluster model to hypernuclei

Many of interesting light nuclei have two- or three-cluster structure. Therefore, if we inject one or two hyperons into them, we have interesting light hypernuclei with 3-, 4-, or 5-cluster structure. Thus, the light hypernucleus is a good place to apply our cluster model and few-body calculation method and make predictive suggestions to the hypernuclear experiment that is in a quickly developing stage. Strategy for studying hypernuclear physics is presented in Ref. [20].

Along this line, Hiyama has developed GEM so that it is applicable to precise 4- and 5-body calculations [3]. She has studied a lot of hypernuclei (reviewed in Refs. [21, 22]) such as i) single-$\Lambda$ hypernuclei (${}^{4}_{\Lambda}$H, ${}^{4}_{\Lambda}$He, ${}^{6}_{\Lambda}$He, ${}^{6}_{\Lambda}$Li, ${}^{7}_{\Lambda}$He, ${}^{7}_{\Lambda}$Li, ${}^{7}_{\Lambda}$Be, ${}^{8}_{\Lambda}$Li, ${}^{8}_{\Lambda}$Be, ${}^{9}_{\Lambda}$Be, ${}^{10}_{\Lambda}$Be, ${}^{10}_{\Lambda}$B, ${}^{13}_{\Lambda}$C), ii) double-$\Lambda$ hypernuclei (${}^{7}_{\Lambda\Lambda}$He, ${}^{7}_{\Lambda\Lambda}$Li, ${}^{8}_{\Lambda\Lambda}$Li, ${}^{9}_{\Lambda\Lambda}$Li, ${}^{9}_{\Lambda\Lambda}$Be, ${}^{10}_{\Lambda\Lambda}$Be, ${}^{11}_{\Lambda\Lambda}$Be) and iii) single-$\Xi^-$ hypernuclei (${}^{5}_{\Xi^-}$H, ${}^{7}_{\Xi^-}$H, ${}^{9}_{\Xi^-}$Li, ${}^{10}_{\Xi^-}$Li, ${}^{12}_{\Xi^-}$Be). Among those hypernuclei, I briefly introduce three types of her predictions compared with the experiment.

a) **Shrinkage of hypernuclei**

One of the interesting issues in hypernuclear physics is to investigate dynamical change of nuclear structure induced by a glue-like role of an added $\Lambda$ hyperon. In 1999, Hiyama *et al.*[23] proposed to measure specifically the $B$(E2) value of the transition $5/2^+_1 \to 1/2^+_1$ in ${}^{7}_{\Lambda}$Li and suggested how to derive the size of the ground state of ${}^{7}_{\Lambda}$Li from that $B(E2)$ value. Taking ${}^{5}_{\Lambda}{\rm He} + N + N$ three-body model, they obtained $B(E2; 5/2^+_1 \to 1/2^+_1) = 2.42e^2{\rm fm}^4$ and therefore predicted that the size is shrinked by 25% due to the addition of $\Lambda$ particle. Later, in Ref.[24](2001), with the use of more precise $\alpha + \Lambda + n + p$ four-body model, they recalculated the $B$(E2) value and obtained $2.85e^2{\rm fm}^4$ which means that the shrinkage is 22%.

The first observation of the hypernuclear $B$(E2) strength was made in the KEK-E419 experiment for $B(E2; 5/2^+_1 \to 1/2^+_1)$ in ${}^{7}_{\Lambda}$Li. The strength was reported in 2001 to be $3.6 \pm 0.5^{+0.5}_{-0.4} e^2{\rm fm}^4$ [25]. From this, the shrinkage was estimated to be $19 \pm 4\%$, which was consistent with our prediction. This confirmed, for the first time, the shrinkage of the nuclear size induced by the $\Lambda$ particle. Our prediction about the shrinkage of ${}^{12}$C was given in [26, 27].

b) **Spin-orbit splitting in hypernuclei**

The strong $NN$ spin-orbit force is one of the essential ingredients of nuclear physics. In hypernuclei, spin-orbit splitting energy was first precisely calculated in 2000 by Hiyama *et al.* [27] for the $5/2^+_1$-$3/2^+_1$ doublet states in ${}^{9}_{\Lambda}{\rm Be}(= 2\alpha + \Lambda)$ and the $3/2^-_1$-$1/2^-_1$ states in ${}^{13}_{\Lambda}{\rm C}(= 3\alpha + \Lambda)$; a very small splitting was predicted quite differently from the splitting in the ordinary nucleus when the quark-model based two-body $\Lambda N$ spin-orbit force was taken. The calculated splitting was $\Delta E_{\rm CAL}(5/2^+_1$-$3/2^+_1) = 35 - 40$ keV in ${}^{9}_{\Lambda}$Be and $\Delta E_{\rm CAL}(3/2^-_1$-$1/2^-_1) = 150 - 200$ keV in ${}^{13}_{\Lambda}$C.

Afterwards, experimental values were reported as $\Delta E_{\rm EXP}(5/2^+_1 - 3/2^+_1) = 31.4^{+2.5}_{-3.6}$ keV in ${}^{9}_{\Lambda}$Be by BNL-E930 [28] in 2002 and $\Delta E_{\rm CAL}(3/2^-_1 - 1/2^-_1) = 150 \pm 54 \pm 36$ keV in ${}^{13}_{\Lambda}$C by BNL-E929 [29] in 2001, which is consistent with our prediction. The very weak spin-orbit component of the $\Lambda N$ interaction compared with that of the $NN$ interaction was confirmed.

c) **Double $\Lambda$ hypernuclei**

The double strangeness ($S = -2$) nucleus is an entry to multi-strangeness hadronic systems (including the core of a neutron star). Since the present experimental information on the $S = -2$ system is very limited, study of the structure of double $\Lambda$ hypernuclei is absolutely important.

In 2002 by Hiyama *et al.* [30], energy levels of ${}^{7}_{\Lambda\Lambda}$He, ${}^{7}_{\Lambda\Lambda}$Li, ${}^{8}_{\Lambda\Lambda}$Li, ${}^{9}_{\Lambda\Lambda}$Li, ${}^{9}_{\Lambda\Lambda}$Be, and ${}^{10}_{\Lambda\Lambda}$Be were predicted with the framework of $\alpha x \Lambda\Lambda$ four-body model, where $x = n, p, d, t$, ${}^{3}$He, and $\alpha$, respectively. Also, energy levels of the double $\Lambda$ hypernucleus, ${}^{11}_{\Lambda\Lambda}$Be were calculated within the framework of a $\alpha\alpha n\Lambda\Lambda$ five-body model in Ref. [31] in 2010.

As for ${}^{10}_{\Lambda\Lambda}$Be, we predicted $B^{\rm cal}_{\Lambda\Lambda}(2^+_1) = 11.88$ MeV and $B^{\rm cal}_{\Lambda\Lambda}(0^+_1) = 14.74$ MeV. The Demachi-Yanagi event, identified as ${}^{10}_{\Lambda\Lambda}$Be and $B^{\rm exp}_{\Lambda\Lambda} = 11.90 \pm 0.13$ MeV [32], was interpreted as an observation of its $2^+$ excited state. The Hida event, observed in KEK-E373 experiment as $B^{\rm exp}_{\Lambda\Lambda} = 20.83 \pm 1.27$ MeV for ${}^{11}_{\Lambda\Lambda}$Be [32], was interpreted as an observation of the ground state of ${}^{11}_{\Lambda\Lambda}$Be since our calculation gave $B^{\rm cal}_{\Lambda\Lambda}(3/2^-_1, {\rm g.s.}) = 18.23$ MeV and $B^{\rm cal}_{\Lambda\Lambda}(5/2^-_1) = 15.48$ MeV (the experimental error is large in this case; further experiment at J-PARC is waited for).

Since the emulsion experiment itself cannot specify the spin-parity of the observed state and cannot tell whether the state is the ground state or an excited state, theoretical calculation is quite helpful in determining the property of the observed state.

## 5. Concluding remarks

Predictive power of theory requires good models and accurate calculation methods. It is desirable for theoreticians to make predictive suggestions to experiments in quickly developing fields. Here, we have shown some examples of the efforts, along this line, done by our Kyushu group taking the cluster model and calculation methods of CDCC and GEM.

## References

[1] M. Kamimura *et al.*, Prog. Theor. Phys. Suppl. No.**89**, 1 (1986).
[2] M. Kamimura, Phys. Rev. **A38**, 621 (1988).
[3] E. Hiyama, Y. Kino, and M. Kamimura, Prog. Part. Nucl. Phys. **51** (2003) 223.
[4] N. Austern, Y. Yahiro and M. Kawai, Phys. Rev. Lett. **63**, 2649(2007); Phys. Rev. C **53** (1996), 394.
[5] A. Deltuva *et al.*, Phys. Rev. C **76**, 064602(2007).
[6] K. Ikeda *et al.*, Prog. Theor. Phys. Suppl. No.**52** (1971) 1.
[7] S. Saito *et al.*, Prog. Theor. Phys. Suppl. No.**62** (1977) 1.
[8] K. Ikeda *et al.*, Prog. Theor. Phys. Suppl. No.**68** (1980) 1.
[9] Y. Fukushima and M. Kamimura, J. Phys. Soc. Japan **57** (1978) 706.
[10] M. Kamimura, Nucl. Phys. A **351** (1981) 456.
[11] A. Lenhardt *et al.*, Nucl. Instrum. Method Phys. Res., Sect. A **562** (2006) 320.
[12] Y. Funaki *et al.*, Eur. Phys. J. A. **28** (2006) 259.
[13] M. Kamimura and Y. Fukushima, Proceedings of INS Intererernational Symposium on Nuclear Direct reaction Mechanism, Fukuoka, 1978, p.409.
[14] M. Itoh *et al.*, Nucl. Phys. A **738** (2004) 268.
[15] Y. Sakuragi, private communication (1983).
[16] K. Katori *et al.*, Nucl. Phys. A**480** (1988) 323.
[17] Y. Sakuragi, M. Kamimura and K. Katori, Phys. Lett. B **205** (1988) 204.
[18] Y. Sakuragi, private communication (1988).
[19] Y. Yamagata, Y. Sakuragi, *et al.*, Phys. Rev. C **39** (1989) 873.
[20] E. Hiyama, *et al.*, Prog. Theor. Phys. Suppl. No.**185** (2010), 1.
[21] E. Hiyama, *et al.*, Prog. Theor. Phys. Suppl. No.**185** (2010), 106.
[22] E. Hiyama, *et al.*, Prog. Theor. Phys. Suppl. No.**185** (2010), 152.
[23] E. Hiyama, *et al.*, Phys. Rev. C **59**, 2351 (1999).
[24] E. Hiyama, *et al.*, Nucl. Phys. A**684** 227c.
[25] K. Tanida *et al.*, Phys. Rev. Lett. **86**, 1982 (2001).
[26] E. Hiyama, *et al.*, Prog. Theor. Phys. **97**, 881 (1997).
[27] E. Hiyama, *et al.*, Phys. Rev. Lett. **85**, 270 (2000).
[28] H. Akikawa *et al.*, Phys. Rev. Lett. **88**, 82501 (2002).
[29] S. Ajimura *et al.*, Phys. Rev. Lett. **86**, 4225 (2001).
[30] E. Hiyama, *et al.*, Phys. Rev. C **66**, 024007 (2002).
[31] E. Hiyama *et al.*, Phys. Rev. Lett. **104**, 212502(2010).
[32] K. Nakazawa and H. Takahashi, Prog. Theor. Phys. Suppl. No.**185** (2010), 335.
[33] M. Hori *et al.* Phys. Rev. Letters **87** (2001) 093401.

# Spin observable in proton elastic scattering of $^6$He and its relevance to cluster structure

**Tomohiro Uesaka**
Center for Nuclear Study, University of Tokyo, Tokyo 113-0033, Japan
E-mail: `uesaka@cns.s.u-tokyo.ac.jp`

**Satoshi Sakaguchi**
RIKEN Nishina Center, Saitama 351-0198, Japan
E-mail: `satoshi@ribf.riken.jp`

**Abstract.** Vector analyzing power for the proton elastic scattering of $^6$He has been measured at 71 MeV/nucleon by using a polarized proton solid target. Through comparisons with cluster-folding model calculations, a role played by the $\alpha$-core in $^6$He is discussed.

## 1. Introduction

Proton elastic scattering is one of the simplest hadronic reactions and has been used to study various aspects of nuclei. Recently the proton elastic scattering studies are applied for light neutron-rich nuclei[1, 2, 3]. A major goal of the studies is to determine matter distributions, more specifically neutron distributions, in the neutron-rich nuclei and to argue existence of halo and/or skin structures.

The aim of this article is to discuss how the cluster formation in a nucleus affects spin observable in proton elastic scattering, taking an example of the $p$-$^6$He elastic scattering. The $^6$He nucleus is known to be well modeled as a weakly bound system of $\alpha$-core and two neutrons. In addition, because of the small two-neutron separation energy of 1.86 MeV, $^6$He is considered to have a very extended neutron distribution and to be a *halo* nucleus.

Vector analyzing power $A_y$ in the proton elastic scattering originates from a spin-orbit coupling between the proton and the target nucleus. The spin-orbit coupling is usually modeled by a spin-orbit part of optical potential which has a functional form of Thomas-type as

$$V_{\rm LS}(r) \sim \left(\frac{\hbar}{m_\pi c^2}\right)^2 \frac{1}{r}\frac{d}{dr}\rho(r)\,. \qquad (1)$$

One can expect that distinctive neutron distribution in $^6$He cause possible modifications in magnitude and radial shape of the spin-orbit potential.

## 2. Experiments

The polarized proton target working in a low magnetic field of 0.1 Tesla and at high temperature of 100 K was used in the vector analyzing power measurement for the $p$-$^6$He elastic scattering experiment[4, 5]. The target material was a crystal of naphthalene with a small amount ($\sim$ 0.005 mol%) of pentacene as a dopant. The dimensions of the target were 1 mm in thickness and 14 mm in diameter. The proton polarization was 20% at maximum and 14% on average.

**Figure 1.** Setup of the $p$-$^6$He elastic scattering experiment at RIPS.

A radioactive $^6$He beam with an energy of 70.6±1.4 MeV/nucleon was produced via the projectile fragmentation reaction of a primary $^{12}$C beam on a 1.39 g/cm$^2$ beryllium target. Intensity and purity of the $^6$He beam was $3.0\times10^5$ cps and 95%, respectively, after separation in RIPS[6].

The $^6$He particles scattered at forward angles in the laboratory frame were detected by a multi-wire drift chamber (MWDC) and plastic scintillators placed about 1 m downstream of the target. Since the $^6$He nucleus has no bound excited state, particle identification by $\Delta E$-$E$ in the scintillator is sufficient, in principle, for identifying elastic scattering events. The MWDC provides the $^6$He scattering angle and the reaction point on the target.

Two counter telescopes to detect recoiled protons were placed left and right with respect to the beam axis. Each telescope consisted of a single-wire drift chamber (SWDC) for a position measurement and a CsI(Tl) scintillator for a total energy measurement. They covered an angular range of $\theta_{\text{c.m.}} = 35^\circ - 90^\circ$ in the center of mass system. It should be emphasized that the low-magnetic field (0.1 Tesla) operation of the polarized proton target is inevitable for detection of the low-energy recoil protons.

## 3. Results and discussions

Figure 2 shows the measured angular distributions of differential cross section ($d\sigma/d\Omega$) and vector analyzing power ($A_y$) for the $p$-$^6$He elastic scattering by filled circles. Only statistical uncertainties are shown in the figure. Systematic uncertainty in $A_y$, mainly due to uncertainty in absolute normalization of proton polarization, is 19% independent of scattering angles. Open circles and open squares are $p$-$^6$He data by Korsheninnikov [7] and $p$-$^4$He data by Burzynski[8] at the same incident energy. It should be noted that the $^4$He data are plotted at the angle where momentum transfer for $p$-$^6$He is the same as that for the corresponding $p$-$^4$He data.

The steeper angular distribution of $d\sigma/d\Omega$ for the $^6$He differs considerably from the gradual

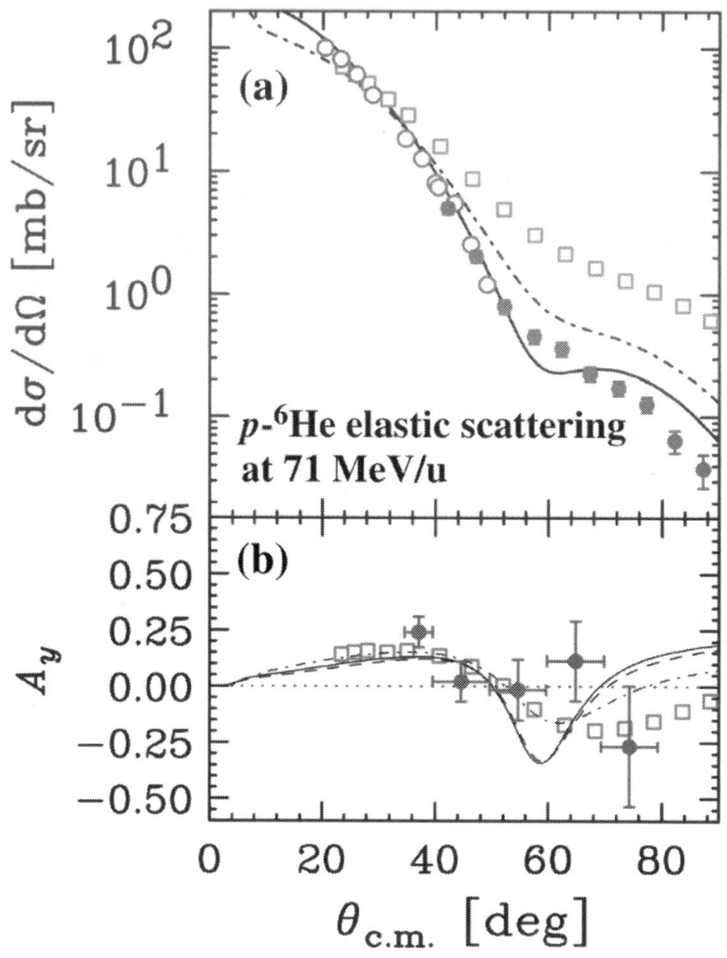


**Figure 2.** Cross section and vector analyzing power for the $p$-$^6$He elastic scattering experiment at 71 MeV/u.

one for the $^4$He, which is a natural consequence of a larger size of $^6$He. On the other hand, the $A_y$ data are similar to each other within the experimental error bars.

Based on an assumption that $^6$He is well modeled by an $\alpha$-$n$-$n$ cluster structure, we compare the data with a cluster-folding (CF) calculations. The CF optical potential can be written as

$$U_{\mathrm{CF}} = \sum_{i=1,2} \int V_{pn_i} \rho_n(r_i) dr_i + \int V_{p\alpha} \rho_\alpha(r_\alpha) dr_\alpha, \tag{2}$$

where $V_{pX}$ includes both central and spin-orbit parts. As the $p$-$\alpha$, a phenomenological optical potential which reproduces the $p$-$^4$He elastic scattering data at 72 MeV/nucleon [8] is used. The phenomenological optical potential includes complicated effects in the $p$-$\alpha$ scattering, such as non-locality due to exchange process, at least in part and can serve as a reasonable model of the $p$-$\alpha$ interaction. The complex effective interaction (CEG)[9] is adopted as the $p$-$n$ interaction. The interactions are folded with the $\alpha$-$n$-$n$ cluster distributions obtained by the Gaussian expansion method [10]. Details of the calculation will be reported elsewhere [5].

In Fig. 2, results of the full CF calculation (solid lines) are compared with the $^6$He data. The CF calculation reproduces both of $d\sigma/d\Omega$ and $A_y$ reasonably, in particular at $\theta_{\mathrm{c.m.}} \sim 35°$–$60°$.

On the other hand, folding model calculations without the cluster assumption can not reproduce the data as well within the present framework[5]. This supports the $\alpha$-$n$-$n$ cluster assumption introduced in the CF calculation. Reasonable reproduction without the cluster assumption requires detailed calculations which explicitly treat non-locality in the scattering[11].

To separate the valence neutron and the $\alpha$-core contributions, calculations without the central part of the $p$-$n$ interaction (dashed lines) and without all the $p$-$n$ interactions(dot-dashed lines) are carried out. The latter corresponds to extraction of a "pure" $\alpha$-core contribution. It is interesting to note that the calculation for the pure $\alpha$-core contribution sits in between the $^6$He data and the $^4$He data. This indicate that a matter distribution of the $\alpha$-core part in $^6$He is wider than that of a bare $^4$He nucleus due to the $\alpha$-core motion in $^6$He, while it is, by definition, narrower than that of $^6$He as a whole. The spin-orbit interaction $V_{pn;\ell s}$ gives negligible effects on $d\sigma/d\Omega$ and $A_y$, which is consistent with predictions in Ref. [12].

From the comparisons with CF, it is concluded that the spin-orbit coupling, which brings about a finite value of vector analyzing power, does not originate from interaction with the valence neutron but from the interaction with the $\alpha$-core. The $\alpha$-core distribution in $^6$He, which is largely affected by recoil of the valence neutrons, seems to be a possible key to understand the behavior of $A_y$. The $\alpha$-core motion in $^6$He make the $\alpha$-core distribution extended and gradual, which in turn makes the spin-orbit coupling between proton and $^6$He weaker. This conclusion is consistent with the result of phenomenological optical model analyses[5].

## 4. Summary

Cross section and vector analyzing power for the proton elastic scattering of $^6$He has been measured at 71 MeV/nucleon by using a polarized proton solid target. Low-magnetic field operation of the target is inevitable for detection of low-energy recoil protons. Data are compared with cluster-folding model calculations. It is concluded that the $\alpha$-core in $^6$He gives dominant contribution to the spin-orbit coupling in the proton elastic scattering of $^6$He and that the $\alpha$-core distribution in $^6$He is a possible key to understand the $p$-$^6$He elastic scattering.

## Acknowledgements

The authors thank the RIKEN and CNS staffs for operation of accelerators during the measurement. One of the authors (S.S.) expresses his gratitude for financial support by a Grant-in-Aid for Japan Society for the Promotion of Science (JSPS) Fellows (No. 18-11398). The authors are also grateful to the JSPS Core-to-Core Program. This work was supported by the Grant-in-Aid No. 17684005 of the Ministry of Education, Culture, Sports, Science, and Technology of Japan.

## References

[1] A. Korsheninnikov *et al.*, Physics Letters B **316**, 38 (1993).
[2] M. D. Cortina-Gil *et al.*, Physics Letters B **401**, 9 (1997).
[3] P. Engels et al., Progress in Particle and Nuclear Physics **46**, 307 (2001).
[4] T. Uesaka, S. Sakaguchi *et al.*, Phys. Rev. C **82**, 021602(R) (2010).
[5] S. Sakaguchi, Y. Iseri *et al.*, *submitted to* Physical Review C.
[6] T. Kubo *et al.*, Nuclear Instruments and Methods in Physics Research B **70**, 309 (1992).
[7] A. Korsheninnikov *et al.*, Nuclear Physics A **617**, 45 (1997).
[8] S. Burzynski *et al.*, Physical Review C **39**, 56 (1989).
[9] N. Yamaguchi, S. Nagata, and T. Matsuda, Progress of Theoretical Physics **70**, 459 (1983).
[10] E. Hiyama, Y. Kino, and M. Kamimura, Progress in Particle and Nuclear Physics **51**, 223 (2003).
[11] S. Karataglidis, P. J. Dortmans, K. Amos, and C. Bennhold, Physical Review C **61**, 024319 (2000).
[12] R. Crespo and A. Moro, Physical Review C **76**, 054607 (2007).

Journal of Physics: Conference Series **321** (2011) 012012 doi:10.1088/1742-6596/321/1/012012

# Search for alpha inelastic condensed state in $^{24}$Mg

**T. Kawabata[1], T. Adachi[2], M. Fujiwara[2], K. Hatanaka[2], Y. Ishiguro[1], M. Itoh[3], Y. Maeda[4], H. Matsubara[5], H. Miyasako[4], Y. Nozawa[1], T. Saito[4], S. Sakaguchi[6], Y. Sasamoto[5], Y. Shimizu[6], T. Takahashi[3], A. Tamii[2], S. Terashima[6], H. Tokieda[5], N. Tomida[1], T. Uesaka[6], M. Uchida[7], Y. Yasuda[2], N. Yokota[1], H. P. Yoshida[3], J. Zenihiro[6]**

[1]Department of Physics, Kyoto University
[2]Research Center for Nuclear Physics, Osaka University
[3]Cyclotron and Radioisotope Center, Tohoku University
[4]Faculty of Engineering, University of Miyazaki
[5]Center for Nuclear Study, University of Tokyo
[6]RIKEN (The Institute for Physical and Chemical Research)
[7]Department of Physics, Tokyo Institute of Technology

E-mail: `kawabata@scphys.kyoto-u.ac.jp`

**Abstract.** The alpha inelastic scattering from $^{24}$Mg was measured to obtain the isoscalar natural-parity excitation strengths and to search for the $\alpha$-condensed states. The multipole decomposition analysis for the measured cross sections was performed. The strength distributions for the $\Delta L = 0$–$3$ were successfully obtained and the possible candidates for the $\alpha$-condensed states around the $^{16}$O core were found.

## 1. Introduction

Alpha particle clustering is an important concept in nuclear physics. On the basis of the Ikeda diagram [1], the $\alpha$ cluster structures are expected to emerge near the $\alpha$-decay threshold energies. For instance, it has been suggested that the 7.65-MeV $0_2^+$ state in $^{12}$C, which locates at an excitation energy higher than the $3\alpha$-decay threshold by 0.39 MeV, has a spatially well-developed $3\alpha$-cluster structure.

This $0_2^+$ state is theoretically described by introducing a novel concept of the nuclear structure, *i.e.*, this state has a dilute-gas-like structure where three $\alpha$ clusters are weakly interacting and are condensed into the lowest $s$-orbit [2]. The next natural question addressed is whether such $\alpha$ condensed states exist in heavier self-conjugate $4n$ nuclei.

Such $\alpha$ condensed states are theoretically predicted up to $n = 10$ [3]. The energy of the $n\alpha$ condensed state relative to the $n\alpha$-decay threshold increases with $n$ due to the short-range nature of the attractive force between $\alpha$ clusters and the long-rage nature of the Coulomb repulsion. Finally $n\alpha$ condensed state becomes unstable beyond $n = 10$.

A recent theoretical work proposed a new conformation of the $\alpha$ condensed state where $\alpha$ clusters are condensed into the lowest $s$-orbit around a core nucleus [4]. Attractive potential for $\alpha$ clusters provided by the core nucleus stabilizes the $\alpha$ condensed state around the core nucleus. Thus, such $\alpha$ condensed states around core nuclei are expected to appear at excitation energies lower than the corresponding cluster-decay threshold energies.

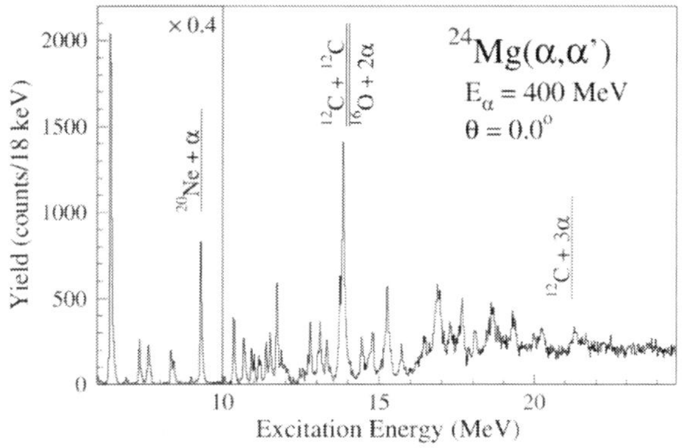


**Figure 1.** Typical spectrum for the $^{24}$Mg($\alpha,\alpha'$) reaction measured at 0°. The excitation spectrum below $E_x = 10$ MeV is downscaled by a factor of 0.4.

From the experimental point of view, the $3\alpha$ and $4\alpha$ condensed states in $^{12}$C and $^{16}$O were extensively studied in the last decade [5, 6, 7, 8, 9]. However, the other nuclei heavier than $^{16}$O are still unexplored due to experimental difficulties. Since the fully condensed states in $^{20}$Ne and $^{24}$Mg are expected to appear about 5 MeV above the $n\alpha$-decay threshold, these states are obscured by continuum states in the highly excited region. Although the $\alpha$ condensed states around the core nuclei should appear at relatively low excitation energies, the level densities near these condensed states are still high and the experimental identifications are not easy.

In our previous works [10, 11], it was demonstrated that spatially well-developed cluster states in light nuclei, whose spins and parities are the same as in the ground state, are strongly excited by isoscalar monopole transitions. The $0_2^+$ state in $^{12}$C, a typical $3\alpha$ cluster state, is strongly excited with an isoscalar monopole strength of $121 \pm 9$ fm$^4$, which is about 3 times larger than the single particle estimation. Besides the $0_2^+$ state in $^{12}$C, the $3/2_3^-$ state in $^{11}$B with the well-developed $2\alpha + t$ structure is also strongly excited by an isoscalar monopole transition. These facts are theoretically explained on the basis of the Bayman-Bohr theorem and the ground-state correlation [12]. Since ground-state wave functions in light nuclei inherently posses the clustering degrees of freedom, it is naturally expected that spatially developed cluster states are excited by stimulating relative motion between the clusters with the monopole operator $r^2$. Therefore, the isoscalar monopole strengths are a key observable to search for the $\alpha$-condensed states.

The alpha inelastic scattering at intermediate energies and at forward angles is one of the most useful probes to measure the isoscalar monopole strengths because its reaction mechanism is simple. It has a selectivity for isoscalar natural-parity transitions and there is a good linear relation between the reaction cross sections and the relevant nuclear transition matrix elements. The alpha inelastic scattering from $^{24}$Mg was previously measured at Texas A&M University [13, 14]. However, these measurements were devoted to study giant resonances in $^{24}$Mg and the excitation-energy resolution was not good enough to separate the $\alpha$-condensed states from the other states.

In the present work, we performed the high resolution measurement of the alpha inelastic scattering from $^{24}$Mg and obtained the isoscalar natural-parity excitation strengths to search for the $\alpha$-condensed states in $^{24}$Mg.

## 2. Experiment

The experiment was performed at the Research Center for Nuclear Physics, Osaka University, using a 400-MeV alpha beam. The halo-free alpha beam extracted from the ring cyclotron was transported to the $^{24}$Mg target. Scattered alpha particles were momentum analyzed by

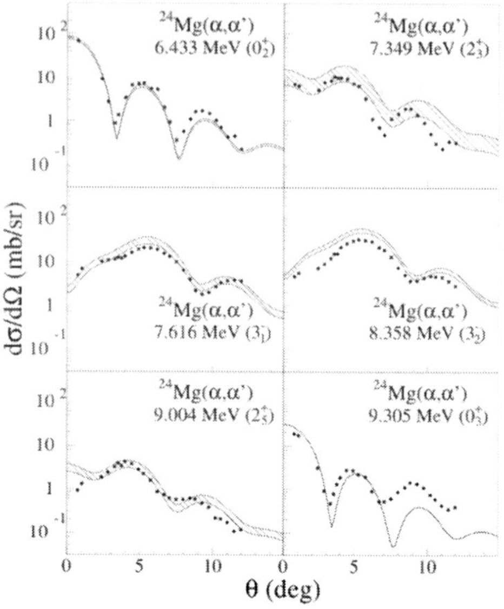


**Figure 2.** Cross sections for the prominent low-lying states in $^{24}$Mg compared with the DWBA calculation. The hatched bands show uncertainties of the calculated cross sections due to the errors of the electromagnetic transition strengths.

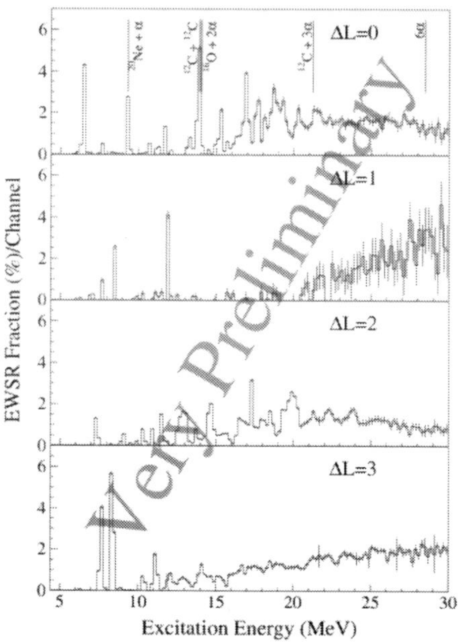


**Figure 3.** Strength distributions for the $\Delta L$ = 0–3 transitions obtained from the multipole decomposition analysis for the $^{24}$Mg($\alpha$,$\alpha'$) reaction.

the high-resolution spectrometer Grand Raiden [15]. The focal-plane detector system of Grand Raiden consisting of two multi-wire drift chambers and plastic scintillation detectors allowed the reconstruction of the scattering angle at the target via ray-tracing techniques.

A typical spectrum for the $^{24}$Mg($\alpha$,$\alpha'$) reaction measured at 0° is shown in Fig. 1. An energy resolution of 80 keV full width at half maximum was obtained.

## 3. Result and Discussion

The measured cross sections for the $^{24}$Mg($\alpha$,$\alpha'$) reaction exciting the prominent low-lying states are compared with the distorted-wave Born-approximation (DWBA) calculation in Fig. 2. The transition potentials used in the DWBA calculation were obtained by folding the macroscopic form factors with the phenomenological $\alpha N$ interaction $V_{\alpha N}$ given by $V_{\alpha N}(r) = -V\exp(-r^2/\alpha_V) - iW\exp(-r^2/\alpha_W)$. The interaction strengths and range parameters of $V$ = 13.1 MeV, $W$ = 8.8 MeV, and $\alpha_V = \alpha_W$ = 5.0 fm were determined to reproduce the cross section for the $\alpha$+$^{24}$Mg elastic scattering.

The normalization factors for the macroscopic form factors were determined to give the known electromagnetic transition strengths except for the $0_3^+$ state at $E_x$ = 9.305 MeV. The hatched bands in Fig. 2 show uncertainties of the calculated cross sections due to the errors in the electromagnetic transition strengths. The normalization factor for the $0_3^+$ state was determined to fit the present data. As seen in Fig. 2, the DWBA calculation reasonably well-describes the measured cross sections. It should be noted that all the parameters in the present DWBA calculation are determined from the previous experimental data and no parameters are tuned to reproduce the present result except for the normalization factor for the $0_3^+$ state.

Since the angular distribution of the cross section for each multipole transition depends on its transferred angular momentum, it is possible to decompose the cross section into each multipole component by fitting the measured angular distribution. In the fitting procedure, the multipole contributions up to $\Delta L = 15$ were taken into account. The strength distributions for the $\Delta L = 0$–3 transitions obtained from the multipole decomposition analysis are shown in Fig. 3. All the prominent low-lying states shown in Fig. 2 are successfully decomposed into the correct multipole components. This proves the validity of the present multipole decomposition analysis.

It is remarkable that the fine structures are observed in the $\Delta L = 0$ strength distribution and some of them are located near the $n\alpha$-decay threshold energies. Although no prominent structure is found near the $6\alpha$-decay threshold, a narrow peak at $E_x = 13.8$ MeV and a broad bump at $E_x = 21.5$ MeV are very close to the $^{16}$O + $2\alpha$ and $^{12}$C + $3\alpha$ threshold energies, respectively. These states are considered to be candidates of the $\alpha$-condensed states around the core $^{16}$O and $^{12}$C nuclei predicted in Ref. [4]. However, it is difficult to conclude that these states are indeed the $\alpha$-condensed states around the core nuclei because the alpha inelastic scattering excites not only $\alpha$-cluster states but also single-particle states such as giant monopole resonances (GMR). These candidates, thus, might be attributed to the GMR instead of the $\alpha$-condensed states, and further information is needed to distinguish between the two possibilities.

In order to clarify the microscopic structure of the excited states, we measured decay particles from the excited states in coincidence with the inelastically scattered $\alpha$ particles. If the excited states above the particle-decay thresholds have spatially well-developed cluster structures, these states should dominantly decay into $\alpha$-emission channels rather than into proton- or neutron-emission channels. On the other hand, single-particle states such as GMR should prefer proton- or neutron-emission channels to $\alpha$-emission channels. The decay branching ratio of the $\alpha$-emission channels to the proton-emission channels, thus, provide complementary information to identify the $\alpha$ cluster states. The analysis of the decay particle is still in progress. The results will be presented soon elsewhere.

## Acknowledgments

The authors acknowledge the effort of the RCNP cyclotron crew for providing the stable and clean beam. This research was supported in part by the Nogami memorial fund for young nuclear physicists, the Grant-in-Aids for Scientific Research No. 21740171 and for the Global COE Program "The Next Generation of Physics, Spun from Universality and Emergence" from the Japan Ministry of Education, Sports, Culture, Science, and Technology.

## References

[1] Ikeda K, Takigawa N and Horiuchi H 1968 *Prog. Theor. Phys. Suppl., Extra Number* 464–475
[2] Tohsaki A, Horiuchi H, Schuck P and Röpke G 2001 *Phys. Rev. Lett.* **87** 192501
[3] Yamada T and Schuck P 2004 *Phys. Rev. C* **69** 024309
[4] Itagaki N, Kimura M, Kurokawa C, Ito M and von Oertzen W 2007 *Phys. Rev. C* **75** 037303
[5] Itoh M *et al.* 2004 *Nucl. Phys. A* **738** 268–272
[6] Chernykh M, Feldmeier H, Neff T, von Neumann-Cosel P and Richter A 2007 *Phys. Rev. Lett.* **98** 032501
[7] Freer M *et al.* 2009 *Phys. Rev. C* **80** 041303
[8] Belyaeva T L *et al.* 2010 *Phys. Rev. C* **82** 054618
[9] Wakasa T *et al.* 2007 *Phys. Lett. B* 173–177
[10] Kawabata T *et al.* 2007 *Phys. Lett. B* **646** 6–11
[11] Kawabata T *et al.* 2004 *Phys. Rev. C* **70** 034318
[12] Yamada T, Funaki Y, Horiuchi H, Ikeda K and Tohsaki A 2008 *Prog. Theor. Phys.* **120** 1139–1167
[13] Youngblood D H, Lui Y W, Chen X F and Clark H L 2009 *Phys. Rev. C* **80** 064318
[14] Youngblood D H, Lui Y W and Clark H L 1999 *Phys. Rev. C* **60** 014304
[15] Fujiwara M *et al.* 1999 *Nucl. Instrum. Methods Phys. Res. A* **422** 484–488

# $^{12}$C spectroscopy above the $3\alpha$ threshold

**Pierre Descouvemont**

Physique Nucléaire Théorique et Physique Mathématique, CP229
Université Libre de Bruxelles (ULB), B1050 Brussels, Belgium

E-mail: pdesc@ulb.ac.be

**Abstract.** We investigate three-$\alpha$ continuum states in the hyperspherical formalism for $J = 0^+$ and $J = 2^+$. We use the shallow Ali-Bodmer potential and the deep potential of Buck *et al.* We determine the $3\alpha$ phase shifts up to $E = 6$ MeV, in parallel with an analysis of resonances in the framework of the Complex Scaling method. We show that shallow potentials provide additional narrow resonances, in contrast with experimental data. Deep potentials, however, only give rise to broad resonances, and are more consistent with the data.

## 1. Introduction

The $^{12}$C spectroscopy below the $3\alpha$ threshold is well known since many years (see Ref. [1] and references therein). The $0^+$ ground state and the $2^+$ first excited state ($E_x = 4.44$ MeV) have been investigated in many experimental and theoretical works, and their properties are now well established. The situation, however, is very different above the $3\alpha$ threshold. In the continuum region, the second $0^+$ state at $E_x = 7.65$ MeV plays a very important role in nuclear physics and in astrophysics. This state cannot be reproduced in standard [2, 3] or in no-core [4] shell model calculations. The $0_2^+$ state is known to have a well-developed cluster structure and has been successfully described by $\alpha$ cluster models [5, 6, 7, 8, 9]. Fermionic molecular dynamics (FMD) calculations [10] also reproduce the $0_2^+$ resonance provided that $\alpha$-cluster structures are introduced in the FMD basis.

Historically, the interest for the $0_2^+$ state started with the prediction of Hoyle [11] that a resonance with zero angular momentum should exist just above the $\alpha$+$^8$Be threshold to explain the $^{12}$C abundance in the Universe. In nuclear physics the interest for the $0_2^+$ structure was recently revived by its interpretation as a dilute cluster gas state of three weakly interacting $\alpha$ particles [12].

The experimental search for a $2^+$ broad resonance is recent. Due to its specific structure, the observation of such resonance is extremely difficult. Some evidence for a $2^+$ broad level was reported by Itoh *et al.* [13] near $E_\alpha \approx 9 - 10$ MeV, though not widely accepted. More recently, the situation was clarified, and a possible $2^+$ resonance at $E_\alpha = 9.6 \pm 0.1$ MeV with $\Gamma = 0.6$ MeV was observed by Freer *et al.* [14]. These properties are in fair agreement with predictions of microscopic cluster models [9, 15].

The aim of the present work (see Ref. [16] for details) is to investigate the $3\alpha$ continuum in a non-microscopic approach. We use the Complex Scaling Method (CSM, see Ref. [17]) complemented by the triple $\alpha$ phase shifts. Recently, we have extended the hyperspherical formalism [18] to three-body continuum states by implementing the $R$-matrix theory with a

Lagrange basis [19]. Although the hyperspherical theory is known to converge slowly in the continuum [20, 21, 19], it provides genuine 3-body phase shifts.

## 2. Theoretical framework

For 3-body systems, the Schrödinger equation can be solved with the hyperspherical approach [18]. The hyperradius $\rho$ and hyperangle $\alpha$ are deduced from the scaled Jacobi coordinates $\boldsymbol{x}$ and $\boldsymbol{y}$. The wave function in partial wave $J\pi$ is then expanded over hyperspherical harmonics as

$$\Psi^{JM\pi}(\rho,\Omega_5) = \rho^{-5/2} \sum_{\gamma K} \chi^{J\pi}_{\gamma K}(\rho)\ \mathcal{Y}^{JM}_{\gamma K}(\Omega_5), \tag{1}$$

where the hyperradial functions $\chi^{J\pi}_{\gamma K}(\rho)$ have to be determined, and where $\mathcal{Y}^{JM}_{\gamma K}(\Omega_5)$ are known functions, depending on the five angles $\Omega_5 = (\Omega_x, \Omega_y, \alpha)$. The Schrödinger equation is replaced by a system of coupled differential equations involving matrix elements of the 2-body potentials. In Eq. (1), a truncation must be done in the summation over $K$; the maximum $K$ value is denoted as $K_{\max}$. The number of $\gamma K$ components increases rapidly when $K_{\max}$ increases.

The treatment of three-body continuum states ($E > 0$), with exact three-body asymptotic conditions, is recent [22, 19]. It is performed with the $R$-matrix method [23, 24], where the configuration space is divided into two regions: the internal region (with radius $a$) where the nuclear force must be taken into account, and the external region where the potentials have reached their asymptotic (Coulomb) behaviour. Accordingly, the external solution reads, for an entrance channel $\gamma'$,

$$\chi^{J\pi}_{\gamma K,\mathrm{ext}}(\rho) = C^{J\pi}_{\gamma K}\left[H^-_{\gamma K}(k\rho)\delta_{\gamma\gamma'}\delta_{KK'} - U^{J\pi}_{\gamma K,\gamma' K'} H^+_{\gamma K}(k\rho)\right], \tag{2}$$

where $k = \sqrt{2m_N E/\hbar^2}$ is the three-body wave number ($m_N$ is the nucleon mass). In Eq. (2), $C^{J\pi}_{\gamma K}$ is a normalization coefficient, $\boldsymbol{U}^{J\pi}$ is the three-body collision matrix and the incoming and outgoing functions $H^{\pm}_{\gamma K}(x)$ are defined as

$$H^{\pm}_{\gamma K}(x) = G_{K+\frac{3}{2}}(\eta_{\gamma K}, x) \pm iF_{K+\frac{3}{2}}(\eta_{\gamma K}, x), \tag{3}$$

where $\eta_{\gamma K}$ is the Sommerfeld parameters in channel $\gamma K$, and $F$ and $G$ are the regular and irregular Coulomb functions. The internal components $\chi^{J\pi}_{\gamma K,\mathrm{int}}(\rho)$ are expanded over a Lagrange mesh $\varphi_i(\rho)$.

According to the $R$-matrix formalism, matrix elements between basis functions $\varphi_i(\rho)$ must be computed over the internal region. Then the use of the Bloch operator [24] makes the kinetic energy Hermitian, and ensures the continuity of the derivative of the wave function at $\rho = a$. The matching between the internal and external solutions provides the collision matrix $\boldsymbol{U}^{J\pi}$. A strong test of the calculation is that the collision matrix should not depend on the channel radius $a$ and on the number $N$ of basis functions, provided that they are large enough to meet the $R$-matrix requirements. The collision matrix may involve many channels, in particular for large $K_{\max}$ values, and is in general analyzed through its eigenvalues. The number of eigenphases is of course equal to the number of channels $\gamma K$.

The $R$-matrix theory provides 3-body wave functions at any scattering energy. The calculation of the phase shifts is obviously the optimal way to investigate the continuum, and in particular broad resonances. In practice, however, dealing with the continuum is an heavy task, especially for three-body systems. In particular, only shallow local potentials can be currently employed. For these reasons a number of approximate methods have been developed. The common idea is to derive resonance properties (energy and width) from bound-state calculations, which only requires slight modifications of well mastered techniques. In the CSM [25, 26, 17], the space radial coordinate $r$ and the momentum $p$ are rotated by a scaling angle $\theta$. The diagonalization of $H(\theta)$ with square-integrable functions can be done in various bases, such as Gaussian [27] or Lagrange basis [28]. The ABC theorem [25, 26] demonstrates that this diagonalization provides continuum states along straight lines in the complex plane, rotated by an angle $2\theta$ from the positive real axis. On the contrary, resonant states are not modified under variations of $\theta$. The resonance energy $E_R$ and width $\Gamma$ are determined from

$$E(\theta) = E_R - i\Gamma/2, \tag{4}$$

and are independent (at the numerical accuracy of the calculation) of $\theta$.

## 3. Application to $3\alpha$ continuum states

The shallow Ali-Bodmer (AB) potential [29] ($\ell$-dependent potential $d$), and the deep potential of Buck *et al.* (BFW) [30] reproduce fairly well the $\alpha + \alpha$ phase shifts up to $E \approx 20$ MeV, i.e. below the proton threshold in $^8$Be. It is well known that a $3\alpha$ potential must be introduced to reproduce the ground-state energy of $^{12}$C. According to Ref. [31], we choose the three-body interaction as $V_{\alpha\alpha\alpha}(\rho) = v_3 \exp(-(\rho/\rho_3)^2)$, with the range $\rho_3 = 6$ fm. The amplitude $v_3$ is adjusted, for each potential and angular momentum, on the experimental energies.

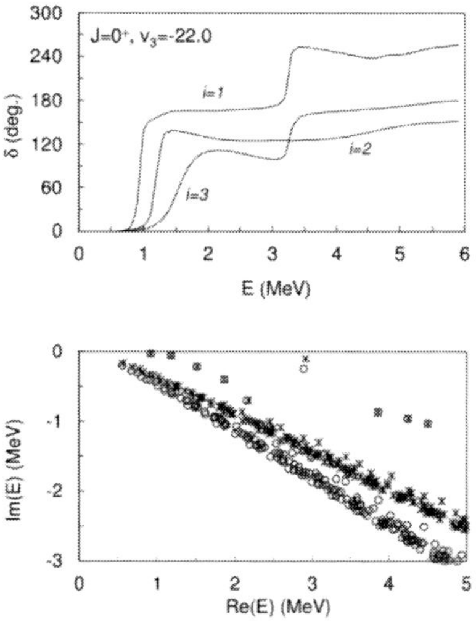


**Figure 1.** Three-$\alpha$ phase shifts (upper panel) and complex energies (lower panel) for the Ali-Bodmer potential ($J = 0^+$), with a 3-body potential $v_3 = -22.0$ MeV. Complex energies are displayed for two angles: $\theta = 0.25$ (crosses), and $\theta = 0.30$ (circles).

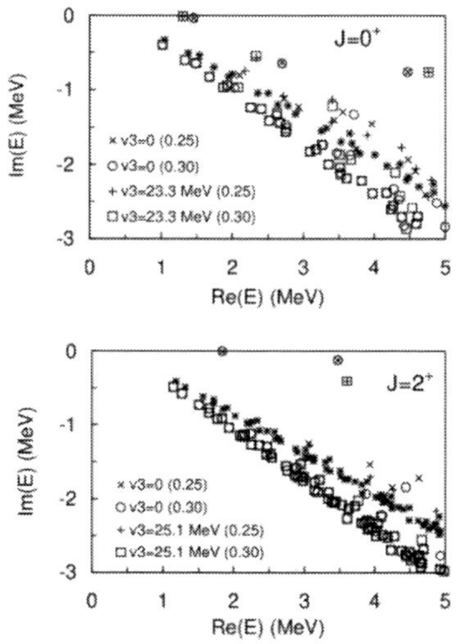


**Figure 2.** Complex energies with the BFW potential for $J = 0^+, 2^+$ with and without a $3\alpha$ potential. Two scaling angles are used: $\theta = 0.25$ and $\theta = 0.30$.

The first three $J = 0^+$ eigenphases (labeled by index $i$) and complex energies are displayed in Fig. 1, for $K_{\max} = 44$, and for two scaling angles: $\theta = 0.25$ and $\theta = 0.30$. Several states are stable and can also be observed in the eigenphases. The second $0^+$ state is located at $-0.12$ MeV but, as for $v_3 = 0$, several additional resonances between 0.5 and 2 MeV are observed in the eigenphases, and confirmed by the CSM method. A further narrow state near 3 MeV ($\Gamma \approx 0.4$ MeV) is obtained.

Journal of Physics: Conference Series **321** (2011) 012013 doi:10.1088/1742-6596/321/1/012013

Figure 2 presents the resonance energies with the BFW potential. With this potential, a small number of resonances is found, in agreement with Ref. [32]. Without the repulsive 3-body potential, the first $2^+$ excited state is unbound, and shows up as a sharp resonance. A resonance is found near $E_x \approx 10.5$ MeV.

## 4. Conclusion

We have shown that the $3\alpha$ continuum is very sensitive to the choice of the $\alpha + \alpha$ potential. This result is not surprising as the bound-state spectrum is also dependent on the potential (see for example Ref. [5]). The AB shallow potential provides several narrow resonances, which do not correspond to experimentally known states. Calculations with the deep BFW potential are currently limited to the CSM method, but provide resonance energies which significantly differ from the AB interaction. In addition to the $0_1^+, 0_2^+$ and $2_1^+$ states, only broad resonances beyond 3 MeV are found. The $2_2^+$ resonance might correspond to the new state observed in recent experiments [14, 33].

## References


[1] Ajzenberg-Selove F 1990 *Nucl. Phys. A* **506** 1
[2] Karataglidis S, Dortmans P J, Amos K and de Swiniarski R 1995 *Phys. Rev. C* **52** 861
[3] Navrátil P and Ormand W E 2003 *Phys. Rev. C* **68** 034305
[4] Navrátil P, Vary J P and Barrett B R 2000 *Phys. Rev. Lett.* **84** 5728
[5] Horiuchi H 1975 *Prog. Theor. Phys.* **53** 447
[6] Uegaki E, Okabe S, Abe Y and Tanaka H 1977 *Prog. Theor. Phys.* **57** 1262
[7] Fujiwara Y, Horiuchi H, Ikeda K, Kamimura M, Katō K, Suzuki Y and Uegaki E 1980 *Prog. Theor. Phys. Suppl.* **68** 29
[8] Kamimura M 1981 *Nucl. Phys. A* **351** 456
[9] Descouvemont P and Baye D 1987 *Nucl. Phys. A* **463** 629
[10] Chernykh M, Feldmeier H, Neff T, von Neumann-Cosel P and Richter A 2007 *Phys. Rev. Lett.* **98** 032501
[11] Hoyle F 1954 *Astrophys. J. Suppl.* **1** 121
[12] Tohsaki A, Horiuchi H, Schuck P and Röpke G 2001 *Phys. Rev. Lett.* **87** 192501
[13] Itoh M, Akimune H, Fujiwara M, Garg U, Hashimoto H, Kawabata T, Kawase K, Kishi S, Murakami T, Nakanishi K, Nakatsugawa Y, Nayak B K, Okumura S, Sakaguchi H, Takeda H, Terashima S, Uchida M, Yasuda Y, Yosoi M and Zenihiro J 2004 *Nucl. Phys. A* **738** 268
[14] Freer M, Fujita H, Buthelezi Z, Carter J, Fearick R W, Förtsch S V, Neveling R, Perez S M, Papka P, Smit F D, Swartz J A and Usman I 2009 *Phys. Rev. C* **80** 041303
[15] Arai K 2006 *Phys. Rev. C* **74** 064311
[16] Descouvemont P 2010 *J. Phys. G* **37** 064010
[17] Ho Y K 1983 *Phys. Rep.* **99** 1
[18] Lin C D 1995 *Phys. Rep.* **257** 1
[19] Descouvemont P, Tursunov E M and Baye D 2006 *Nucl. Phys. A* **765** 370
[20] Fedorov D V and Jensen A S 1996 *Phys. Lett. B* **389** 63
[21] Danilin B V, Thompson I J, Vaagen J S and Zhukov M V 1998 *Nucl. Phys. A* **632** 383
[22] Thompson I J, Danilin B V, Efros V D, Vaagen J S, Bang J M and Zhukov M V 2000 *Phys. Rev. C* **61** 024318
[23] Lane A M and Thomas R G 1958 *Rev. Mod. Phys.* **30** 257
[24] Descouvemont P and Baye D 2010 *Rep. Prog. Phys.* **73** 036301
[25] Aguilar J and Combes J M 1971 *Commun. Math. Phys.* **22** 269
[26] Balslev E and Combes J M 1971 *Commun. Math. Phys.* **22** 280
[27] Tursunov E M 2001 *J. Phys. G* **27** 1381
[28] Descouvemont P, Daniel C and Baye D 2003 *Phys. Rev. C* **67** 044309
[29] Ali S and Bodmer A R 1966 *Nucl. Phys.* **80** 99
[30] Buck B, Friedrich H and Wheatley C 1977 *Nucl. Phys. A* **275** 246
[31] Alvarez-Rodriguez R, Garrido E, Jensen A S, Fedorov D V and Fynbo H O U 2007 *Eur. J. Phys. A* **31** 303
[32] Kurokawa C and Katō K 2007 *Nucl. Phys. A* **792** 87
[33] Diget C A *et al.* 2009 *Phys. Rev. C* **80** 034316

Journal of Physics: Conference Series **321** (2011) 012014 doi:10.1088/1742-6596/321/1/012014

# Many body approach to quartet condensation in nuclear matter

**T Sogo**[1], **P Schuck**[1,2] **and G Röpke**[3]

[1] Institut de Physique Nucléaire, CNRS-IN2P3 and Université Paris-Sud, 91406 Orsay Cedex, France

[2] Laboratoire de Physique et Modélisation des Milieux Condensés, CNRS and Université Joseph Fourier, 25 Avenue des Martyrs, Boîte Postale 166, F-38042 Grenoble Cedex 9, France

[3] Institut für Physik, Universität Rostock, D-18051 Rostock, Germany

E-mail: `sogo@ipno.in2p3.fr`

**Abstract.** A self-consistent theory for quartet condensation in homogeneous infinite nuclear matter is proposed. It goes along similar lines as Gorkov theory for pairing. However, details and physical consequences are very different. For example in the quartet case no sharp quasi-particle pole develops in the single particle propagator. Also quartet condensation only exists in the BEC phase and no extended coherence length develops in weak coupling. The critical temperature for quartet condensation is also investigated in analogy to the Thouless criterion for pairing.

In attractive multi-component fermion systems there exists the possibility of bosonic multi-fermion cluster condensation: for example, in a four component fermion system, like nuclear matter, condensation of a four body bound state, the $\alpha$ particle, can occur because four different nucleons can be bound only with $s$-wave coupling. On the other hand, a four body bound state in a two component system is not energetically favourable because $p$-wave coupling must be involved. In a finite nucleus, the Hoyle state, which is an excited state of $^{12}C$, is considered as the "condensation" state of three $\alpha$ particles in the sense that three $\alpha$-particles occupy a single S-state in the mean field potential [1].

In this work we construct the quartet version in analogy to the BCS pairing theory. The BCS theory describes the condensation of the fermion pairs, where the order parameter is the expectation value of two fermion annihilation operators. We shall develop a BCS-like theory for quartet condensation where the order parameter is represented by the expectation value of four fermion annihilation operators. We apply it to $\alpha$-condensation in symmetric nuclear matter [2].

Let us first briefly recap the usual BCS gap equation. To derive the BCS gap equation, we need two equations. One is the single fermion mass operator: $M_1^{\mathrm{BCS}}(\omega) = \sum_{22'} \frac{1}{2}\bar{v}_{1\bar{1},2\bar{2}} \langle c_{\bar{2}} c_2 \rangle \frac{1}{\omega+(\varepsilon_1-\mu)} \langle c_2^\dagger c_{\bar{2}}^\dagger \rangle \frac{1}{2}\bar{v}_{1\bar{1},2\bar{2}}$, where $c_i, c_i^\dagger$ are the fermion annihilation and creation operators with single particle quantum numbers $i$, i.e. momentum, spin, and isospin in case of nuclear matter, and the single particle kinetic energy $\varepsilon_1 = \frac{k_1^2}{2m}$, the chemical potential $\mu$, antisymmetrised two-body vertex $\bar{v}_{12,34}$, and $\langle ... \rangle$ being a thermal average. Mean field shifts are supposed to be incorporated in an approximate way into the chemical potential. In above expression, $\bar{1}$ means the time reversal conjugate of the state 1. The other equation which needs to be derived for the BCS gap equation is the in-medium two body Schrödinger equation:

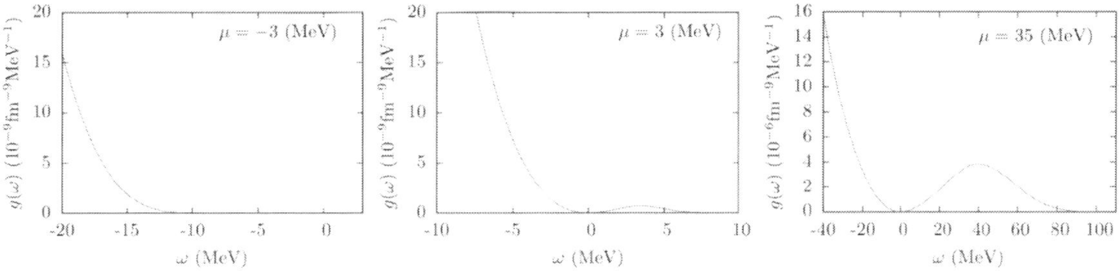


**Figure 1.** Three hole level density $g(\omega)$ at zero temperature for $\mu = -3$, 3, and 35MeV.

$(\varepsilon_1+\varepsilon_2)\langle c_1c_2\rangle+(1-\rho_1-\rho_2)\sum_{1'2'}\frac{1}{2}\bar{v}_{12,1'2'}\langle c_{1'}c_{2'}\rangle = 2\mu\langle c_1c_2\rangle$. where $\rho_1 = \langle c_1^+c_1\rangle$ is an occupation number. The medium effect is included as $(1-\rho_1-\rho_2)$ multiplying the two-body vertex. The occupation number is derived from the one body Green function with the BCS mass operator. With these formulas and considerations, we can derive the BCS gap equation. We need analogous equations for the quartet case.

In analogy to BCS theory, one can show that the single fermion mass operator with quartet condensation has the following form [2]

$$M_1^{\rm quartet}(\omega) = \sum_{2341'2'1''2''}\frac{1}{2}\bar{v}_{12,1'2'}\langle c_4c_3c_{2'}c_{1'}\rangle\frac{\bar{f}_2\bar{f}_3\bar{f}_4+f_2f_3f_4}{\omega+\varepsilon_2+\varepsilon_3+\varepsilon_4-3\mu}\langle c^\dagger_{1''}c^\dagger_{2''}c^\dagger_3c^\dagger_4\rangle\frac{1}{2}\bar{v}_{12,1''2''} \qquad (1)$$

with the Fermi distribution function $f_1 = \frac{1}{e^{(\varepsilon_1-\mu)/T}+1}$ and $\bar{f}_1 = 1-f_1$. It is seen that the quartet mass operator (1) has a similar structure as the BCS mass operator. However, there exist crucial differences. One is the Pauli blocking factor of the three hole Green function in the middle of the quartet mass operator. The BCS mass operator for pairing has just one hole propagator. This has as a consequence that in the quartet mass operator the indices in the three hole Green function are summed, while there is no sum in the BCS mass operator due to momentum conservation. This has strong physical consequences.

Let us consider in detail. Figure 1 shows the three hole level density, which represents the imaginary part of the noninteracting three hole Green function: $g(\omega) = \sum_{234}(\bar{f}_2\bar{f}_3\bar{f}_4 + f_2f_3f_4)\pi\delta(\omega+\varepsilon_2+\varepsilon_3+\varepsilon_4-3\mu)$. Here we consider zero temperature $T = 0$, i.e. the Fermi distribution function becomes the step function, $f_1 = \theta(\mu-\varepsilon_1)$.

For positive chemical potential, see right and middle panels in fig. 1, the level densities become zero at $\omega = 0$, i.e. at the Fermi surface, what stems from the the Pauli blocking factor $(\bar{f}_2\bar{f}_3\bar{f}_4 + f_2f_3f_4)$. This means that there are no correlations in the quartet case for positive chemical potential because the quartet correlations (as the pair correlations) arise at the Fermi surface. Therefore, we can predict that there is no quartet condensation for positive chemical potential. On the contrary, the level density of the one hole Green function in the BCS mass operator has no zero at the Fermi surface. Therefore, the BCS state occurs even for the very small attractive interaction and exists for positive chemical potential. This is a very special case which does not repeat itself for any cluster higher than two body. This is the crucial difference between BCS and quartet cases. For negative chemical potential, the three hole Pauli blocking factor is just one. So in the BEC phase, there is no qualitative difference between pairing and quartetting.

We also derived the in-medium four body Schrödinger equation in analogy to the pair case

$$(\varepsilon_1+\varepsilon_2+\varepsilon_3+\varepsilon_4)\langle c_1c_2c_3c_4\rangle + (1-\rho_1-\rho_2)\sum_{1'2'}\frac{1}{2}\bar{v}_{12,1'2'}\langle c_{1'}c_{2'}c_3c_4\rangle + \text{perm.} = 4\mu\langle c_1c_2c_3c_4\rangle. \quad (2)$$

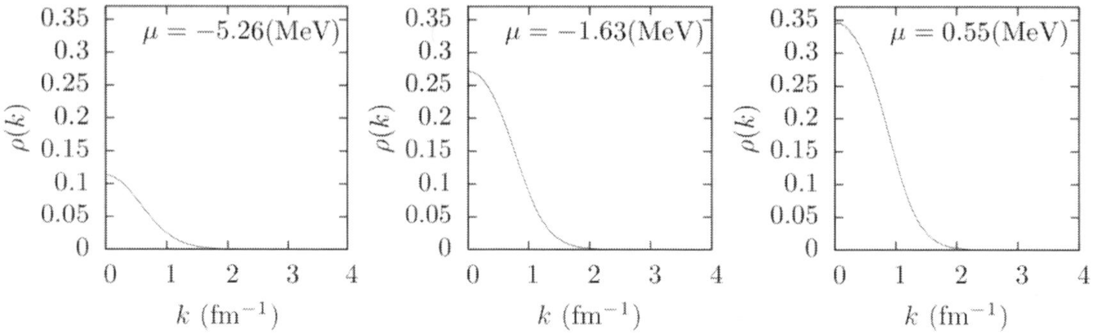


**Figure 2.** Occupation numbers as a function of momentum.

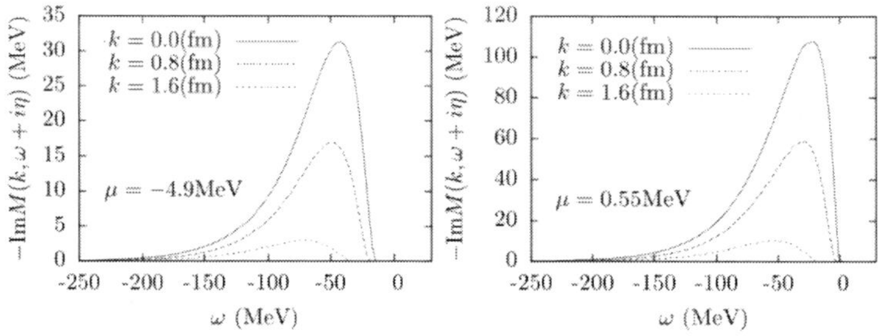


**Figure 3.** Imaginary part of the quartet mass operators of eq. (1) at zero temperature.

where occupation numbers again are obtained from the one body Green function with the quartet mass operator (1). In principle the quartet order parameter is computed iteratively by using eqs. (1) and (2) and the occupation numbers are obtained from the one body Green function in a self-consistent cycle quite similar to the BCS case described above. However, to solve in medium four body equations self consistently is out of reach for the moment. To simplify the numerics, we use the following ansatz for the quartet order parameter which was very successful for the calculation of the critical temperature for $\alpha$ particle condensation (see below): $\langle c_1 c_2 c_3 c_4 \rangle = \phi(k_1)\phi(k_2)\phi(k_3)\phi(k_4)\chi_0(2\pi)^3\delta(\vec{k}_1 + \vec{k}_2 + \vec{k}_3 + \vec{k}_4)$, i.e. it is the Hartree-Fock wave functions projected to zero c.o.m. momentum. $\chi_0$ is the spin-isospin singlet wave function. Besides, we use a separable two body interaction with the parameters adjusted to get the binding energy ($-28.3$MeV) and the rms radius of the free $\alpha$-particle (1.71fm). In this way the problem becomes much simpler because instead of the full four body correlation function, one only has to iterate one single particle S-state wave function.

Figure 2 shows the occupation numbers for various values of the chemical potential at zero temperature. The occupation numbers in fig. 2 are far from step functions but one also get this kind of occupation numbers in BCS theory for negative chemical potential. Note that the order parameter goes to zero above $\mu = 0.55$MeV, i.e. no condensation appears beyond that region where the density is 0.029fm$^{-3}$, i.e. the mean $\alpha$ particle distance is $(0.029)^{-1/3} = 5.2$fm. Therefore, the $\alpha$ condensation breaks down at very low density compared with the rms radius of the free $\alpha$ particle implying that the $\alpha$'s are barely overlapping with their density tails. This demonstrates the dramatic difference with the pairing case. For quartets only the BEC phase exists whereas pairs (at rest) can be entangled as much as one can think of without losing their coherence.

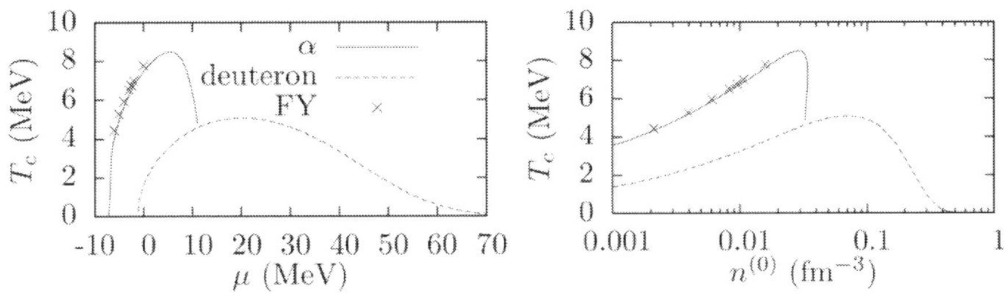


**Figure 4.** The critical temperature for alpha condensation (solid line) and deuteron condensation (dashed line) as functions of the chemical potential (left) and free nucleon density (right). The density is calculated by $n^{(0)} = 4\int d^3kf(k,T_c)/(2\pi)^3$ with $f(k,T_c) = [e^{(k^2/(2m)-\mu)/T_c} + 1]^{-1}$. Crosses are calculated with the Faddeev-Yakubovsky method using a realistic nuclear force.

Figure 3 shows the imaginary part of the mass operator at zero temperature as a function of energy for several momentum values. It is seen that there exists a considerable width of the single particle states which comes from the above mentioned integration over the three hole momenta. Therefore, the quasi-particle picture is not at all valid in the case of quartet condensation. On the other hand, the imaginary part of the BCS mass operator is sharp because the momentum in the one hole Green function in the BCS mass operator is fixed by momentum conservation.

For the calculation of the critical temperature, one has to linearise the in-medium four body Schrödinger equation (2) with respect to the order parameter, i.e. one makes the change $\rho_i \to f_i$ [3]. This is in analogy to the Thouless criterion for pairing.

Figure 4 shows the critical temperature in symmetric nuclear matter. The solid lines are for $\alpha$-condensation calculated with the product ansatz for the order parameter. Crosses are calculated by the Faddeev-Yakubovsky method with a realistic nuclear force. As seen in fig. 4, the agreement with the mean field values are very good for negative chemical potentials, the only domain where the numerical solution of the realistic four body problem was possible. Therefore, the product ansatz is a very valid approximation in the bound state domain. We compared with the case of deuteron condensation. The $\alpha$ condensation has higher critical temperature, but breaks down at much lower density than in the deuteron case. Note that we also investigated the critical temperature in asymmetric nuclear matter [4], where it is revealed that for highly imbalanced nuclear matter the $\alpha$-condensation has higher critical temperature and survives until higher density than in the case of deuteron BEC.

## Acknowledgments
Discussions and collaborations with Y Funaki, H Huriuchi, A Tohsaki and T Yamada are gratefully acknowledged.

## References
[1] Tohsaki A, Horiuchi H, Schuck P and Röpke G 2001 *Phys. Rev. Lett.* **87** 192501
[2] Sogo T, Schuck P and Röpke G 2010 *Phys. Rev.* C **81** 064310
[3] Sogo T, Lazauskas R, Röpke G and Schuck P 2009 *Phys. Rev.* C **79** 051301
[4] Sogo T, Schuck P and Röpke G 2010 *Phys. Rev.* C **82** 034322

Journal of Physics: Conference Series **321** (2011) 012015 doi:10.1088/1742-6596/321/1/012015

# Fission properties of the Barcelona-Catania-Paris energy density functional

**L.M. Robledo**

E-mail: luis.robledo@uam.es
Dep. Física Teórica (Módulo 15), Universidad Autónoma de Madrid, E-28049 Madrid, Spain

**M. Baldo**

Instituto Nazionale di Fisica Nucleare, Sezione di Catania, Via Santa Sofia 64, I-95123 Catania, Italy

**P. Schuck**

Institut de Physique Nucléaire, CNRS, UMR8608, F-91406 Orsay, France

**X. Viñas**

Departament d'Estructura i Constituents de la Matèria and Institut de Ciències del Cosmos, Facultat de Física, Universitat de Barcelona, Diagonal *647*, 08028 Barcelona, Spain

**Abstract.** Fission properties of the Barcelona-Catania-Paris (BCP) energy density functional are explored by performing constrained mean field Hartree-Fock-Bogoliubov (HFB) calculations along the fission path. These calculations provide us with the quantities required to estimate the spontaneous fission half lives and fragment mass distribution. The results obtained are compared to experimental data and other calculations.

## 1. Introduction

The Barcelona-Catania-Paris (BCP) energy density functional (EDF) has been proposed [1] as an alternative to more traditional approaches based on Gogny or Skyrme type functionals (see Ref [2] for a review) in the calculation of nuclear properties all over the Nuclide Chart. The definition of the BCP EDF conceptually differs from the underlying philosophy of the Skyrme or Gogny EDF's in its use of realistic nuclear matter equations of state to inspire a finite nuclei functional by means of the Local Density Approximation (LDA). The suitability of the BCP EDF to describe finite nuclei was tested first in [1] where masses and radii of spherical nuclei were used to define the parameters of the functional (called BCP1 and BCP2). Next, the ability of BCP1 and BCP2 to reproduce quadrupole deformation properties was considered in [3]. Next in the multipole parametrization of the nuclear shape is the octupole degree of freedom, responsible of the breaking of parity symmetry in several rare earth and actinide isotopes. In [4] we studied the suitability of BCP1 and BCP2 to describe octupole deformation properties in the context of the collective Schrödinger equation (CSE). We concluded that BCP was good in reproducing experimental quantities related to octupole deformation. The results of [4] not only indicated the validity of the potential energy curves (PEC) but also the suitability of the collective inertias required in the CSE. Both in [3] and [4] the fission barrier of $^{240}$Pu was considered as an example of a mixed case where both quadrupole and octupole deformation play a role. The conclusion reached in both cases was that the BCP1 and BCP2 fission barriers were both rather close to the one of the Gogny D1S.

Given the relevant role of fission in establishing the very limits of existence of heavy nuclei, including the famous "island of stability" [5] it is timely to address in a more systematic way the description of fission with the BCP functional. Fission involves a delicate balance between a variety of very different kinds of shapes that makes it a perfect testing ground to complement previous studies of deformation properties with BCP.

In this paper we present the results of fission calculations in three characteristic nuclei, two actinides ($^{246}$Cm and $^{252}$ No) and a superheavy ($^{270}$Sg). Fission barrier heights and spontaneous fission half lives are given. The role played by the collective mass in the description of half lives is discussed. The differences with the results obtained with the Gogny force [6] (D1S parametrization [7]) are also discussed.

## 2. Methodology

The BCP energy density functional is made of a bulk part which is inspired by fully microscopic and realistic calculations of symmetric and neutron matter equations of state [8]. The two equations of state (symmetric and neutron matter) given as a function of the nuclear density are parametrized using a fitting procedure in terms of polynomials of the densities. To account for finite size effects related to the surface energy, a phenomenological finite range interaction of gaussian type is included. In addition, the Coulomb interaction and the spin-orbit term are taken exactly as in the Skyrme or Gogny forces. To deal with open-shell nuclei we include in the BCP functionals a zero range density-dependent pairing interaction fitted to reproduce the nuclear matter gaps obtained with the Gogny force [9]. The only free parameters of these functionals are the isospin like $V_L$ and unlike $V_U$ strengths of the surface term, the range of the Gaussian form factor of the surface term $r_0$ and the strength of the spin-orbit interaction $W_0$ [1]. These free parameters are adjusted in the usual way to reproduce the ground state energy and charge radii of some selected spherical nuclei. With these ingredients, the BCP functionals give an excellent description of 161 even-even spherical nuclei with rms values for the ground-state energies and charge radii comparable with the most performing of Skyrme, Gogny and Relativistic mean field type. Apart from the advantages already mentioned in [1], the BCP functionals are advantageous in their application to finite nuclei as a consequence of a reduced computational cost as compared to Gogny or even Skyrme. Recently, open questions about BCP and ways to improve the EDF have been discussed in [10].

To describe the path to fission we use mean field configurations given in terms of HFB wave functions that define the standard and pairing densities required by the BCP functional. The amplitudes of the Bogoliubov transformation defining those HFB wave functions are computed as the solution of the HFB equation. In our calculations we have recast the HFB equation in terms of the minimization of the energy. The minimization process uses the gradient method as described in [11, 12]. The most evident advantage of this method is its handling of constraints, which allows a larger number of them to be treated at once. For the calculation of fission barriers and to reduce the computational effort we have restricted the calculation to axial symmetry preserving configurations (but reflection symmetry breaking is allowed). The quasiparticle operators are expanded in an axially symmetric harmonic oscillator basis.

Dynamical parameters like quadrupole collective masses and moments of inertia are computed by recurring to the usual approximations of neglecting the interaction in the stability matrix (cranking approximation [13]).

The spontaneous fission half life is computed with the standard WKB formalism where $T_{\rm sf}$

is given (in seconds) by

$$T_{\rm sf} = 2.86\,10^{-21}(1 + \exp(2S)).$$

The action along the $Q_2$ constrained path is

$$S = \int_a^b dQ_2 \sqrt{2B(Q_2)(V(Q_2) - E_0)}.$$

For the collective quadrupole inertia $B(Q_2)$ we have used the Adiabatic Time Dependent HFB (ATDHFB) expression computed in the "cranking" approximation [13]. The quantity $V(Q_2)$ entering the action is the HFB energy minus the zero point energy (ZPE) correction $\epsilon_0(Q_2)$ associated with the quadrupole motion and minus the rotational energy correction given by the rotational approximation with the Yoccoz moment of inertia (see [14, 15] for further details). Finally, an additional parameter $E_0$ is introduced in the action. It is meant to represent the quantal ground state energy obtained after considering quantal fluctuations in the quadrupole degree of freedom. This quantity is often taken as a free parameter or kept fixed to some reasonable value. We have followed the latter approach with $E_0 = 0.5$ MeV.

## 3. Results

Fission barrier calculations for three heavy nuclei, namely $^{246}$Cm (Z=96), $^{252}$No (Z=102), and $^{270}$Sg (Z=106) are discussed. We have compared the results of the calculations with the BCP1 interaction with those obtained by using the Gogny force with the standard D1S parametrization. The reason for this comparison is the success of the Gogny D1S in describing a variety of fission related phenomena [14, 15, 16, 17, 18, 19, 20]. The BCP2 parametrization has not been considered as it gives results which are very close to the BCP1 ones. As mentioned in the previous section, axial symmetry is preserved in the calculations and therefore the first fission barrier won't be reduced as a consequence of triaxiallity. This implies that our results for the spontaneous fission (SF) half life are overestimated.

### *3.1. Curium*

In Fig 1 several properties relevant to fission are depicted for the $^{246}$Cm nucleus and the two EDF considered. On the left panel the HFB energy is given as a function of axial quadrupole moment $Q_2$. Two fission barriers are observed in both cases. Both the inner and outer barriers are lower for BCP1 than for D1S. Please remember that the inner barrier height is very sensitive to triaxial effects not considered in this paper. The excitation energy of the fission isomer is a couple of MeV lower for BCP than for D1S. The lower energies obtained with BCP could be a consequence of the lower surface energy for BCP as compared to D1S [3]. On the right panels and from bottom to top we have the particle-particle correlation energies ($E_{pp} = {\rm Tr}\Delta\kappa$) for protons and neutrons. They are systematically lower for BCP than for D1S indicating weaker pairing correlations in BCP. In the middle panel the octupole $Q_3$ and hexadecapole $Q_4$ moments are depicted. They are very similar in both calculations indicating that the shapes along the fission path are also very similar. Octupole deformation starts to develop at $Q_2 = 50$b (at the fission isomer position) and grows as the nucleus evolves to scission. Finally, in the upper panel the inertia $B(Q_2)$ entering the action in the WKB formula is depicted. The BCP inertia is systematically higher than the D1S one, probably as a consequence of the weaker pairing correlations in BCP (the inertia is roughly proportional to the inverse of the pairing gap).

With some of these quantities (and others not depicted, like the rotational and zero point energy corrections) we can compute the spontaneous fission (SF) half life for this isotope. We obtain the values $4.3\,10^{21}$ s for D1S and $6.2\,10^{17}$ s for BCP1. The shorter half life given by BCP1, as compared to D1S, is the consequence of the balance between the barrier heights which are lower in BCP and favor shorter half lives and the collective inertia that is higher in BCP1

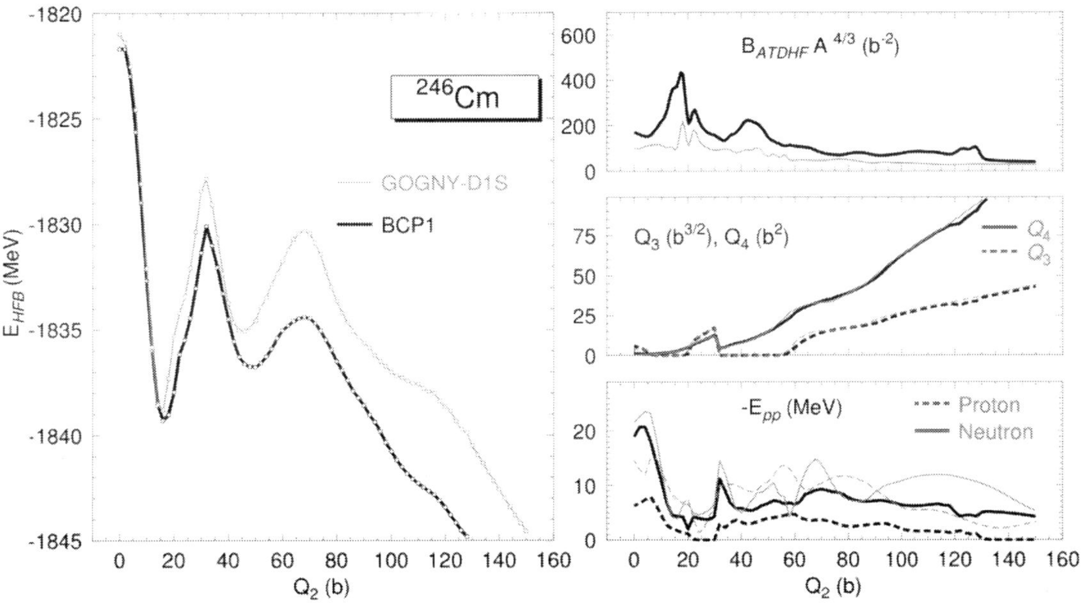


**Figure 1.** Results for the $^{246}$Cm nucleus. On the left panel, the HFB energy as a function of the axial quadrupole moment $Q_2$ obtained with the Gogny D1S EDF (red, thin curve) and the BCP one (black, thick curve). On the right panels, from bottom to top, the particle-particle (pairing) correlation energies for protons (dashed) and neutrons (full), red thin curves for D1S and black thick ones for BCP1. In the middle panel the octupole $Q_3$ and hexadecapole $Q_4$ moments. In the top panel, the collective inertia (see text for details).

what favors longer half lives. Both results should be compared with the experimental value 1.8 $10^7$ yr (5.7 $10^{14}$ s) [21]. Two comments are in order, first the reduction of the first barrier due to triaxiality can lower the half lives although its impact can be softened by the increase of the collective inertias due to triaxiality [16]. The second comment has to do with the sensitivity of half lives to the $E_0$ parameter, by changing it to 1.0 MeV we obtain a half life of 1.5 $10^{13}$ s with BCP, meaning a reduction of four orders of magnitude in the half life.

Finally, by looking at the density matter distribution for very elongated shapes we can identify the nascent fragments and compute their mass contents. The results correspond to an asymmetric mass distribution ($Q_3$ is different from zero at large deformations) with fragments $^{140}$Xe (Z=54, N=86) spherical and $^{106}$Mo (Z=42, N=64) deformed. This result is in good agreement with experiment [22].

### *3.2. Nobelium*

The results for the $^{252}$No isotope are depicted in Fig 2 where we observe very similar behaviors as the ones encountered in the previous subsection for the Curium isotope. The only relevant differences are the severe reduction of the second fission barrier height that makes very unlikely the existence of a fission isomer and the octupole moment which is substantially smaller. The computed half lives are 4.2 $10^5$ s for BCP1 and 3.2 $10^{10}$ s for D1S. The $^{252}$No isotope has a half life of 2.44 s with an $\alpha$-decay branching ratio of 66.7%. It decays via SF with a branching ratio of 32.2% and the remaining is electron capture [23]. Therefore the experimental SF half life is 7.2 s which is several orders of magnitude shorter than the computed values. As discussed in the previous section this is not a severe limitation as the impact of triaxiality has not been considered and the $E_0$ parameter has not been used for fine tuning of the SF half lives. What

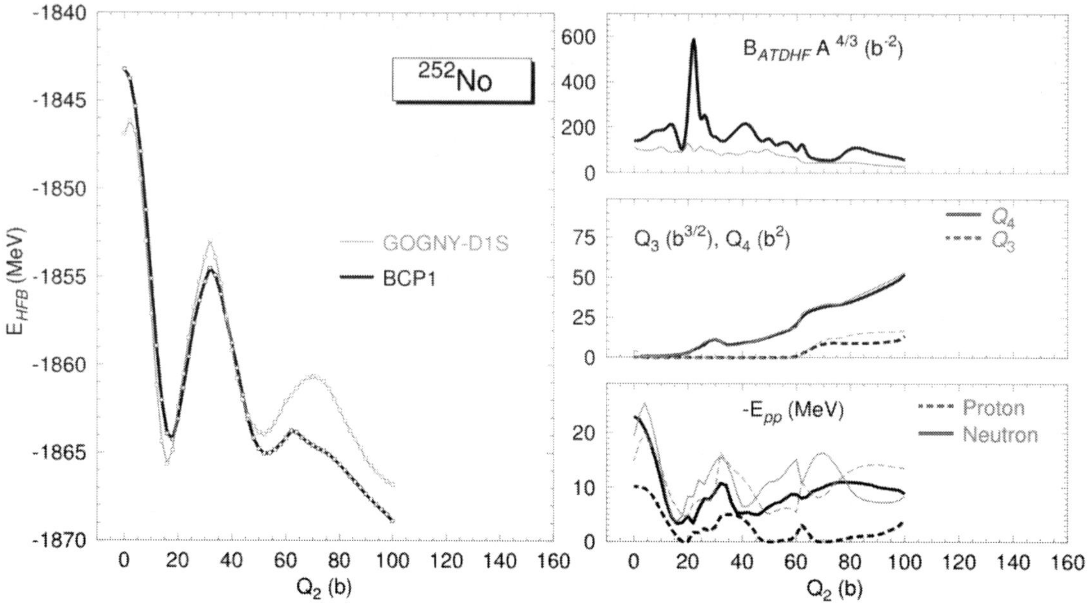


**Figure 2.** Same as in Fig 1 but for the isotope $^{252}$No.

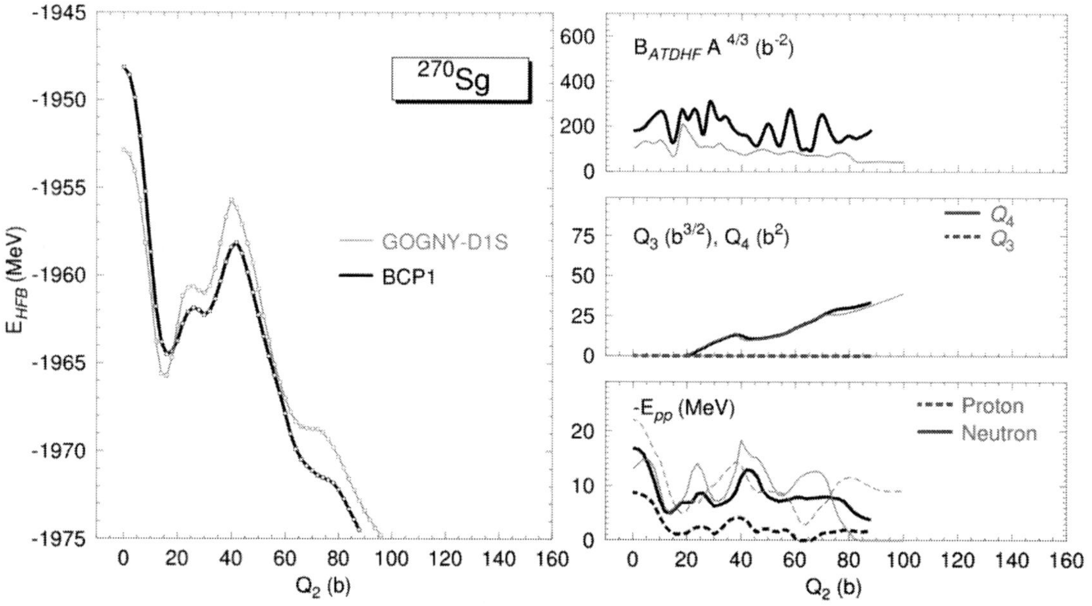


**Figure 3.** Same as in Fig 1 but for the isotope $^{270}$Sg.

is relevant is that under the same circumstances, the reduction in the theoretical predictions in going from $^{246}$Cm to $^{252}$No agree very well with the reduction observed in the experimental data. The mass distribution of the fragments corresponds to the isotopes $^{136}$Xe (Spherical) and $^{116}$Cd (slightly deformed) and agrees well with the experimental indications.

*3.3. Seaborgium*

The results for the $^{270}$Sg isotope are depicted in Fig 3 where we observe very similar behaviors as the ones encountered in the previous subsections for the Curium and Nobelium isotopes. For this heavy nucleus the second fission barrier does not exist. The $^{270}$Sg isotope has Z=106 and N=164 and therefore a symmetric split in two $^{135}$I will be favored [24] as it implies ending up with magic neutron number N=82 in the fragments. We will not study here multimodal fission suggested by the experimental fragment mass distribution and fragment's kinetic energies. We observe that the octupole moment obtained is zero for the whole quadrupole moment range leading to a symmetric split. The values for the half lives are 5.3 s for BCP1 and $1.2\,10^5$ s for D1S. Again, the Gogny D1S half lives are longer than the ones of BCP1.

## 4. Conclusions

Fission properties of the BCP EDF have been analyzed in three characteristic examples. The behavior with mass number of the results obtained with BCP1 is similar to the one of the experimental data and other EDFs. We conclude that the BCP EDF is also well suited for the description of fission making broader its range of applicability.

## Acknowledgments

Work supported in part by MICINN grants Nos. FPA2009-08958, FIS2009-07277, and FPA2008-03865-E/IN2P3 and by the Consolider-Ingenio 2010 program CPAN CSD2007-00042 and MULTIDARK CSD2009-00064. X. V. also acknowledges the support from FIS2008-01661 (Spain and FEDER) and 2009SGR-1289 (Spain). Support by CompStar, a Research Networking Programme of the European Science Foundation is also acknowledged.

## References

[1] M. Baldo, P. Schuck, and X. Vias, Phys. Lett. B **663**, 390 (2008)
[2] M. Bender, P.-H. Heenen, and P.-G. Reinhard, Rev. Mod. Phys. **75**, 121 (2003)
[3] L.M. Robledo, M. Baldo, P. Schuck and X. Viñas, Phys. Rev. **C77**, 051301(R) (2008)
[4] L.M. Robledo, M. Baldo, P. Schuck and X. Viñas, Phys. Rev. **C81**, 034315 (2010)
[5] Hans J. Specht, Rev. Mod. Phys. **46**, 773 (1974); S. Bjrnholm and J. E. Lynn, Rev. Mod. Phys. **52**, 725 (1980); "Fifty Years of Fission", Nucl. Phys. **A502** (1989).
[6] J. Decharge and D. Gogny, Phys. Rev. C**21**, 1568 (1980)
[7] J.F. Berger, M. Girod and D. Gogny. Nucl. Phys. **A428**, 23c (1984)
[8] M. Baldo, C. Maieron, P. Schuck, and X. Vias, Nucl. Phys. **A736**, 241 (2004)
[9] E. Garrido, P. Sarriguren, E. Moya de Guerra, and P. Schuck, Phys. Rev. C **60**, 064312 (1999)
[10] M. Baldo, L.M. Robledo, P. Schuck, and X. Viñas J. of Phys. **G37** 064015 (2010)
[11] P. Ring and P. Shuck, *The Nuclear Many Body Problem* (Springer–Verlag Edt. Berlin, 1980)
[12] J.L. Egido, J. Lessing, V. Martin, and L.M. Robledo Nucl. Phys. **A594** 70 (1995)
[13] M.J. Giannoni and P. Quentin, Phys. Rev. **C21** (1980) 2060; *ibid* 2076
[14] J.L. Egido and L.M. Robledo, Phys. Rev. Lett. **85**, 1198 (2000)
[15] M. Warda, J.L. Egido, L.M. Robledo and K. Pomorski, Phys. Rev. C**66**, 014310 (2002); Intl. J. of Mod. Phys. E**13**, 169 (2004)
[16] J.-P. Delaroche, M. Girod, H. Goutte and J. Libert, Nucl. Phys. A 771 (2006) 103
[17] N. Dubray, H. Goutte, and J.-P. Delaroche, Phys. Rev. C**77**, 014310 (2008)
[18] V. Martin and L.M. Robledo, Int. J. Mod. Phys. **E 18**, 788 (2009).
[19] S. Perez-Martin and L.M. Robledo, Int. J. Mod. Phys. **E 18**, 861 (2009).
[20] W. Younes and D. Gogny, Phys. Rev. C 80, 054313 (2009)
[21] C. Thompson, Phys. Rev C2, 763 (1970)
[22] F. Pleasonton, R. L. Ferguson, F. Plasil, and C. E. Bemis, Jr. Phys. Rev. **C 8**, 1018 (1973).
[23] C. E. Bemis, et al. Phys. Rev. **C 15**, 705 (1977).
[24] M.G. Itkis, et al. Nucl. Phys. **A654**, 870c (1999).

Journal of Physics: Conference Series **321** (2011) 012016 doi:10.1088/1742-6596/321/1/012016

# A new covariant density functional with point-coupling and its application

**J. Meng[1,2], P. W. Zhao[1], L. L. Li[1], Z. P. Li[3], H. Z. Liang[1], J. M. Yao[3]**

[1]State Key Laboratory of Nuclear Physics and Technology, School of Physics, Peking University, Beijing 100871, China

[2]School of Physics and Nuclear Energy Engineering, Beihang University, Beijing 100191, China

[3]School of Physical Science and Technology, Southwest University, Chongqing 400715, China

E-mail: `mengj@pku.edu.cn`

**Abstract.** The covariant density functional theory with a few number of parameters allows a very successful phenomenological description of ground state properties of nuclei all over the nuclear chart. The newly proposed point-coupling parametrization PC-PK1 [1] and its application is reported. In particular, the description of the nuclear ground state properties and the neutron single-particle spectrum in $^{132}$Sn is discussed in details.

## 1. Introduction

The density functional theory (DFT) is well known for its numerous applications to the description of nuclear ground and excited states. In particular, the covariant version of DFT takes the Lorentz symmetry into account in a self-consistent way. During the past years, the covariant DFT has received wide attention due to its successful description of lots of nuclear phenomena [2, 3].

There exist a number of attractive features in the covariant DFT, especially in its practical applications in the self-consistent relativistic mean-field (RMF) framework. The most obvious one is the natural inclusion of the spin degree of freedom and the resulting spin-orbit potential that emerges automatically with the empirical strength in a covariant way. The relativistic effects are responsible for the existence of approximate pseudospin symmetry in the nuclear single-particle spectra [4, 5, 6, 7] and spin symmetry in the anti-nucleon spectra [8]. Moreover, a covariant treatment of nuclear matter provides a distinction between scalar and four-vector nucleon self-energies, leading to a natural saturation mechanism.

The traditional RMF framework is based on the meson-exchange representation, i.e., the Dirac nucleons interact with each other via the exchange of mesons. In recent years, the RMF model with point-coupling interaction (RMF-PC) [9], in which the zero-range point-coupling interaction is used instead of the meson exchange, has attracted more and more attentions owing to the following advantages. First, it avoids the possible physical constraints introduced by explicit usage of the Klein-Gordon equation to describe mean meson fields, especially the fictitious $\sigma$ meson. Second, it is possible to study the role of naturalness in effective theories for nuclear-structure-related problems [10]. Third, it provides more opportunities to investigate its relationship to the nonrelativistic approaches. Finally, it is relatively easy to study the effects beyond the mean-field for nuclear low-lying collective excited states [11].

In practical application of the RMF-PC model, the most widely used parametrizations include PC-LA [9], PC-F1 [12], and DD-PC1 [13]. However, the isospin dependence of the binding energy given by both PC-LA and PC-F1 along either the isotopic or the isotonic chain deviates from the data remarkably. The density-dependent parametrization DD-PC1 was proposed from the equation of state for nuclear matter and the masses of 64 deformed nuclei. Although it reproduces the binding energies, deformations, and charge radii of deformed nuclei quite well, the differences between the predicted binding energies and the corresponding data are somewhat large for spherical nuclei.

In the symposium, we presented the newly proposed point-coupling parametrization PC-PK1 [1] for covariant DFT and its application including the description of

- the nuclear ground state properties [1];
- the neutron single-particle spectrum in $^{132}$Sn [14];
- the magnetic rotation bands in $^{60}$Ni [15];
- the nuclear low-lying excitation states [1].

In this report, only the description of the nuclear ground state properties and the neutron single-particle spectrum in $^{132}$Sn will be discussed in details due to the limitation in length.

## 2. PC-PK1 parametrization

The parametrization of PC-PK1 is determined by a multiparameter fitting to both the binding energies and charge radii for 60 selected spherical nuclei with the Levenberg-Marquardt method. Meanwhile, the empirical neutron pairing gaps for 6 nuclei obtained with the five-point formula are also employed to constrain the pairing strengths. With the experimental observables and the calculated values, the ensemble of parameters can be obtained by minimizing the square deviation. More details for the fitting procedure and coupling constants of the parameter set PC-PK1 can be found in Ref. [1].

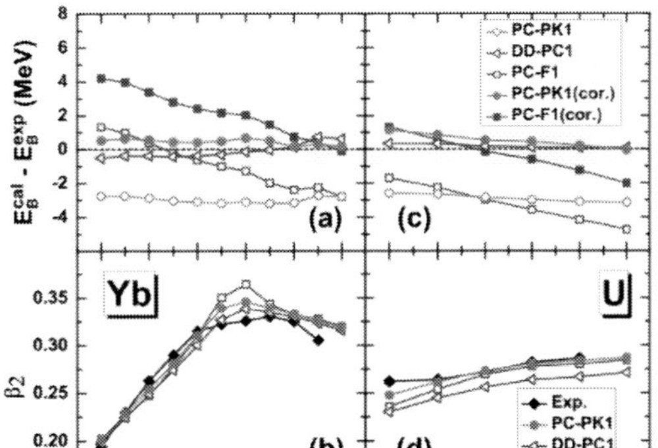


**Figure 1.** Deviations of the calculated binding energies from the data [16] for Yb and U isotopes obtained by PC-PK1, DD-PC1, and PC-F1 (upper panel) as well as the corresponding calculated ground state deformations in comparison with data [17] (lower panel). The filled circles and squares in the upper panels correspond to the rotational corrected values given by PC-PK1 and PC-F1, respectively. Taken from Ref. [1].

## 3. Nuclear ground state properties

Taking Yb and U isotopes as examples, the binding energies and quadrupole deformations of the ground states are investigated. In the upper panels of Fig. 1, the deviations of the calculated binding energies with PC-PK1, DD-PC1, and PC-F1 from the data [16] are shown. When the rotational corrections are not taken into account, a systematic underestimation of the binding energies around 3 MeV for both Yb and U isotopes is found for PC-PK1. For PC-F1, the difference between the calculated and the observed binding energy decreases monotonically with the isospin values. As almost all the isotopes shown in Fig. 1 are used to adjust the parameters, the predictions with DD-PC1 are in good agreement with the data (within 1 MeV).

After taking into account the rotational correction energies with the cranking approximation [18], the calculated results by PC-PK1 reproduce the data quite well (within 1 MeV) for both Yb and U isotopes, while the differences between the corrected binding energies given by PC-F1 and data are still large. Since DD-PC1 is adjusted to the binding energies of well-deformed nuclei, the rotational correction energy is not considered in the corresponding calculations.

The calculated ground state deformations are shown in the lower panels of Fig. 1 in comparison with the data available [17]. It is found that PC-PK1 can provide also good performance in the description of the deformations as well as their corresponding evolutions with neutron number for both Yb and U isotopes.

## 4. Neutron single-particle spectrum in $^{132}$Sn

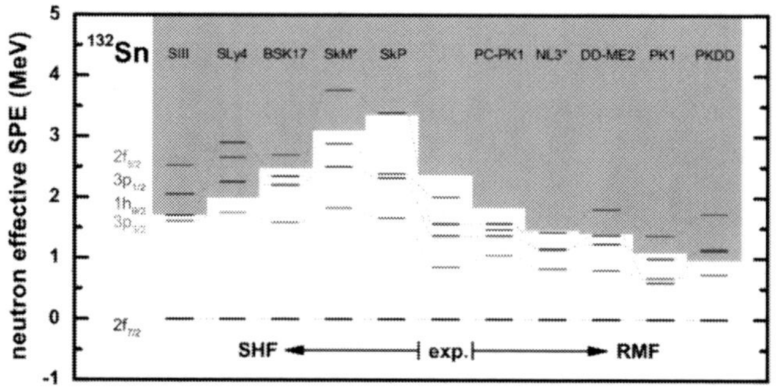


**Figure 2.** Neutron-effective single-particle energies with respect to the orbital $2f_{7/2}$ in $^{132}$Sn calculated by the RMF theory with effective interactions PC-PK1, NL3*, DD-ME2, PK1, and PKDD. The experimental data and the results calculated by the Skyrme-Hartree-Fock theory with SkP, SkM*, BSK17, SLy4, and SIII are also shown for comparison. The shaded areas indicate the area beyond the neutron threshold. Taken from Ref. [14].

The concept of magic numbers is one of the most fundamental ingredients for understanding the nature of atomic nuclei. Recently, by using the $^{132}\mathrm{Sn(d,p)}^{133}\mathrm{Sn}$ reaction, the experiment performed at Oak Ridge National Laboratory revealed for the first time that the spectroscopic factors of the neutron single-particle states $3p_{1/2}$, $3p_{3/2}$, $2f_{5/2}$, and $2f_{7/2}$ outside the $N = 82$ core are consistent with $S \approx 1$ [19]. This is one of the critical pieces of evidence to conclude that $^{132}$Sn is a perfectly radioactive doubly magic nuclei, even better than the stable $^{208}$Pb.

In Fig. 2, the neutron-effective single-particle energies (SPE) with respect to the orbital $2f_{7/2}$ in $^{132}$Sn calculated by the RMF theory with PC-PK1, NL3*, DD-ME2, PK1, and PKDD are compared with the experimental data [19] and results calculated by the Skyrme-Hartree-Fock theory with typical effective interactions SkP, SkM*, BSK17, SLy4, and SIII. At the present mean-field level, it is clearly shown that the overall structure of the neutron-effective SPE for such a neutron-rich doubly magic nucleus can be well reproduced in the relativistic framework, where both the spin-orbit and orbit-orbit potentials are described in a self-consistent way. In contrast, the nonrelativistic results overestimate nearly twice the single-particle energy spacing between the orbitals $3p_{3/2}$ and $2f_{7/2}$, where the reduction of the orbit-orbit potential plays the most important role.

Furthermore, one can extract the Nilsson spin-orbit parameter $C$ and orbit-orbit parameter $D$ by fitting to the single-particle energies. It is found that the RMF results agree quite well with the extracted Nilsson spin-orbit parameter $C$ and orbit-orbit parameter $D$ from experimental SPE of $^{132}$Sn, but remarkably differ from the traditional Nilsson parameters.

## 5. Conclusion

In summary, the newly proposed point-coupling parametrization PC-PK1 [1] for covariant DFT and its application are briefly reported. In particular, the description of the nuclear ground state properties with PC-PK1 is discussed in details. PC-PK1 is determined by fitting to the observables of 60 selected spherical nuclei, including the binding energies, charge radii, and empirical pairing gaps. It is found that PC-PK1 can provide very good descriptions for the binding energies as well as its isospin dependence along either the isotopic or the isotonic chains, which makes it reliable for application in exotic nuclei.

Investigations have also been applied to the single-particle spectrum of the neutron-rich doubly magic nucleus $^{132}$Sn. It is found that the RMF results agree quite well with the experimental single-particle spectrum in $^{132}$Sn as well as the Nilsson spin-orbit parameter $C$ and orbit-orbit parameter $D$ thus extracted, but remarkably differ from the traditional Nilsson parameters.

## Acknowledgements

We thank P. Ring and T. Nikšić for helpful discussions. This work was partly supported by the Major State 973 Program 2007CB815000, the NSFC under Grants No. 10947013, and No. 10975008, China Postdoctoral Science Foundation Grant No. 20100480149, and Southwest University Initial Research Foundation Grant (No. SWU109011 and No. SWU110039) as well as the Fundamental Research Funds for the Central Universities (XDJK2010B007).

## References

[1] Zhao P W, Li Z P, Yao J M and Meng J 2010 *Phys. Rev. C* **82** 054319
[2] Ring P 1996 *Prog. Part. Nucl. Phys.* **37** 193 – 263
[3] Meng J, Toki H, Zhou S, Zhang S, Long W and Geng L 2006 *Prog. Part. Nucl. Phys.* **57** 470–563
[4] Meng J, Sugawara-Tanabe K, Yamaji S, Ring P and Arima A 1998 *Phys. Rev. C* **58** R628
[5] Meng J, Sugawara-Tanabe K, Yamaji S and Arima A 1999 *Phys. Rev. C* **59** 154
[6] Ginocchio J N 2005 *Phys. Rep.* **414** 165
[7] Liang H, Zhao P, Zhang Y, Meng J and Giai N V 2011 *Phys. Rev. C* **83** 041301
[8] Zhou S G, Meng J and Ring P 2003 *Phys. Rev. Lett.* **91** 262501
[9] Nikolaus B A, Hoch T and Madland D G 1992 *Phys. Rev. C* **46** 1757–1781
[10] Friar J L, Madland D G and Lynn B W 1996 *Phys. Rev. C* **53** 3085–3087
[11] Yao J M, Meng J, Ring P and Pena Arteaga D 2009 *Phys. Rev. C* **79** 044312
[12] Bürvenich T, Madland D G, Maruhn J A and Reinhard P G 2002 *Phys. Rev. C* **65** 044308
[13] Nikšić T, Vretenar D and Ring P 2008 *Phys. Rev. C* **78** 034318
[14] Liang H, Zhao P, Li L and Meng J 2011 *Phys. Rev. C* **83** 011302
[15] Zhao P, Zhang S, Peng J, Liang H, Ring P and Meng J 2011 *Phys. Lett. B* **699** 181–186
[16] Audi G, Wapstra A H and Thibault C 2003 *Nucl. Phys. A* **729** 337 – 676
[17] Raman S, Nestor C and Tikkanen P 2001 *At. Data Nucl. Data Tables* **78** 1 – 128
[18] Girod M and Grammaticos B 1979 *Nucl. Phys. A* **330** 40 – 52
[19] Jones K L *et al.* 2010 *Nature* **465** 454

Many-body correlations from dilute to dense nuclear systems (MBC 2011) IOP Publishing
Journal of Physics: Conference Series **321** (2011) 012017 doi:10.1088/1742-6596/321/1/012017

# Skyrme energy-density functional approach to collective dynamics

**Kenichi Yoshida**[1,2], **Nobuo Hinohara**[2], **Takashi Nakatsukasa**[2]

[1] Department of Physics, Graduate School of Science and Technology, Niigata University, Niigata 950-2181, Japan

[2] RIKEN Nishina Center for Accelerator-Based Science, Wako, Saitama 351-0198, Japan

E-mail: kyoshida@nt.sc.niigata-u.ac.jp

**Abstract.** Our recent developments for the microscopic description of nuclear collective dynamics in the framework of the time-dependent density functional theory is presented. It is shown that the quasiparticle random-phase approximation is ready for the systematic investigation of the giant resonances in the entire mass region of nuclear chart and that the collective Hamiltonian approach gives the quantitative description of the low-lying states in transitional nuclei with the Skyrme and pairing energy density functionals as a microscopic input. We put emphasis on necessity of the massive use of high performance computers to carry out such microscopic calculations.

## 1. Introduction

Density functional theory has been widely used to describe a variety of quantum many-body systems, including nuclear many-body systems [1]. Ground-state properties including nuclear deformation are described well by the modern nuclear energy-density functional (EDF) method. Experimental evidences of the nuclear shape change are related to the low-lying quadrupole collectivity, such as the ratio of the excitation energy of $2_1^+$ and $4_1^+$ states, the reduced transition probability $B(E2; 2_1^+ \to 0_1^+)$, etc. Furthermore, it is known that nuclear deformation also affects the high-frequency collective modes of excitation, giant resonances (GRs) [2, 3]. For instance, the peak splitting of the giant dipole resonance (GDR), which is caused by the different frequencies of oscillation along the long and short axes, has been observed in experiments [4]. A typical example of the shape phase transition from spherical to deformed ground states and the evolution of deformation splitting in GDRs have been observed in Nd and Sm isotopes [2, 3, 4, 5, 6].

Magic number or shell closure is an essential concept in understanding the stability against deformation. With the advent of RI beam technology, physics of neutron-rich nuclei has been one of the main current subjects in nuclear physics. It has been revealed that the conventional magic numbers disappear and new magic numbers appear instead in neutron-rich nuclei. Quite recently, the collectivity around $N = 40$ has been extensively studied both experimentally and theoretically. The observed small excitation energy of the $2_1^+$ state in neutron-rich Cr isotopes indicates that the deformation develops toward $N = 40$ [7, 8, 9, 10, 11].

In this paper, we demonstrate that the Skyrme-EDF method describes well the GDR and low-lying collective states in nuclei undergoing the shape phase transition.

## 2. Skyrme-EDF approach for collective excitations using parallel computers

To describe the nuclear deformation and the pairing correlations simultaneously in good account of the continuum, we solve the Hartree-Fock-Bogoliubov (HFB) equations [12] in the coordinate space using cylindrical coordinates. The Skyrme and the density-dependent contact interactions are employed for the particle-hole and particle-particle channels, respectively. We assume axial and reflection symmetries in the ground state. Since the parity ($\pi$) and the magnetic quantum number ($\Omega$) are good quantum numbers, the HFB Hamiltonian takes a block diagonal form with respect to each ($\Omega^\pi, q$) sector, where $q$ stands for neutron or proton. The HFB equations for each sector are solved independently using parallel computers. Then, the densities and HFB Hamiltonian are updated, which requires the communication among processors. The modified Broyden's method [13] is utilized to calculate new densities. The qp states are truncated according to the qp energy cutoff at $E_\alpha \leq 60$ MeV.

We introduce the additional truncation for the quasiparticle-random-phase approximation (QRPA) calculation in terms of the two-quasiparticle (2qp) energy as $E_\alpha + E_\beta \leq 60$ MeV to reduce the number of 2qp states. The calculation of the QRPA matrix elements in the qp basis is performed in parallel computers. In the present calculation, all the matrix elements are real. For diagonalization of the matrix, we use the ScaLAPACK PDSYEV subroutine. Details of the setup for the HFB+QRPA calculation are given in Ref. [14].

To describe the evolution of the transitional quadrupole collectivity, we need to go beyond the QRPA because the dynamical change of the mean-field potential associated with the large-amplitude collective motion should be taken into account. We thus take the quadrupole collective Hamiltonian approach with the local QRPA (LQRPA) method [15] in the present paper. For evaluation of the collective masses, the LQRPA equations are solved on the constrained HFB states along the collective coordinate. As a collective coordinate, we take the axially symmetric quadrupole deformation $\beta$. The collective coordinate is discretized into 20-30 mesh points in the present calculation. The LQRPA equations are solved independently on the discretized points of the collective coordinate. We confirm that the physical quantities calculated by using 20 discretized points get already converged. Details of the model employed and the calculation scheme are given in Ref. [16].

## 3. Photoabsorption cross sections of heavy deformed nuclei: Small-amplitude dynamics

Figure 1 shows the photoabsorption cross sections of $^{238}$U. We have a deformation splitting associated with the prolately deformed ground state. We compare it with the experimental data, and we can see that the peak energies are well reproduced. To calculate the QRPA matrix elements and to diagonalize the matrix, it takes about 580 CPU hours and 330 CPU hours, respectively for the $K^\pi = 0^-$ state. Employing 512 processors, we can calculate the total cross sections in 3.5 hours. Our new QRPA calculation code is ready for the systematic investigation of the GRs in all the nuclei.

We apply this new QRPA calculation code to the GDR in Nd and Sm isotopes undergoing the shape-phase transition [14]. The Skyrme-HFB calculation reproduces well the evolution of the ground-state deformation; onset of deformation between $N = 84$ and 86, and gradual increase toward $N = 92$. Associated with the ground-state deformation, the isotopic dependence of the peak broadening is well reproduced, surprisingly, even for the transitional nuclei. The width for $N = 82$ and 84 is calculated as $\Gamma = 4.5$ MeV, and it gradually increases to about 6 MeV for $N = 88$, then the splitting becomes visible for $N \geq 90$. Here, the width $\Gamma$ is evaluated by fitting the calculated cross section with a Lorentz line. This broadening effect is commonly interpreted as the mode-mode coupling effects to the low-lying collective modes. In the present QRPA calculation, the mode coupling is not explicitly taken into account. However, the QRPA based on the deformed HFB state may implicitly include a part of the coupling effect.

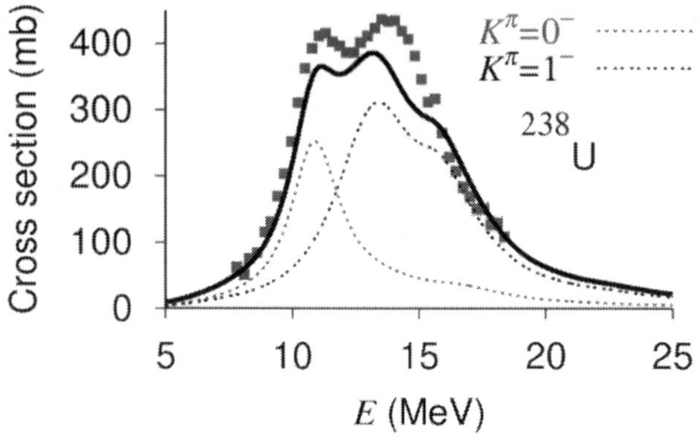


**Figure 1.** Photoabsorption cross sections of $^{238}$U calculated by employing the Skyrme SkM* and the pairing functionals [17]. Experimental data [18] are denoted by filled squares.

The GDR peak energies agree well with experimental values. However, a small increase of the width from $^{142}$Nd to $^{144}$Nd observed in the experiment cannot be fully reproduced in the calculation. Since the HFB calculation produces the spherical ground state for $^{144}$Nd, this requires an explicit higher-order calculation beyond the QRPA. We may notice another small disagreement in the peak shape: The calculated GDR peak has a shoulder in the spherical nuclei, and this shoulder is becoming the third peak in the deformed nuclei. This is due to the Landau fragmentation, however, this feature is not clearly observed in the experiments. We found that detailed properties of the Landau fragmentation depend on the choice of the Skyrme EDF. For instance, the fragmentation effect becomes weaker with the SkP functional, to give a better agreement with the experiments.

## 4. Low-lying states in Cr isotopes: Large-amplitude dynamics

We investigated the deformation properties in neutron-rich Cr isotopes with $N = 34 - 44$ employing the constrained HFB method. In $^{58}$Cr and $^{60}$Cr, the collective potentials are soft against the $\beta$ deformation even we get the HFB minima at $\beta = 0$. Beyond $N = 38$, the prolate minimum gradually develops up to $N = 44$. We can see a shoulder or a shallow minimum at the oblately deformed region as well.

To investigate the effects of the softness of the collective potential on properties of low-lying states in Cr isotopes, the collective Hamiltonian approach with the LQRPA method is employed [16]. When the neutron number increases from $^{58}$Cr, the excitation energy of the $2_1^+$ state drops toward $^{62}$Cr. The ratio of the $E_{4^+}$ to $E_{2^+}$ ($R_{4/2}$) and the transition probability $B(E2; 2_1^+ \to 0_1^+)$ grow up at the same time. These features are consistent with the experimental results [9, 10, 11], and it clearly shows the onset of deformation at $N \sim 38$.

Evolution of deformation is also seen in the matter radii. In the deformed systems, the increase of radius is proportional to $\beta^2$. Thus, beyond $N = 38$ the radius is constantly larger than the systematics of $A^{1/3}$. Furthermore, we can see that the calculated radius is larger than the HFB result in $^{58}$Cr and $^{60}$Cr associated with the large-amplitude fluctuation. Investigating the collective wave functions, we can discuss more about the dynamics microscopically. In $^{58}$Cr, the wave function is distributed around $\beta = 0$. But the wave function is not sharply localized at the spherical minimum of the potential, which reflects the softness of the collective potential against deformation. When two neutrons are added to $^{58}$Cr, broadening of the wave function can be seen. When two more neutrons are added, the wave function moves toward a prolately deformed minimum with a large spreading. From this analysis, we can say that $^{60}$Cr is located close to the critical point of the shape-phase transition in neuron-rich Cr isotopes and the large-amplitude dynamics plays a dominant role.

## 5. Summary

We have investigated effects of the shape transition on $E1$ strength distribution in the rare-earth nuclei, using the newly developed parallelized HFB+QRPA calculation code with the Skyrme EDF. This enables us to simultaneously study both high-energy GRs and low-energy collective and non-collective states. The typical characteristics of the GDR for shape-phase transition from spherical to deformed nuclei, especially the isotopic dependence of broadening and splitting of the GDRs, are extremely well reproduced in the calculation.

We have furthermore developed a new framework of the microscopic model for the large-amplitude collective motion based on the nuclear EDF method. The collective masses and the potential appearing in the quadrupole collective Hamiltonian are evaluated by employing the constrained HFB and local QRPA approach, where the time-odd components of the mean field are fully taken into account. We applied this new framework to the shape-phase transition in neutron-rich Cr isotopes around $N = 40$. The present calculation gives consistent results for the low-lying excited states with the observations and the other theoretical approaches. Investigating the collective wave functions, we reach a conclusion that $^{60}$Cr is located close to the critical point of the shape-phase transition, and the onset of deformation takes place at $N = 38$. The large-amplitude dynamics plays a dominant role in the shape changes in neutron-rich Cr isotopes.

Systematic calculations with the HFB+QRPA and the collective Hamiltonian approach with the LQRPA method employing the massively paralleled computers for spherical-to-deformed and light-to-heavy nuclei help us not only to understand and to predict new types of collective modes of excitation, but also to shed light on the nuclear EDF of new generations.

## Acknowledgments

The work is supported by Grant-in-Aid for Scientific Research (Nos. 23740223, 21340073 and 20105003) and by the Joint Research Program at Center for Computational Sciences, University of Tsukuba. The numerical calculations were performed on RIKEN Integrated Cluster of Clusters (RICC) and T2K-Tsukuba.

## References

[1] Bender M, Heenen P-H and Reinhard P-G 2003 *Rev. Mod. Phys.* **75** 121
[2] Bohr A and Mottelson B R 1975 *Nuclear Structure*, vol. II (Amsterdam: Benjamin)
[3] Harakeh M N and van der Woude A 2001 *Giant Resonances* (Oxford)
[4] Berman B L and Fultz S C 1975 *Rev. Mod. Phys.* **47** 713
[5] Carlos P, Beil H, Bergère R, Leprêtre A and Veyssière A 1971 *Nucl. Phys.* A **172** 437
[6] Carlos P, Beil H, Bergère R, Leprêtre A, De Miniac A and Veyssière A 1974 *Nucl. Phys.* A **225** 171
[7] Sorlin O *et al.* 2003 *Eur. Phys. J.* A **16** 55
[8] Bürger A *et al.* 2005 *Phys. Lett.* B **622** 29
[9] Zhu S *et al.* 2006 *Phys. Rev.* C **74** 064315
[10] Aoi N *et al.* 2009 *Phys. Rev. Lett.* **102** 012502
[11] Gade A *et al.* 2010 *Phys. Rev.* C **81** 051304R
[12] Dobaczewski J, Flocard H and Treiner J 1984 *Nucl. Phys.* A **422** 103
[13] Baran A, Bulgac A, Forbes M M, Hagen G, Nazarewicz W, Schunck N and Stoitsov M V 2008 *Phys. Rev.* C **78** 014318
[14] Yoshida K and Nakatsukasa T 2011 *Phys. Rev.* C **83** 021304R
[15] Hinohara N, Sato K, Nakatsukasa T, Matsuo M and Matsuyanagi K 2010 *Phys. Rev.* C **82** 064313
[16] Yoshida K and Hinohara N 2011 *Preprint* arXiv:1103.5270
[17] Yamagami M, Shimizu Y R and Nakatsukasa T 2009 *Phys. Rev.* C **80** 064301
[18] Veyssière A, Beil H, Bergère R, Carlos P, Leprêtre A and Kernbach K 1973 *Nucl. Phys.* A **199** 45

Journal of Physics: Conference Series **321** (2011) 012018 doi:10.1088/1742-6596/321/1/012018

# Microscopic theory of particle-vibration coupling

**Gianluca Colò, Pier Francesco Bortignon**

Dipartimento di Fisica, Università degli Studi di Milano and INFN, Sez. di Milano, via Celoria 16, 20133 Milano (Italy)

E-mail: colo@mi.infn.it

**Hiroyuki Sagawa**

Center for Mathematics and Physics, University of Aizu, Aizu-Wakamatsu, Fukushima 965-8560 (Japan)

**Kassem Moghrabi, Marcella Grasso, Nguyen Van Giai**

Institut de Physique Nucléaire, Université Paris-Sud, IN2P3-CNRS, 91406 Orsay Cedex (France)

**Abstract.** Some recent microscopic implementations of the particle-vibration coupling (PVC) theory for atomic nuclei are briefly reviewed. Within the nonrelativistic framework, the results seem to point to the necessity of fitting new effective interactions that can work beyond mean field. In keeping with this, the divergences which arise must be cured. A method is proposed, and the future perspectives that are opened are addressed.

## 1. Introduction

The fact that nucleons can be viewed as moving independently in an average potential, lies at the basis of the validity of self-consistent mean field (SCMF) models. These models, either in the nonrelativistic or covariant framework, are now rather sophisticated and produce good results for bulk nuclear properties like masses, radii, or the trends of deformations. A review of SCMF can be found in Ref. [1]. At the same time, the SCMF models do not account well for the single-particle spectroscopic data.

This fact can be considered to be consistent with the idea that SCMF is an implementation of the Density Functional Theory (DFT) for atomic nuclei, and single-particle states may not be within the DFT framework. However, the consequence is that we still lack a fully microscopic model that is able to describe single-particle properties along the isotopic table (in keeping with the limited applicability of *ab-initio* methods, and of the microscopic shell model). This issue has been recently addressed in Ref. [2].

In this reference, results for the single-particle states obtained from a recent implementation of the DFT are compared with those coming from the particle-vibration coupling (PVC) approach. In the case of PVC, results obtained within either a nonrelativistic or relativistic framework are discussed. PVC is supposed to take care of most of the dynamical correlations missing in the static SCMF models, in keeping with the fact that collective vibrations are the most important low-lying states of spherical nuclei. However, the conclusion of Ref. [2] is that it is hard to extract a clear understanding why different theories predict different results - even in the case of a standard nucleus like $^{208}$Pb. Moreover, none of the theories is very accurate in reproducing the experimental data for the centroid (or main peak) of the single-particle strength distributions.

Experimentally, the single-particle strength displays also a non-negligible fragmentation, and theory should account for it. The fragmentation lies definitely outside the framework of DFT. This is perhaps the main motivation to try to generalize this framework, in which the basic variable is the static density, to a many-body theory where dynamical correlations are included. A simple version of this, is the PVC model (also called "dynamical" shell model). There are of course other motivations: for instance, the damping width of single-particle and collective states (like the nuclear giant resonances) can be described only by means of theories beyond SCMF or present day DFT.

If theories beyond mean field are necessary, we should expect to be obliged to re-fit at that level the effective forces that, so far, have been fitted at the mean field level. This has still to be done. The calculations discussed below, namely the recent implementations of PVC both in a nonrelativistic [3] and in a covariant [4] scheme, employ self-consistently the same interactions used for the mean field case like SLy5 or NL3.

In the case of zero-range forces, like the Skyrme interactions that have been used in Ref. [3] as well as in previous works, divergences arise when they are used beyond mean field. These divergences are associated with the use of a large model space. In the practical calculations, the model space is always truncated. For future applications, however, it is desirable to dispose of a regularization procedure.

A first step in this direction has been taken in Ref. [5]. In it, at least for a simplified Skyrme force applied to the case of uniform matter, it has been shown that the second-order divergence can be removed. In fact, for each upper cutoff in the model space, one can re-fit the interaction parameters so that the physical results are unchanged. The purpose of this contribution is to show these new results together, and discuss the perspectives that are open as a consequence.

## 2. Formalism of the particle-vibration coupling approach

Using standard many-body techniques, one can introduce the energy-dependent part $\Sigma(\omega)$ of the self-energy. It enters the well-known Dyson equation,

$$G(\omega) = G_0(\omega) + G_0(\omega)\Sigma(\omega)G(\omega), \tag{1}$$

that allows determining the Green's function $G$ in terms of the unperturbed one $G_0$. In the present context, $G_0$ is the mean-field (i.e., Hartree-Fock) Green's function. The $\Sigma(\omega)$ term is associated with the coupling to (mainly density) vibrations. We shall work in the so-called diagonal approximation, as defined in Ref. [6]: the quantities appearing in the Dyson equation (1) are written in the basis in which the mean-field is diagonal (that is, the HF basis), and only the diagonal matrix elements of $\Sigma(\omega)$ are considered. If the indices $i$, $p$ and $h$ label, respectively, a generic single-nucleon state, a particle state above the Fermi energy and a hole state below the Fermi energy, the diagonal matrix element $\Sigma_i$ reads

$$\Sigma_i(\omega) = \frac{1}{2j_i+1}\left(\sum_{nL,\ p>F}\frac{|\langle i||V||p,nL\rangle|^2}{\omega-\varepsilon_p-\omega_{nL}+i\eta}+\sum_{nL,\ h<F}\frac{|\langle i||V||h,nL\rangle|^2}{\omega-\varepsilon_h+\omega_{nL}-i\eta}\right). \tag{2}$$

The energies $\omega_{nL}$ are the energies of the vibrational states (phonons) with angular momentum $L$, labeled by the index $n$. A (small) imaginary part $\eta$ is added in the denominators (it is set at 0.05 MeV in the calculations presented below). The numerators contain the squared modulus of a reduced matrix element called PVC vertex (see below).

Journal of Physics: Conference Series **321** (2011) 012018 doi:10.1088/1742-6596/321/1/012018

The solution of the Dyson equation in the diagonal approximation is equivalent to the diagonalization of the matrix

$$\begin{pmatrix} \varepsilon_i & \frac{\langle i||V||k_1,n_1L_1\rangle}{\sqrt{2j_i+1}} & \frac{\langle i||V||k_2,n_2L_2\rangle}{\sqrt{2j_i+1}} & \dots \\ \frac{\langle i||V||k_1,n_1L_1\rangle}{\sqrt{2j_i+1}} & \varepsilon_{k_1}+\hbar\omega_{n_1L_1} & 0 & \dots \\ \frac{\langle i||V||k_2,n_2L_2\rangle}{\sqrt{2j_i+1}} & 0 & \varepsilon_{k_2}+\hbar\omega_{n_2L_2} & \dots \\ \dots & \dots & \dots & \dots \end{pmatrix}. \quad (3)$$

In this paper, we will show results obtained by using this method. They will be compared with those obtained by means of perturbation theory, that is, by calculating dressed single-particle energies defined by the equation

$$\tilde{\varepsilon}_i = \varepsilon_i + \Sigma_i(\omega)|_{\omega=\varepsilon_i}. \quad (4)$$

## 3. Results

Indeed, most of the calculations performed in the past have employed second-order perturbation theory, that is, Eq. (4). In the review paper [7], several calculations performed in the 80's are compared and discussed. Qualitatively, they all demonstrate the relevance of the PVC to produce the single-particle strength fragmentation, as well as the increase of the level density (that is, the enhancement of the effective mass $m^*$) close to the Fermi energy which is empirically observed. However, none of them is consistent and it is hard to extract quantitative conclusions due to the many approximations involved.

Only recently, consistent calculations have been carried out. In the calculations presented below, based on the Skyrme interaction, first we solve the Hartree-Fock (HF) equations in coordinate space with the Skyrme parameter set SLy5 [8]. With the same parameter set we also perform fully self-consistent RPA calculations according to the formalism developed in Ref. [9, 10]. Using single-particle and phonon states, we can write and diagonalize the matrix (3). The interaction $V$ is the particle-hole interaction associated with the SLy5 functional. Both the velocity-dependent and velocity-independent parts are taken into account (only the residual two-body spin-orbit is dropped). All details, in particular the choice of the model space and the numerical parameters, are exactly as in Ref. [3], except for the fact that the perturbative expression was used there.

The results are shown in Table 1. They are compared with those obtained in Ref. [3] and the differences turn out to be small. In fact the r.m.s. difference between theory and experiment, which was quoted as 0.62 MeV in [3], is 0.59 MeV in the present work.

In Ref. [3], not only $^{40}$Ca but also $^{208}$Pb has been studied. In the latter case the r.m.s. between theory and experiment is 1.21 MeV: this number is obtained using perturbation theory, but we have seen that a full diagonalization is unlikely to change it signifcantly.

In the case of the relativistic mean field (RMF) plus PVC calculations of Ref. [4], the results seem qualitatively similar but quantitatively better. Perhaps this is due to the fact that the low-lying states display more collectivity in the case of the calculation of Ref. [4], than by employing the Skyrme set SLy5 of the present work. To check this point, new HF plus PVC calculations using one of the Skyrme sets that include tensor terms [13] (for instance using the set T44 that reproduces the values of the electromagnetic transition probabilities of the low-lying states within 10-20% [14]) should be envisaged. In any event, re-fitting the force would be certainly more satisfactory from a conceptual point of view.

## 4. The problem of the divergences beyond mean field

As recalled in the Introduction, the problem of re-fitting a Skyrme force by including contributions beyond mean field is hindered by the presence of divergences. Generally speaking,

**Table 1.** Results for the single-particle neutron energies in $^{40}$Ca, obtained within HF by employing the Skyrme force SLy5 (including the tensor force of Ref. [11] that has a negligible effect in this case), are shown in the second column. In the third and fourth columns we display the dressed single-particle energies obtained either from the perturbative formula (4), as in Ref. [3], or from the diagonalization of the matrix shown in Eq. (3). Finally, in the last column the experimental energies are shown (from Ref. [12]). All energies are in MeV.

| | HF energies SLy5 + tensor | HF plus PVC perturbation theory | HF plus PVC diagonalization | Experimental energies |
|---|---|---|---|---|
| $1f_{5/2}$ | -1.26 | -2.81 | -2.64 | -3.38 |
| $2p_{1/2}$ | -3.11 | -5.16 | -4.93 | -4.76 |
| $2p_{3/2}$ | -5.28 | -7.43 | -7.17 | -6.76 |
| $1f_{7/2}$ | -9.69 | -10.64 | -10.55 | -8.62 |
| $1d_{3/2}$ | -15.17 | -15.80 | -15.76 | -15.64 |
| $2s_{1/2}$ | -17.26 | -18.32 | -18.39 | -18.19 |
| $1d_{5/2}$ | -22.10 | -22.39 | -22.37 | -22.39 |

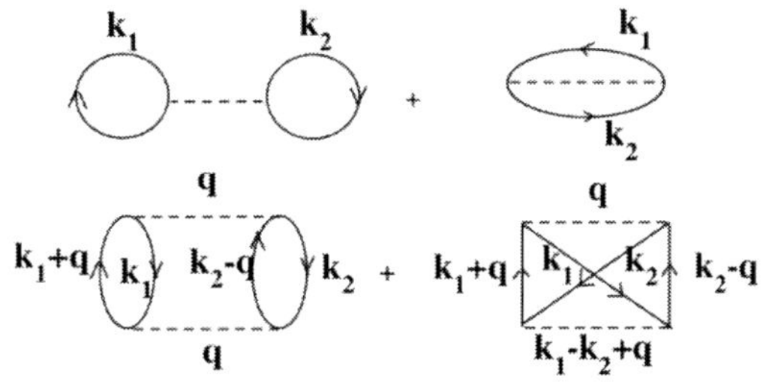


**Figure 1.** First- and second-order diagrams for the total energy in uniform fermionic matter. Labels refer to momentum states.

these divergences prevent from assessing unambiguously the results of calculations beyond mean field. Usually, these calculations are done by truncating the model space in some arbitrary way and this is by far an unsatisfactory procedure.

The self-energy of Eq. (2) belongs to the class of quantities that have an untraviolet divergence in the case of a zero-range interaction. The same can be said for the corrections beyond mean field to the total energy or to other observables like the nuclear radius. Solving this problem is of general interest. Zero-range interactions are also employed for the study of dilute atomic gases [15]. In nuclear physics, the finite-range Gogny force has, nonetheless, a zero-range density-dependent term.

For all these reasons, the problem of ultraviolet divergences has been attacked in Ref. [5]. In this work, the zero-range interaction has been taken as a simplified Skyrme force with only velocity-independent $t_0$ and $t_3$ terms. We have studied the case of the second-order contributions to the total energy (see the bottom row of Fig. 1), in the case of a uniform Fermi system. A simple power counting argument shows the presence of the ultraviolet divergence. Whereas $k_1$ and $k_2$ are limited by the Fermi momentum $k_F$, $q$ can be arbitrarily large and for large values of this momentum transfer the second-order diagrams behave like

$$\sim \int d^3q \, \frac{v^2(q)}{q^2}. \tag{5}$$

For a zero-range force the matrix elements $v$ are independent of $q$ and a linear divergence shows up.

In Ref. [5] the problem has been solved in the following way. The values of the second-order diagrams displayed in Fig. 1 have been calculated analytically as a function of the density $\rho$ of the system, and of the upper limit of the $\vec{q}$-integral defined as $\Lambda$ (the momentum cutoff). These second-order diagrams are found to provide a correction to the total energy $\Delta E$ that can be written as

$$\frac{\Delta E(\rho,\Lambda)}{A} = \chi(\rho) I(\rho,\Lambda). \tag{6}$$

$A$ is the particle number, the value of $\chi$ is

$$\chi(\rho) \equiv -\frac{3}{4\pi^6}\frac{mk_F^7 g^2}{\hbar^2 \rho}, \tag{7}$$

where $m$ is the fermion mass and $g$ is the strength of the zero-range force ($g = t_0 + \frac{1}{6}t_3\rho^\alpha$), and the divergent term reads

$$\begin{aligned} I(\rho,\Lambda) &= \frac{1}{105}(43 - 46\ln 2) - \frac{18}{35} + \frac{\Lambda}{35k_F} + \frac{11\Lambda^3}{210k_F^3} + \frac{\Lambda^5}{840k_F^5} + \frac{16\ln 2}{35} \\ &+ \left(\frac{\Lambda^5}{60k_F^5} - \frac{\Lambda^7}{1680k_F^7}\right)\ln\left(\frac{\Lambda}{k_F}\right) + \left(\frac{1}{35} - \frac{\Lambda^2}{30k_F^2} + \frac{\Lambda^4}{48k_F^4} - \frac{\Lambda^5}{120k_F^5} + \frac{\Lambda^7}{3360k_F^7}\right)\ln\left(-2 + \frac{\Lambda}{k_F}\right) \\ &- \left(\frac{1}{35} - \frac{\Lambda^2}{30k_F^2} + \frac{\Lambda^4}{48k_F^4} + \frac{\Lambda^5}{120k_F^5} - \frac{\Lambda^7}{3360k_F^7}\right)\ln\left(2 + \frac{\Lambda}{k_F}\right). \end{aligned} \tag{8}$$

By making appropriate expansions it is possible to show that this expression is linearly divergent for large values of $\Lambda$, as expected from the above power counting argument.

The next step has been to include the momentum cutoff $\Lambda$ among the parameters of the interaction, and to show numerically that for every value of $\Lambda$ the remaining parameters ($t_0$, $t_3$ and $\alpha$) can be determined in such a way that the total energy of the system remains always the same. Usually, the Skyrme forces are fitted so to reproduce the empirical saturation point of symmetric matter. We have assumed that the entire equation of state (EoS) of nuclear matter obtained with the Skyrme force SkP [16] at the mean field (HF) level, can be taken as a benchmark. This curve can be perfectly reproduced after including the second-order contributions to the energy, for every value of $\Lambda$, by using the renormalized parameters (or, in other words, by adopting a renormalized interaction that can be labeled as $\text{SkP}_\Lambda$). The good reproduction of the EoS can be viewed in Fig. 2, for different realistic values of $\Lambda$.

The procedure discussed in this Section realizes the effective field theory (EFT) program in a simple, yet realistic case. Of course, the cutoff regularization might be replaced by other methods such as dimensional regularization if this turned out to be useful.

## 5. Conclusion

Although the success of mean field methods or, more generally, of DFT-based methods is undeniable, there exist situations where a many-body description in terms of dynamical quantities, like the Green's function and the mass operator, is preferable. In nuclear physics, one of the most important dynamical processes that affect the single-particle properties is the coupling with low-lying, density vibrations. This is the rationale behind the well-known PVC theory. While in the past the PVC calculations contained uncontrolled approximations and phenomenological ingredients, microscopic PVC calculations are nowadays available. In this contribution, we have shown results of a recent Skyrme-HF implementation of PVC in the case

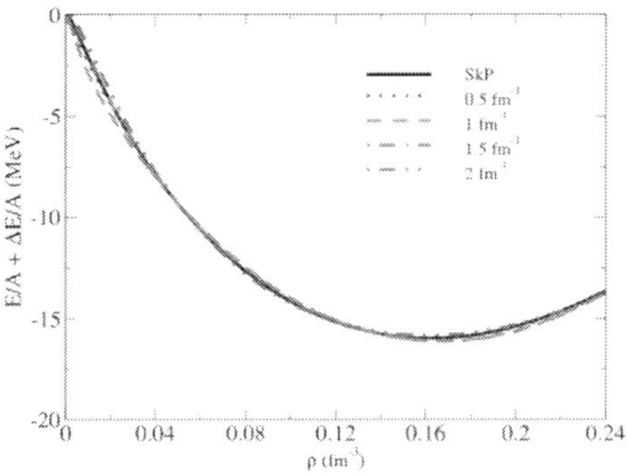


**Figure 2.** Second-order-corrected equations of state compared with the reference equation of state (SkP at mean field level), for different values of the cutoff $\Lambda$.

of $^{40}$Ca, and compared in detail the energy of the neutron states obtained from perturbation theory and from matrix diagonalization. They are quite similar and in better agreement with experiment than the Skyrme-HF results.

At this stage, more systematic calculations should be envisaged, with the purpose of understanding to which extent the accuracy of PVC results can be improved, and whether higher order processes than those included in the PVC model of Sec. 2 are important. In any event, an effective force that has been fitted at the mean field level should be re-fitted when used beyond that level of approximation. In the case of zero-range forces, this is hindered by ultraviolet divergences.

We have shown in Sec. 4 that it is possible to develop strategies for the regularization of these divergences. We have indeed shown this explicitly in the case of the second-order contributions to the total energy in uinform symmetric matter. This paves the way to many new possible steps. Not only PVC models, but models where 2 particle-2 hole, or $n$ particle-$n$ hole couplings with zero-range forces are introduced, can benefit from further refinements of our technique. Using zero-range forces also the RPA correlation energy diverges; one should be able, eventually, to cure this divergence as well.

## References

[1] Bender M, Heenen P H and Reinhard P G 2003 *Rev. Mod. Phys.* **75** 121
[2] Bortignon P F, Colò G, and Sagawa H 2010 *J. Phys. G:* Nucl. Part. Phys. **37** 064013
[3] Colò G, Bortignon P F and Sagawa H 2010 *Phys. Rev.* C **82** 064307
[4] Litvinova E and Ring P 2006 *Phys. Rev.* C **73** 044328
[5] Moghrabi K, Grasso M, Colò G and Giai N V 2010 *Phys. Rev. Lett.* **105** 262501
[6] Ring P and Werner E 1973 *Nucl. Phys.* A **211** 198
[7] Mahaux C, Bortignon P F, Broglia R A and Dasso C 1985 *Phys. Rep.* **120** 1
[8] Chabanat E, Bonche P, Haensel P and Meyer J 1998 *Nucl. Phys.* A **635** 231
[9] Colò G, Bortignon P F, Fracasso S and Giai N V 2007 *Nucl. Phys.* A **788** 173c
[10] Colò G, Cao L, Giai N V and Capelli L 2011 *Comp. Phys. Comm.* (submitted)
[11] Colò G, Sagawa H, Fracasso S and Bortignon P F 2007 *Phys. Lett.* B **646** 227
[12] Oros A 1996 Ph.D. thesis University of Koeln
[13] Lesinski T, Bender M, Bennaceur K, Duguet T and Meyer J 2007 *Phys. Rev.* C **76** 044319
[14] Cao L, Sagawa H and Colò G 2011 *Phys. Rev.* C **83** 034324
[15] Bruun G, Castin Y, Dum R and Burnett K 1999 *Eur. Phys. J.* D **7** 333
[16] Dobaczewski J, Flocard H and Treiner J 1984 *Nucl. Phys.* A **422** 103

# Nuclear surface dynamics and pairing correlations

**F Barranco**[1], **R A Broglia**[2,3,4], **A. Idini**[2,3], **E. Vigezzi**[3]
[1]Escuela Técnica Superior de Ingenieros, Universidad de Sevilla, Spain
[2]Dipartimento di Fisica, Università di Milano, Italy
[3]INFN, Sezione di Milano, Italy
[4]The Niels Bohr Institute, University of Copenhagen, Denmark

E-mail: barranco@us.es

**Abstract.** We solve the Nambu-Gor'kov equations in $^{120}$Sn, demonstrating that the coupling of quasiparticles with collective surface vibrations renormalizes the low-lying part of the nuclear spectrum and influences pairing correlations in a substantial way.

## 1. Introduction

It is well known that single-particle properties in closed shell nuclei are strongly renormalized by the coupling to collective modes [1, 2]. In this paper we study renormalization effects in superfluid nuclei [3, 4, 5]. We shall present results of calculations carried out within the framework of Nambu-Gor'kov equations. The formalism is consistent with the basic rules of Nuclear Field Theory (NFT) which have been extensively applied to the case of normal nuclei (see [6] and refs. therein; see also [7]). It takes into account the essential many-body processes which renormalize the self-energy of the quasiparticles and fragment their strength, increasing the effective mass and the level density close to the Fermi energy, and affecting pairing correlations in a substantial way.

## 2. Formalism

A convenient formalism for the calculation of the properties of quasiparticles in superfluid nuclei taking into account effects beyond mean field was given by Van der Sluys et al. [8]. The derivation of the formalism was based on the equation of motion method, which has a very close relation to the Green's functions formalism used to derive the Nambu-Gor'kov equations, commonly used to study superconductivity in condensed matter physics [9]. We start from a Hartree-Fock (HF) +BCS calculation. The HF basis is calculated making use of an effective interaction, while the BCS calculation is performed with a nucleon-nucleon bare pairing force. In this way one obtains unperturbed quasiparticle energies $E_a$ and amplitudes $u_a, v_a$, associated with a given spherical orbital $a \equiv \{nlj\}$. We then add the coupling between quasiparticles and vibrations, by solving the energy-dependent Nambu-Gor'kov equations

$$\begin{pmatrix} (\epsilon_a - \epsilon_F) + \tilde{\Sigma}^{11}_{a(n)} & \tilde{\Sigma}^{12}_{a(n)} \\ \tilde{\Sigma}^{12}_{a(n)} & -(\epsilon_a - \epsilon_F) + \tilde{\Sigma}^{22}_{a(n)} \end{pmatrix} \begin{pmatrix} \tilde{u}_{a(n)} \\ \tilde{v}_{a(n)} \end{pmatrix} = \tilde{E}_{a(n)} \begin{pmatrix} \tilde{u}_{a(n)} \\ \tilde{v}_{a(n)} \end{pmatrix}, \quad (1)$$

where $\epsilon_a$ and $\epsilon_F$ denote the HF single-particle and the Fermi energy. Eq. (1) has several solutions, which are labeled by the index $n$, leading to a fragmentation of the original

Journal of Physics: Conference Series **321** (2011) 012019 doi:10.1088/1742-6596/321/1/012019

quasiparticle strength; the renormalized energies and amplitudes associated with a given fragment are denoted by $\tilde{E}_{a(n)}$ and by $\tilde{u}_{a(n)}$ and $\tilde{v}_{a(n)}$, while $\tilde{\Sigma}^{11}_{a(n)}$ and $\tilde{\Sigma}^{12}_{a(n)}$ denote the energy-dependent normal and anomalous self-energies. In the following we shall focus on the anomalous self-energy. It can be separated into two terms, writing $\tilde{\Sigma}^{12}_{a(n)} = \Delta^{BCS}_a + \tilde{\Sigma}^{12,pho}_{a(n)}$.

The first term, $\Delta^{BCS}_a$, is the pairing gap associated with the bare interaction obtained in the previous HF+BCS calculation, while the second term is associated with the phonon-induced interaction and is given by

$$\tilde{\Sigma}^{12,pho}_{a(n)} = -\sum_{b,m} \frac{(2j_b+1)}{2} V_{ind}(a(n)b(m))\tilde{u}_{b(m)}\tilde{v}_{b(m)}, \tag{2}$$

where we have introduced the induced pairing interaction:

$$V_{ind}(a(n)b(m)) = \sum_{\lambda,\nu} \frac{2h^2(ab\lambda\nu)}{(2j_b+1)} \times \left[\frac{1}{\tilde{E}_a(n) - \tilde{E}_b(m) - \hbar\omega_{\lambda\nu}} - \frac{1}{\tilde{E}_a(n) + \tilde{E}_b(m) + \hbar\omega_{\lambda\nu}}\right]. \tag{3}$$

The term $h(ab\lambda\nu)$ denotes the basic particle-vibration coupling vertex, which will be calculated following the collective model by Bohr and Mottelson (cf. [10], Eq. (6.209)):

$$h(ab\lambda\nu) = -(-1)^{j_a-j_b}\chi_\lambda\beta_\lambda < a|r_1\frac{\partial U}{\partial r_1}|b >< j_b||Y_\lambda||j_a > \left[\frac{1}{(2j_a+1)(2\lambda+1)}\right]^{1/2}. \tag{4}$$

The coefficient $\chi_\lambda$ multiplies the self-consistent value of the coupling constant introduced in [10], in order to reproduce the experimental deformation parameters $\beta_\lambda$ and the energies $\hbar\omega_{\lambda\nu}$ of the lowest $2^+, 3^-, 4^+$ and $5^-$ modes in our QRPA calculation with a separable force; typically $\chi_\lambda \approx 0.9$.

Furthermore, we can symmetrize the matrix (1) in order to get a $2\times 2$ eigenvalue equation which is formally identical to the BCS eigenvalue equation. This can be achieved multiplying Eq. (1) by the $Z_{a(n)}$ energy-dependent function, $[Z_{a(n)}]^{-1} = 1 - \tilde{\Sigma}^{odd}_{a(n)}/\tilde{E}_{a(n)}$, where $\tilde{\Sigma}^{odd}_{a(n)}$ is the odd part of $\tilde{\Sigma}^{11}_{a(n)}$. It is then possible to rewrite Eq. (1) as

$$\begin{pmatrix} \tilde{\epsilon}_{a(n)} - e_F & \tilde{\Delta}_{a(n)} \\ \tilde{\Delta}_{a(n)} & -(\tilde{\epsilon}_{a(n)} - e_F) \end{pmatrix} \begin{pmatrix} \tilde{u}_{a(n)} \\ \tilde{v}_{a(n)} \end{pmatrix} = \tilde{E}_{a(n)} \begin{pmatrix} \tilde{u}_{a(n)} \\ \tilde{v}_{a(n)} \end{pmatrix} \tag{5}$$

where $\tilde{\epsilon}_{a(n)} - e_F = Z_{a(n)}\left[(\epsilon_a - e_F) + \tilde{\Sigma}^{even}_{a(n)}\right]$, and where $\tilde{\Sigma}^{even}_{a(n)}$ is the even part of $\tilde{\Sigma}^{11}_{a(n)}$. The term $\tilde{\epsilon}_{a(n)}$ in Eq. (5) represents the renormalized single-particle energy, and one can now identify the pairing gap with the term $\tilde{\Delta}_{a(n)}$ [9]:

$$\tilde{\Delta}_{a(n)} = Z_{a(n)}\tilde{\Sigma}^{12}_{a(n)} = Z_{a(n)}\left(\Delta^{BCS}_a + \tilde{\Sigma}^{12,pho}_{a(n)}\right) \equiv \tilde{\Delta}^{bare}_{a(n)} + \tilde{\Delta}^{pho}_{a(n)}. \tag{6}$$

## 3. Results

Summarizing, our calculations are carried out in the following steps:

(a) We perform a HF calculation to obtain the single-particle levels which are going to be renormalized. We make use of the effective interaction SLy4, whose effective mass $m_k \approx 0.7m$ leads to a single-particle density in reasonable agreement with experiment, after renormalization effects are taken into account.

(b) We work out a BCS calculation with the bare nucleon-nucleon interaction Argonne $v_{14}$ in the HF basis. This calculation takes into account the coupling with orbitals lying at very high

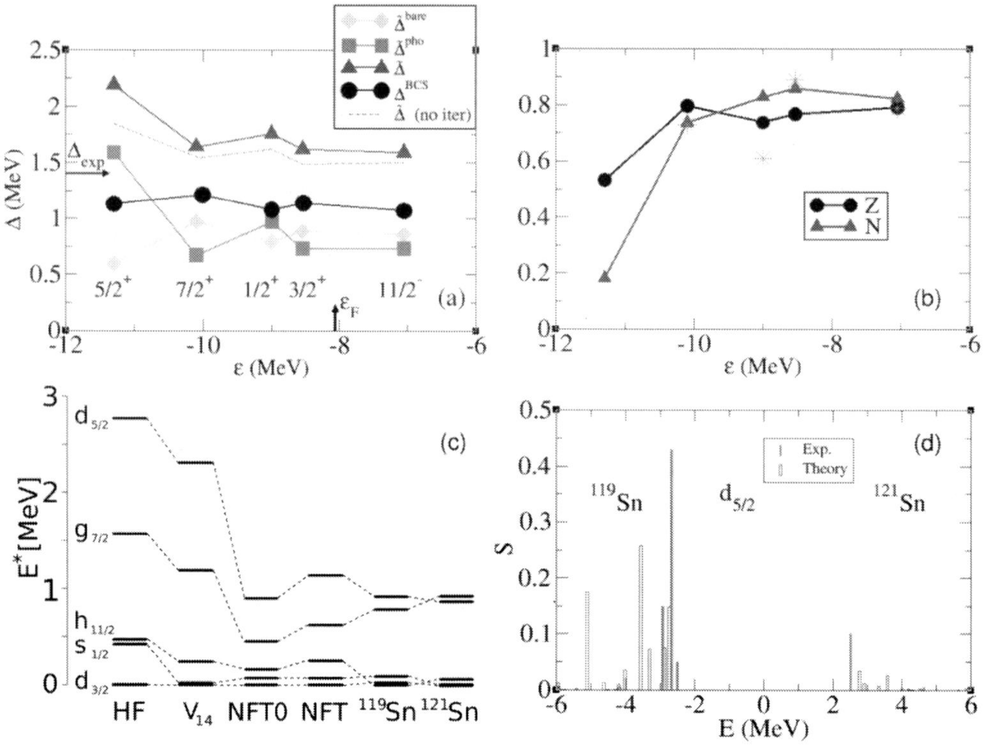


**Figure 1.** Results obtained solving the Nambu-Gor'kov equations in $^{120}$Sn. (a) State-dependent neutron paring gaps (see the text). The values of the Fermi energy $\epsilon_F$ and of the gap obtained from the experimental odd-even mass difference $\Delta_{exp}$ are also indicated. (b) $Z-$ factor and quasiparticle strength $N$ associated with the lowest quasiparticle peaks. The values of the experimental quasiparticle strength [13] are indicated by stars, except for the $d_{5/2}$ orbital which shows a pronounced fragmentation. (c) The theoretical quasiparticle spectra obtained at the various steps of the calculation are compared to the experimental data (see the text). (d) Theoretical and experimental strength functions associated with the $d_{5/2}$ orbital.

energy (about 1 GeV) due to the effect of the hard core, but we will be interested in the values of the quasiparticle energies and amplitudes of the few valence orbitals close to the Fermi energy. We remark that the absolute value of the resulting pairing gap is sensitive (within 20-30%) to the momentum dependence of the effective mass at large momenta, which for the time being represents an open theoretical problem [11, 12].

(c) We do a QRPA calculation obtaining a phonon spectrum which will be coupled to the quasiparticles obtained in point (b).

(d) We solve by iteration the Nambu-Gor'kov equations (1), obtaining a renormalized low-lying spectrum, and a redistribution of the quasiparticle strength. At the same time, we obtain the renormalized value of the pairing gap $\tilde{\Delta}$.

The main results are shown in Fig. 1. In Fig. 1(a) we show the pairing gap $\Delta^{BCS}$ obtained at the BCS level (filled dots), which displays a very limited state-dependence and is close to 1.1 MeV. The gap is shown as a function of the unperturbed single-particle energy in the HF field for the five orbitals $d_{5/2}$, $g_{7/2}$, $s_{1/2}$, $d_{3/2}$ and $h_{11/2}$. The value of the BCS gap is diminished by the spectroscopic $Z$−factor shown in Fig. 1(b) ($Z \approx 0.8$), leading to a value $\tilde{\Delta}^{bare} = Z\Delta^{BCS} \approx 0.8$ MeV (diamonds). We also show in Fig. 1(b) the quasiparticle strength $N$ associated with the lowest fragments for each orbital, compared to the experimental findings. The total state-dependent gap (triangles) is obtained by summing $\tilde{\Delta}^{bare}$ with the phonon-induced part $\tilde{\Delta}^{pho} \approx 0.8$ MeV (squares). We then obtain a total gap $\tilde{\Delta} \approx 1.6$ MeV, that can be compared with the typical estimate of the gap derived from the odd-even mass difference, which is in the range 1.35-1.45 MeV, depending on the particular formula used. Also displayed in Fig. 1(a) (dashed line) is the gap obtained after the first iteration of the Nambu-Gor'kov equations. The low-lying quasiparticle spectrum obtained at the various stages of the calculation is displayed in Fig. 1(c), where the positions of the lowest fragment associated with each of the five valence orbitals are indicated. In the first column (labeled HF) we show the quantities $|\epsilon_a - \epsilon_F|$ resulting from the HF calculation, referred to the value for the level $d_{3/2}$, lying closest to the Fermi energy. In the second column ($v_{14}$) we report the quasiparticle energies obtained from the BCS calculation. In the third and fourth column we give the quasiparticle energies obtained taking into account the renormalization processes by solving the Nambu-Gor'kov equation, after the first iteration (NFT0) and at convergence (NFT). The theoretical results are compared to the experimental spectra in the last two columns. It is seen that the particle-vibration coupling increases the level density, leading to a much better agreement with experiment. While the quasiparticle strength of the three orbitals lying close to the Fermi energy is well described by the BCS quasiparticle picture, the $d_{5/2}$ orbital becomes strongly fragmented (cf. Fig. 1(d)).

## 4. Conclusions and outlook

One can conclude that it is possible, at least in the case of $^{120}$Sn, to draw a picture that is consistent with the available experimental data concerning the quasiparticle properties close to Fermi energy. However, several elements should be further investigated, before one can reach firm quantitative theoretical results, in particular concerning the absolute value of the total pairing gap. On the one hand, the mean field should be either derived microscopically, or at least refitted comparing experiment with theoretical results taking into account renormalization effects; on the other hand, contributions which are expected to provide a repulsive contribution to the pairing interaction, associated with three-body effects and with the influence of spin modes, should also be considered in detail. As suggested by P. Schuck at this Conference, it would be interesting to investigate renormalization effects in finite nuclei in the case of $T = 0$ pairing gap, which has been found to take very large values in nuclear matter [14, 15].

## References

[1] Mahaux C, Bortignon PF, Broglia RA and Dasso CH 1985 *Phys. Rep.* **120** 1
[2] Soloviev VG 1992 *Theory of atomic nuclei. Quasiparticles and nuclei* (Institute of Physics Publishing)
[3] Terasaki J, Barranco F, Broglia RA, Vigezzi E and Bortignon PF 2002 *Nucl. Phys. A* **697** 127
[4] Barranco F, Broglia RA, Colò G, Vigezzi E and Bortignon PF 2004 *Eur. Jou. Phys. A* **21** 57
[5] Idini A, Barranco F and Vigezzi E 2011 *nucl-th*
[6] Bortignon PF, Broglia RA, Bes DR and Liotta R 1977 *Phys. Rep.* **30C** 305
[7] Bes DR and Kurchan J (1990) *The treatment of collective coordinates in many-body systems. An application of the BRST invariance* (World Scientific)
[8] Van der Sluys V, Van Neck D, Waroquier M and Ryckebusch J 1993 *Nucl. Phys. A* **551** 210
[9] Schrieffer JR 1964 *Superconductivity* (Benjamin)
[10] Bohr A and Mottelson B R 1975 *Nuclear Structure, Vol II* (Benjamin)
[11] Hebeler K, Duguet T, Lesinski T and Schwenk A 2009 *Phys. Rev. C* **80** 044321
[12] Baldo M, Lombardo U, Pankratov SS and Saperstein EE 2010 *J. Phys. G* **37** 064016
[13] Dickey SA, Kraushaar JJ, Ristinen R and Rumore MA 1982 *Nucl. Phys. A* **377** 137
[14] Garrido E, Sarriguren P, Moya de Guerra E, Lombardo U, Schuck P and Schulze HJ 2001 *Phys. Rev. C* **63** 037304
[15] Cao LG, Lombardo U and Schuck P 2006 *Phys. Rev. C* **74** 064301

Many-body correlations from dilute to dense nuclear systems (MBC 2011) IOP Publishing
Journal of Physics: Conference Series **321** (2011) 012020 doi:10.1088/1742-6596/321/1/012020

# Finite-size analysis of the Fermi liquid properties of the homogeneous electron gas

**Markus Holzmann**[a,b*], **Bernard Bernu**[a], **and David M. Ceperley**[c]
[a]LPTMC, UMR 7600 of CNRS, Université Pierre et Marie Curie, 4 Place Jussieu, 75005 Paris, France
[b]Univ. Grenoble 1/CNRS, LPMMC UMR 5493, Maison des Magistères, 38042 Grenoble, France
[c]Dept. of Physics and NCSA, U. of Illinois at Urbana-Champaign, Urbana, IL 61801, USA
E-mail: *markus@lptmc.jussieu.fr

**Abstract.** We analyze the extrapolation to the thermodynamic limit of Fermi liquid properties of the homogeneous electron gas in two and three dimensions. Using field theory, we explicitly calculate finite-size effects of the total energy, the renormalization factor, and the effective mass at the Fermi surface within the random phase approximation (RPA) and discuss the validity for general metallic systems.

## 1. Introduction

We consider the Fermi liquid phase of the (unpolarized) homogeneous electron gas in two and three dimensions ($d = 2, 3$) at zero temperature, $T = 0$. Quantum Monte Carlo (QMC) calculations [1, 2] have not only provided the most precise evaluations of the ground state energy, $E$, as a function of density, $n$, but also give access to spectral properties, *e.g.* the momentum distribution, $n_k$, the effective mass, $m^*$, of quasiparticle excitations, and the corresponding spectral weight, the renormalization factor $Z$ [2, 3, 4, 5, 6, 7, 8, 9, 10]. Whereas QMC provides a powerful strategy to quantify correlations in Fermi systems, it still suffers from two fundamental limitations: the fixed-node error, and the extrapolation to the thermodynamic limit. As shown recently, calculations of the Fermi liquid parameters are drastically affected by size effects [4, 7], since typical QMC computations are done with $N \lesssim 10^3$ electrons, but extrapolation schemes based on analytical knowledge of underlying properties significantly reduce the finite size error [11, 12, 4, 7]. Here, to provide further insight, we address the issue of extrapolating calculations of finite (small) systems to the infinite volume limit within common language of field theory. Strategies of improving the many-body wave function to reduce the fixed-node error in QMC are discussed elsewhere [13, 14, 15].

At high densities, $r_s \to 0$, the random phase approximations (RPA) is expected to be valid, and we explicitly calculate leading order size corrections within this approximation, and discuss their general validity. Here, $r_s = a/a_B$ with $a = [2(d-1)\pi n/d]^{-1/d}$ is the mean inter particle distance, $a_B = \hbar^2/me^2$ the Bohr radius, and $e$ and $m$ are the charge and mass of the electrons. Our results agree with the leading order corrections in [11, 4, 6, 7] where finite size corrections have been derived from a formulation directly based on the many-body wave function. In the following, we explicitly concentrate on three fundamental properties, the total energy, $E$, the renormalization factor, $Z = n_{k_F^-} - n_{k_F^+}$, and the effective mass, $m^*$, at the Fermi surface, $k_F$.

## 2. Formalism

From the dielectric function, $\epsilon(q,\omega)$, the ground state energy per particle, $E$, inside a box of volume $V$ with periodic boundary conditions, and $n = N/V$, can be obtained via coupling constant integration [16]

$$E = E_0 + \frac{1}{2V}\sum_{\mathbf{q}\neq 0}\left(-v_q - \frac{\hbar}{n}\int_0^1 d\lambda \int_0^\infty \frac{d\omega}{\pi}\left[\frac{1}{\epsilon(q,i\omega;\lambda)} - 1\right]\right) \tag{1}$$

where $v_q = 2(d-1)\pi e^2/q^{d-1}$ is the Coulomb potential and

$$E_0 = \frac{2}{N}\sum_{\mathbf{k}\leq k_F} \varepsilon_k \tag{2}$$

is the ideal gas kinetic energy with $\varepsilon_k = \hbar^2 k^2/2m$. In the following we set $\hbar = 1$. Within the RPA [17, 18], the dielectric function is given in terms of the ideal-gas density-density response function, $\chi_0(k,\omega)$,

$$\epsilon_{RPA}(k,\omega) = 1 - v_k \chi_0(k,\omega) \tag{3}$$

and the coupling constant integration can be done explicitly

$$E_{RPA} = E_0 + \frac{1}{N}\sum_{\mathbf{q}\neq 0}\left[-\frac{v_q}{2n} + \int_0^\infty \frac{d\omega}{2\pi}\log \epsilon_{RPA}(\mathbf{q},i\omega)\right] \tag{4}$$

Spectral properties are obtained from the single particle Green's function, $G(k,\omega)$, which equals

$$G^{-1}(k,\omega) = \omega + \mu - \varepsilon_k - \Sigma(k,\omega) \tag{5}$$

where the self energy, $\Sigma(k,\omega)$ accounts for interaction effects, and $\mu$ is the chemical potential which fixes the Fermi energy, $\varepsilon_F = \mu$, via $\varepsilon_F = \varepsilon_{k_F} + \Sigma(k_F, 0; \mu \equiv \varepsilon_F)$. From a frequency integration over the Green's function, the momentum distribution, $n_k$, can be obtained, and the magnitude of the jump at the Fermi surface, $Z = n_{k_F^-} - n_{k_F^+}$, is directly related to the residual of $G$ at ($k_F$,$\varepsilon_F$). In general, poles of the Green's function give rise to quasi-particle excitations, whose energies are given in terms of an effective mass $m^*$ at the Fermi surface.

Within the RPA, the self energy equals [19, 20, 21]

$$\begin{aligned}\Sigma_{RPA}(k,\omega) \quad = \quad & -\frac{1}{V}\sum_{\mathbf{q}\neq 0}\left\{\frac{v_q}{2} + \frac{v_q}{\epsilon(q,\varepsilon_{\mathbf{k+q}}-\omega)}\left[\theta(\varepsilon_F - \varepsilon_{\mathbf{k+q}}) - \theta(\omega - \varepsilon_{\mathbf{k+q}})\right]\right. \\ & \left. + \int_{-\infty}^{\infty}\frac{d\nu}{(2\pi)}\frac{v_q}{\epsilon(q,i\nu)}\frac{1}{i\nu + \omega - \varepsilon_{\mathbf{k+q}}}\right\}\end{aligned} \tag{6}$$

where $\varepsilon_F = k_F^2/2m$ is the unperturbed Fermi energy. From the self-energy, the renormalization factor, $Z$, at $k = k_F$ is given by

$$Z^{-1} = 1 - \frac{\partial \Sigma(k_F,\omega)}{\partial\omega}\Big|_{\omega=\varepsilon_F} \tag{7}$$

and the effective mass, $m_\sigma^*$, which characterizes the quasi-particle dispersion at the Fermi surface, is then obtained via

$$\frac{m}{m^*} = Z\left(1 + \frac{m}{k_F}\frac{\partial\Sigma(k,\varepsilon_F)}{\partial k}\Big|_{k=k_F}\right) = \frac{1 + \frac{m}{k_F}\frac{\partial\Sigma(k,\varepsilon_F)}{\partial k}\Big|_{k=k_F}}{1 - \frac{\partial\Sigma(k_F,\omega)}{\partial\omega}\Big|_{\omega=\varepsilon_F}} \tag{8}$$

Formally, the RPA contains the leading order corrections beyond the independent particle Hartree-Fock approximation, and its validity is restricted to the high density region, $r_s \to 0$. In this limit, it is consistent to expand Eq. (8). This leads to the so-called on-shell approximation (OSA) of the effective mass,

$$\frac{m}{m^*_{OSA}} = 1 + \frac{m}{k_F}\frac{\partial\Sigma(k,\varepsilon_F)}{\partial k}\Big|_{k=k_F} + \frac{\partial\Sigma(k_F,\omega)}{\partial\omega}\Big|_{\omega=\varepsilon_F}. \tag{9}$$

In the literature, there exists a long-standing controversy [19, 22, 23, 24, 25], whether Eq. (8) or Eq. (9) should be used to determine the effective mass. Strictly speaking, the validity of the RPA is confined to the high density region where both formulas essentially agree, $r_s \lesssim 1$, and the validity of RPA or RPA-like results at lower densities essentially relies on cancelation of errors. However, given the exact self-energy beyond RPA, only Eq. (8) provides the effective mass, and our finite-size analysis below is based on Eq. (8) , since, as we will see, our results are not restricted to the RPA.

## 3. Thermodynamic limit extrapolation of finite system results

All formulas of the previous section are valid for systems of finite size, $L = V^{1/d}$, where wave vectors are discrete. Perturbative expressions for the Green's function and the dielectric functions, as *e.g.* the RPA expressions, are based on the linked-cluster theorem, valid in a grand-canonical ensemble of fixed volume introducing small number fluctuations. Since the non-interacting Green's function does not depend explicitly on the system size, the extrapolation from the finite to the infinite system corresponds to the conversions of all underlying discrete sums to integrals, $(2\pi/L)^d \sum_{\mathbf{q}} \ldots \to \int d\mathbf{q} \ldots$, and finite size corrections are equivalent to discretization errors in quadrature. On this basis, the elimination of the size error in the second term on the rhs of the total energy expression, Eq. (1), is straightforward, leading to the Madelung constant

$$v_M = \sum_{\mathbf{q}} \frac{v_q}{2V} - \int \frac{d\mathbf{q}}{(2\pi)^d}\frac{v_q}{2}. \tag{10}$$

In the following, we show that a similar analysis of the quadrature error can be used to reduce the finite size error in other quantities.

If the integrand is regular, the discretization error is rapidly vanishing (of order $1/N$ or higher). The main contribution of order $N^{-\alpha}$ with $\alpha < 1$ comes from non-analytical points in the integrand. From the general structure of the perturbation expansion there are only certain discrete points where non-analyticities can develop: in the long wavelength limit, $k = 0$, where the Coulomb potential is singular, and for integer multiples of the Fermi wave-vector $k = j\,k_F$ ($j = 1, 2, \ldots$) due to the sharp Fermi surface of the underlying wave function.

Explicitly, the sharp Fermi surface at $k_F$ leads to well-known shell oscillations in the energy, already present in the energy of the ideal Fermi gas, Eq. (2). Twisted boundary conditions strongly reduce these oscillations [26], and grand-canonical twist averaging (GTABC) completely eliminates them [11]. Within GTABC, we have

$$E_0^{GTABC} = \frac{2}{N}\int_{|\vartheta_\alpha|\le\pi/L}\frac{d\vartheta}{(2\pi/L)^d}\sum_{\mathbf{k}_\alpha=2\pi m/L}\theta(\varepsilon_F - \varepsilon_{\mathbf{k}+\vartheta})\varepsilon_{\mathbf{k}+\vartheta} = \frac{2}{n}\int\frac{d\mathbf{k}}{(2\pi)^d}\theta(\varepsilon_F - \varepsilon_{\mathbf{k}})\varepsilon_{\mathbf{k}} \tag{11}$$

and $E_0^{GTABC} \equiv E_0(N \to \infty)$. Similarly, GTABC also eliminates size effects of other non-interacting properties, *e.g.* in the ideal gas density response, $\chi_0^{GTABC}(q,\omega) \equiv \chi_0(q,\omega; N \to \infty)$, as can be seen from the spectral representation

$$\chi_0^{GTABC}(q,\omega) = \frac{2}{V} \int_{|\theta_\alpha| \leq \pi/L} \frac{d\vartheta}{(2\pi/L)^d} \sum_{\mathbf{k}_\alpha = 2\pi m/L} \frac{\theta(\varepsilon_F - \varepsilon_{\mathbf{k}+\vartheta}) - \theta(\varepsilon_F - \varepsilon_{\mathbf{k}+\vartheta+\mathbf{q}})}{\omega + \varepsilon_{\mathbf{k}+\vartheta} - \varepsilon_{\mathbf{k}+\vartheta+\mathbf{q}} + i\eta} \tag{12}$$

although $\chi_0$ is still only given at discrete wave vectors ($\mathbf{q}_\alpha = 2\pi j/L$ with integer $j$). Therefore, twisted boundary conditions as well as GTABC do not modify the discrete sum in the exchange-correlation energy on the rhs of Eq. (1). However, GATBC lead to an accelerated convergence of the integrand towards the thermodynamic limit on all allowed wave vectors.

Within GTABC, we assume that size effects in the integrands of all expressions can be neglected, and focus on the presumably largest contributions from the Coulomb singularity at $k = 0$. For the total energy, Eq. (1), we expect that the leading order size corrections of the exchange-correlation energy, $\delta E$, is given by

$$\delta E \simeq \int_{|\mathbf{q}_\alpha| \leq \pi/L} \frac{d\mathbf{q}}{(2\pi)^d} \int_0^1 d\lambda \int_0^\infty \frac{d\omega}{2\pi n} \left[ \frac{1}{\epsilon(q, i\omega; \lambda)} - 1 \right]. \tag{13}$$

Similarly, the effect of the thermodynamic limit extrapolation on the renormalization factor and the effective mass can be obtained from explicit (approximate) expressions of the self-energy, Eq. (6), together with Eq. (7) and Eq. (8).

## 4. Results

The Lindhard function, $\chi_0(q,\omega)$, has the following limiting form in the long wave length limit

$$\chi_0(k,\omega) \approx nk^2/m\omega^2, \quad k \to 0 \tag{14}$$

and we have

$$\epsilon(k,\omega) \approx 1 - \omega_p^2(k)/\omega^2, \quad k \to 0 \tag{15}$$

where the plasma frequency is given by $\omega_p(k) = (nv_k k^2/m)^{1/2}$.

### *4.1. Total energy*

Inserting the limiting form of the dielectric function in Eq. (4), we obtain

$$\delta E = \frac{1}{n} \int_{|\mathbf{q}_\alpha| \leq \pi/L} \frac{d\mathbf{q}}{(2\pi)^d} \int \frac{d\omega}{2\pi} \log(1 + \omega_p^2(q)/\omega^2) = \frac{1}{n} \int_{|\mathbf{q}_\alpha| \leq \pi/L} \frac{d\mathbf{q}}{(2\pi)^d} \frac{\omega_p(q)}{2} \tag{16}$$

and size corrections in the energy have a simple interpretation in terms of zero point energy of long wavelength plasmon modes. Explicitly, in three dimensions, we have

$$\delta E_{3d} = \frac{\hbar^2}{2ma_B^2} \sqrt{\frac{3}{r_s^3}} \frac{1}{N} \tag{17}$$

in agreement with Ref. [11], and

$$\delta E_{2d} = s_2 \frac{\pi}{5} (4\pi)^{1/4} \frac{1}{r_s^{3/2}} \frac{1}{N^{5/4}} \tag{18}$$

in two dimensions [4] where $s_2 = \frac{5}{4\pi} \int_{|q_\alpha|<1} |q|^{1/2} = \frac{4}{\pi} \int_0^1 du\, (1 + u^2)^{1/4} \simeq 1.362073657$ accounts for the cubic geometry.

### *4.2. Renormalization factor and effective mass*

For the calculation of the renormalization factor and the effective mass, we need the derivatives of the self energy at the Fermi surface. Within the RPA, we have

$$\frac{\partial\Sigma(k_F,\varepsilon_F)}{\partial\omega} = -\frac{1}{V}\sum_{\mathbf{q}\neq 0}\int_{-\infty}^{\infty}\frac{d\nu}{(2\pi)}\left[\frac{1}{\epsilon(q,i\nu)}-\frac{1}{\epsilon(q,0)}\right]\frac{v_q}{[i\nu+\varepsilon_F-\varepsilon_{k_F+\mathbf{q}}]^2} \tag{19}$$

$$\frac{m}{k_F}\frac{\partial\Sigma(k_F,\varepsilon_F)}{\partial k} = \frac{1}{V}\sum_{\mathbf{q}\neq 0}\int_{-\infty}^{\infty}\frac{d\nu}{(2\pi)}\frac{1}{\epsilon(q,i\nu)}\frac{v_q}{[i\nu+\varepsilon_F-\varepsilon_{k_F+\mathbf{q}}]^2}\left[1+\frac{\mathbf{k}_F\cdot\mathbf{q}}{k_F^2}\right] \tag{20}$$

Since $\chi_0(q,0) = -dn/\varepsilon_F$ for $q \to 0$, the static dielectric function diverges in the long wave length limit,

$$\epsilon_{RPA}(q,0) = 1 + dnv_q/\varepsilon_F \sim v_q \sim q^{1-d}, \quad q \to 0 \tag{21}$$

providing perfect screening, and the dominant size correction, $\Delta$, of $\partial\Sigma/\partial\omega$ exactly equals that of $-\partial\Sigma/\partial k$, and is given by

$$\Delta = -\int_{|\mathbf{q}_\alpha|\leq\pi/L}\frac{d\mathbf{q}}{(2\pi)^d}\int_{-\infty}^{\infty}\frac{d\nu}{(2\pi)}\frac{1}{\epsilon(q,i\nu)}\frac{v_q}{[i\nu+\varepsilon_F-\varepsilon_{k_F+\mathbf{q}}]^2} \tag{22}$$

$$\simeq \int_{|\mathbf{q}_\alpha|\leq\pi/L}\frac{d\mathbf{q}}{(2\pi)^d}\int_{-\infty}^{\infty}\frac{d\nu}{(2\pi)}\frac{v_q}{1+\omega_p^2(q)/\nu^2}\frac{1}{\nu^2} \tag{23}$$

We get

$$\Delta = -\int_{|\mathbf{q}_\alpha|\leq\pi/L}\frac{d\mathbf{q}}{(2\pi)^d}\frac{v_q}{4\omega_p(q)} \tag{24}$$

and we have

$$\Delta_{3d} = -c_3\left(\frac{r_s}{3}\right)^{1/2}\left(\frac{3}{4\pi N}\right)^{1/3} \tag{25}$$

in three dimension [7] with $c_3 = \frac{1}{4\pi}\int_{|q_\alpha|<1}|q|^{-2} = \frac{6}{\pi}\int_0^1 du\,\frac{\ln(2+u^2)}{1+u^2} \simeq 1.221374804$, and

$$\Delta_{2d} = -c_2\left(\frac{r_s}{8}\right)^{1/2}\left(\frac{\pi}{N}\right)^{1/4} \tag{26}$$

in two dimensions [4] with $c_2 = \frac{5}{4\pi}\int_{|q_\alpha|<1}|q|^{-3/2} = \frac{4}{\pi}\int_0^1 du\,(1+u^2)^{-3/4} \simeq 1.057929920$. The leading order size corrections for the renormalization factor are then given by

$$Z_\infty^{-1} \simeq Z_N^{-1} + \Delta \tag{27}$$

and, using Eq. (8), the effective mass correction writes

$$\left.\frac{m}{m^*}-1\right|_\infty \simeq \left.\left(\frac{m}{m^*}-1\right)\right|_N(1+\Delta Z) \tag{28}$$

Note, that, within the validity of the RPA, we may put $Z = 1$ on the rhs of Eq. (28). The slow decay of $\Delta$ with increasing system size can lead to important qualitative and quantitative changes of spectral quantities when properly extrapolated to the thermodynamic limit with respect to calculations which rely on linear extrapolation schemes.

## 5. Discussion

We have discussed size effects of the total energy, the renormalization factor, and the effective mass. Particularly, in two dimensions, $Z$ and $m^*$ show an unexpected slowly decaying size effect $\sim N^{-1/4}$. In general, we expect that these size effects are also present in other response quantities, *e.g.* the spin susceptibility, as they are connected to $m^*$ by relations based on Landau's Fermi liquid theory. Our results on the finite size corrections are quite general and do not rely on the validity of the RPA, as they are based on the large wave length behavior of the dielectric function, Eq. (15), which applies to all metallic systems. Further, all results agree with the leading order corrections in [11, 4, 7] where finite size corrections have been derived from a formulation directly based on the many-body wave function.

## Acknowledgements

This research was supported by CNRS-IDRIS, Project IDRIS 061801.

## References

[1] D. M. Ceperley and B. J. Alder, Phys. Rev. Lett. **45**, 566 (1980).
[2] B. Tanatar and D.M. Ceperley, Phys. Rev. B **39**, 5005 (1989).
[3] N. D. Drummond and R. J. Needs, Phys. Rev. B **79**, 085414 (2009).
[4] M. Holzmann, B. Bernu, V. Olevano, R. M. Martin, and D. M. Ceperley, Phys. Rev. B **79**, 041308 (2009).
[5] G. Ortiz and P. Ballone, Phys. Rev. B **50**, 1391 (1994); G. Ortiz and P. Ballone, Phys. Rev. B **56**, 9970 (1997).
[6] S. Huotari, J. A. Soininen, T. Pylkkänen, K. Hämäläinen, A. Issolah, A. Titov, J. McMinis, J. Kim, K. Esler, D.M. Ceperley, M. Holzmann, and V. Olevano, Phys. Rev. Lett. **105**, 086403 (2010).
[7] M. Holzmann, B. Bernu, C. Pierleoni, J. McMinis, D.M. Ceperley, V. Olevano, and L. Delle Site, arXiv:1105.2338 (2011).
[8] Y. Kwon, D.M. Ceperley, and R.M. Martin, Phys. Rev. B **50**, 1684 (1994).
[9] Y. Kwon, D.M. Ceperley, and R.M. Martin, Phys. Rev. B **53**, 7376 (1996).
[10] N. D. Drummond and R. J. Needs, Phys. Rev. B **80**, 245104 (2009).
[11] S. Chiesa, D. M. Ceperley, R. M. Martin, and M. Holzmann, Phys. Rev. Lett. **97**, 076404 (2006).
[12] N.D. Drummond, R.J. Needs, A. Sorouri, and W.M.C. Foulkes, Phys. Rev. B **78**, 125106 (2008).
[13] D.M. Ceperley and B.J. Alder, J. Chem. Phys. **81**, 5833 (1984).
[14] M. Holzmann, B. Bernu, and D.M. Ceperley, Phys. Rev. B **74**, 104510 (2006).
[15] M. Nava, E. Vitali, A. Motta, D.E. Galli, and S. Moroni, arXiv:1103.0915 (2011).
[16] P. Nozières and D. Pines, Phys. Rev. **111**, 442 (1958).
[17] D. Bohm and D. Pines, Phys. Rev. **92**, 609 (1953).
[18] P. Nozières and D. Pines, *The Theory of Quantum Liquids* (Benjamin, New York,1966).
[19] T.M. Rice, Ann. of Phys. **31**, 100 (1965).
[20] J.J. Quinn and R.A. Ferell, Phys. Rev. **112**, 812 (1958).
[21] H.-J. Schulze, P. Schuck, and N. Van Giai, Phys. Rev. **B 61**, 8026 (2000).
[22] C. S. Ting, T. K. Lee, and J. J. Quinn, Phys. Rev. Lett. **34**, 870 (1975); T. K. Lee, C. S. Ting, and J. J. Quinn, Phys. Rev. Lett. **35**, 1048 (1975).
[23] G.E. Santoro and G.F. Giuliani, Phys. Rev. B **39**, 12818 (1989).
[24] Y. Zhang and S. Das Sarma, Phys. Rev. B **71**, 045322 (2005).
[25] R. Asgari, B. Davoudi, M. Polini, G. F. Giuliani, M. P. Tosi, and G. Vignale, Phys. Rev. **B 71**, 045323 (2005).
[26] C. Lin, F.-H. Zong, and D. M. Ceperley, Phys. Rev. E **64**, 016702 (2001).

Journal of Physics: Conference Series **321** (2011) 012021 doi:10.1088/1742-6596/321/1/012021

# Richardson-Gaudin models: the hyperbolic family

**Jorge Dukelsky**[1], **Stefan Rombouts**[1,2] **and Gerardo Ortiz**[3]

[1] Instituto de Estructura de la Materia, CSIC, Serrano 123, 28006 Madrid, Spain

[2] Departamento de Física Aplicada, Universidad de Huelva, 21071 Huelva, Spain

[3] Department of Physics, Indiana University, Bloomington IN 47405, USA

E-mail: j.dukelsky@csic.es

**Abstract.**
We show that the hyperbolic family of Richardson-Gaudin models could describe p-wave superfluids in two dimensional lattices. The analysis of the exact solution reveals an exotic phase diagram with a third order quantum phase transition. Moreover, we propose a separable integrable pairing Hamiltonian within the same family that could be useful to describe exactly superconductivity in heavy nuclei.

## 1. Introduction

The celebrated paper of Bardeen, Cooper and Schrieffer (BCS) of 1957 [1] gave, within a variational approach, the first microscopic description of the superconducting phenomenon. The following year, Bohr, Mottelson and Pines [2] noted that similar physics may underly the large gaps seen in the spectra of even-even atomic nuclei, emphasizing however that finite-size effects would be critical for a proper description of such systems. At roughly the same time that number projected BCS (PBCS) [3] theory was developed, Richardson [4] showed that for a pure pairing Hamiltonian it is possible to *exactly* solve the Schrödinger equation by following closely Cooper's original idea. Years later and from a different perspective, Gaudin introduced an integrable spin model having striking similarities with the Richardson exact solution [5]. However, he couldn't find the explicit relation between both integrable models. In an intensive collaboration with Peter Schuck, we were able to find this relation through a generalization of the Gaudin integrals of motion giving rise to three classes of pairing-like models that we called the Richardson-Gaudin (RG) integrable models, all of which were integrable and all of which could be solved exactly for both fermion and boson systems [6, 7]. During the last decade, the rational family of the RG models was extensively used to describe ultrasmall superconduction grains, heavy nuclei, quantum dots, ultracold atomic gases, etc [8]. More recently, we have found two physical realizations of the hyperbolic family, one for p-wave polarized atomic gases in two dimensional lattices [9, 10], and the other as a potentially useful realistic pairing Hamiltonian for heavy nuclei. In this contribution we will briefly describe the exact solution for $p_x + ip_y$ superfluids, and we will present preliminary results showing how the integrable hyperbolic Hamiltonian could reproduce Gogni Hartree-Fock-Bogoliubov (HFB) gaps in heavy nuclei.

## 2. The Hyperbolic Model

We start with the integrals of motion of the hyperbolic RG model [7], which can be written in a compact form [11] by making the replacements $\sinh(x) = \frac{\eta-1}{2\sqrt{\eta}}$ and $\coth(x) = \frac{\eta-1}{\eta+1}$ as

$$R_i = S_i^z - 2\gamma \sum_{j\neq i} \left[ \frac{\sqrt{\eta_i\eta_j}}{\eta_i - \eta_j} \left(S_i^+ S_j^- + S_i^- S_j^+\right]) + \frac{\eta_i + \eta_j}{\eta_i - \eta_j} S_i^z S_j^z \right], \tag{1}$$

where $S_i^z$, $S_i^\pm$, are the three generators of the $SU(2)_i$ algebra of mode $i$, $i = 1, \cdots, L$, with spin representation $s_i$ such that $\langle S_i^2 \rangle = s_i(s_i + 1)$. We assume that there are $L$ copies of the $SU(2)$ algebra or equivalently $L$ modes. Therefore, the $L$ operators $R_i$ contain $L$ free parameters $\eta_i$ plus the strength of the quadratic term $\gamma$. The integrals of motion (2) commute with the $z$ component of the total spin $S^z = \sum_{i=1}^L S_i^z$.

It is worthwhile to verify that the set of operators $R_i$ commute among themselves, conforming a complete set of integrals of motion. Therefore, they have a complete set of common eigenstates which are parametrized by the ansatz

$$|\Psi\rangle = \prod_{\alpha=1}^{M} S_\alpha^+ |\nu\rangle, \quad S_\alpha^+ = \sum_i \frac{\sqrt{\eta_i}}{\eta_i - E_\alpha} S_i^+, \tag{2}$$

where the $E_\alpha$ are the pair energies or pairons which are to be determined such that the ansatz (2) satisfies the eigenvalue equations $R_i |\Psi\rangle = r_i |\Psi\rangle$.

In the pairing representations each $SU(2)$ copy is associated with a single particle level $i$ and $M$ is the number of active pairs. The vacuum $|\nu\rangle$ is defined by a set of seniorities, $|\nu\rangle = |\nu_1, \nu_2, \ldots, \nu_l\rangle$, where the *seniority* $\nu_i$ is the number of unpaired particles in level $i$ with single particle degeneracy $\Omega_i$, such that $s_i = (\Omega_i - 2\nu_i)/4$.

Although any function of the integrals of motion generates an exactly solvable Hamiltonian, we will restrict ourselves in this presentation to the simple linear combination $H = \sum_i \eta_i R_i$ that after some algebraic manipulations reduces to

$$H = \sum_i \eta_i S_i^z - G \sum_{i,j} \sqrt{\eta_i\eta_j} S_i^+ S_j^- . \tag{3}$$

This separable Hamiltonian has the eigenvectors (2) and the eigenvalues $E = \sum_i < \nu|S_i^z|\nu >$ $+\sum_\alpha E_\alpha$, where the pairons $E_\alpha$ are a solution of the set of non-linear Richardson equations

$$\sum_i \frac{s_i}{\eta_i - E_\alpha} - \sum_{\alpha'(\neq\alpha)} \frac{1}{E_{\alpha'} - E_\alpha} = \frac{Q}{E_\alpha}, \tag{4}$$

with $Q = \frac{1}{4\gamma} - \frac{L_c}{4} + \frac{M-1}{2}$, $L_c = 2\sum_i s_i$, and $M$ is the number of pairons.

Each particular solution of Eq. 4 defines a unique eigenstate. For the remaining discussion we will assume that $\langle\nu|H_h|\nu\rangle = 0$, which amounts to a simple shift in the energy scale, without loss of generality.

## 3. The $p_x + ip_y$ pairing Hamiltonian

In recent years $p$-wave paired superfluids have attracted a lot of attention, in part due to their exotic properties [12]. Of particular interest is the chiral two-dimensional (2D) $p_x+ip_y$ superfluid of spinless fermions, that supports a topological phase with zero energy Majorana modes [13] and, unlike the $s$-wave superfluid, it has a quantum phase transition in the crossover from BCS to BEC whose properties are not yet well understood. Therefore, the derivation of an exactly solvable model could be essential for the understanding of this exotic superfluid.

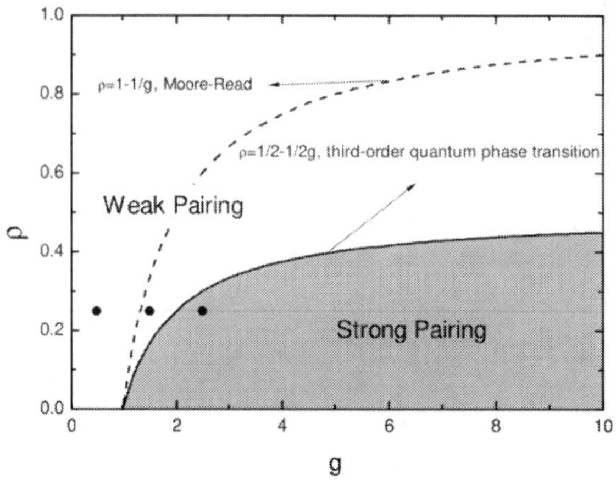


**Figure 1.** Phase diagram of the $p_x+ip_y$ model in terms of the density $\rho$ and the pairing strength $g$. The three circles at quarter filling indicate the configurations studied in the following figure.

In two spatial dimensions, one can define a representation of the $SU(2)$ algebra in terms of creation (annihilation) spinless fermions operators in momentum space, $c_k^\dagger$ ($c_k$). Each pair of states $(k, -k)$ is associated to a single-particle level $\eta_k$, where the index $k$ now refers to the momentum in 2D (in order to avoid double counting we select $k_x > 0$ to label the levels). Furthermore, one can include a phase factor in the definition of $SU(2)$ generators:

$$S_k^z = \frac{1}{2}\left(c_k^\dagger c_k + c_{-k}^\dagger c_{-k} - 1\right), S_k^+ = \frac{k_x + ik_y}{|k|} c_k^\dagger c_{-k}^\dagger, S_k^- = \frac{k_x - ik_y}{|k|} c_{-k} c_k. \tag{5}$$

By taking $\eta_k = k^2$, one obtains the exactly solvable $p_x + ip_y$ model first introduced Ibañez et. al [9]:

$$H_{p_x+ip_y} = \sum_{k,k_x>0} \frac{k^2}{2}\left(c_k^\dagger c_k + c_{-k}^\dagger c_{-k}\right) - G \sum_{k,k_x>0,k',k'_x>0} (k_x + ik_y)(k'_x - ik'_y) c_k^\dagger c_{-k}^\dagger c_{-k'} c_{k'}. \tag{6}$$

Coming back to the Richardson equations (4) that solves the Hamiltonian (6), we recognize two special cases: *case (i)* all are real and negative if $\frac{1}{G} \leq L - 2M + 1$; we will see that the boundary coincides with the phase transition line. *case (ii)* All pairons converge to zero for $\frac{1}{G} = L - M + 1$; this situation determines the so called *Moore-Read* line [14, 9], with interesting properties associated with the fractional quantum Hall effect. Between these two regimes, a fraction of the pairons can converge to zero at integer values of $G^{-1}$. The phase diagram of the $p_x + ip_y$ Hamiltonian (6) depicted in Fig. 1 is characterized by the density $\rho = M/L$ and the scaled pairing strength $g = GL$. The transition between the strong pairing region (BEC) with all pairons real and negative and the weak pairing region (BCS) takes place when one pairon change sign implying that one of the bound molecules in the BEC gets unbounded. For this reason we characterized the transition as a confinement-deconfinement quantum phase transition.

In order to get a more quantitative picture of the pairon distribution in the three regions of the quantum phase diagram, we plot in Fig. 2 the pairon distributions for three representative values of the coupling strength, $g = 0.5, 1.5, 2.5$, at quarter filling for a disk of radius 18 corresponding to a total pair degeneracy $L = 504$. The positions of these points in the quantum phase diagram

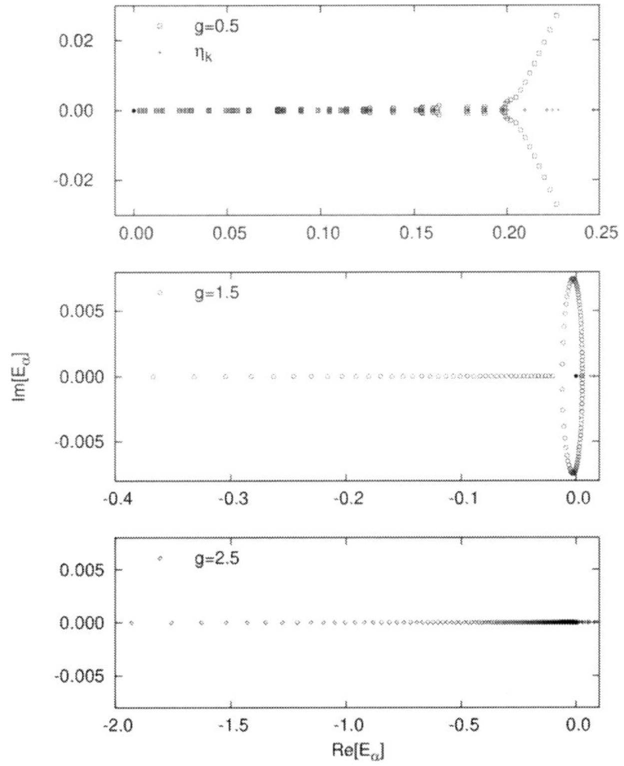


**Figure 2.** Pairon distribution for $L = 504$ at quarter filling, $\rho = \frac{M}{L} = 0.25$, for $g = GL = 0.5, g = 1.0$, and $g = 2.5$.

of Fig. 1 is indicated by the three filled circles. In weak coupling BCS part of the pairons stick to the lower part of the real positive axis, while the remaining pairons form an arc in the complex plane. Approaching the Moore-Read line it looks like the arc is going to close around the origin, but just at the Moore-Read line all pairons collapse to zero, and then a first real negative pairon emerges. In the intermediate weak pairing region a successive series of collapses ensues, at integer values of $Q$, each time producing one more real negative pairon and reducing the size of the arc around the origin. When the last pairon turns real and negative, the system enters the strong pairing phase. From then on the most negative pairon diverges proportional to the interaction strength $G$, while the least negative pairon converges to a finite value.

In order characterize the quantum phase transition, we study the energy density derivatives as describe by the BCS theory which is exact in the thermodynamic limit. As can be seen in Fig. 3, the third derivative shows a discontinuity confirming that the phase transition is third order in the Erhenfest classification.

## 4. The integrable nuclear pairing Hamiltonian

Let us come back to the separable pairing Hamiltonian (3) to note that if we interpret the parameters $\eta_k$ as single particle energies corresponding to a nuclear mean field potential, the pairing interaction has the unphysical behavior of increasing in strength with energy. In order to reverse this unwanted effect we define $\eta_k = 2(\varepsilon_k - \alpha)$, where the free parameter $\alpha$ plays the role of an energy cutoff and $\varepsilon_k$ is the single particle energy in the mean field level $k$. Making

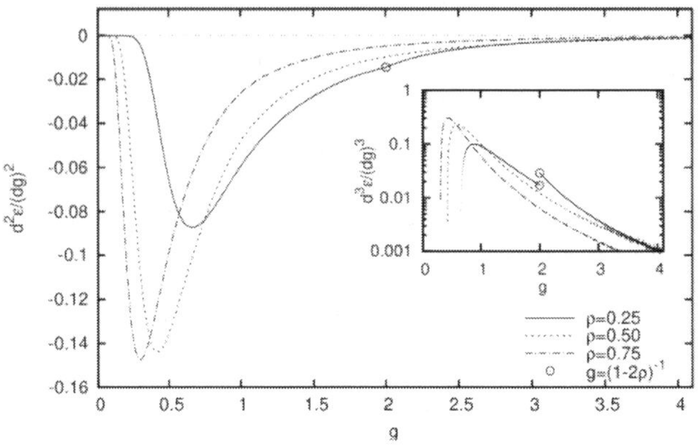


**Figure 3.** Higher order derivatives of the energy density as a function of $g$ for various densities. The open circles mark the transition point at $g = (1-2\rho)^{-1}$.

use of the pair representation of the $SU(2)$, $S_k^+ = c_k^\dagger c_{\bar{k}}^\dagger$, $S_k^z = \frac{1}{2}(c_k^\dagger c_k + c_{\bar{k}}^\dagger c_{\bar{k}} - 1)$, the exactly solvable pairing Hamiltonian (3) reduces to

$$H = \sum_{k>0} \varepsilon_k \left( c_k^\dagger c_k + c_{\bar{k}}^\dagger c_{\bar{k}} \right) - G \sum_{kk',>0} \sqrt{(\alpha - \varepsilon_k)(\alpha - \varepsilon_{k'})} c_k^\dagger c_{\bar{k}}^\dagger c_{\bar{k}'} c_{k'} \tag{7}$$

Our aim is to compare at the BCS level of approximation the results coming from the integrable Hamiltonian (7) with those from a Gogni HFB calculation. As a first step in ascertain the quality of the Hamiltonian (7) to reproduce the superfluid features of heavy nuclei we compare the trend of the energy dependent gaps in both treatments. The energy dependent gaps of the Gogni self-consistent mean field in the canonical basis and those of the integrable pairing Hamiltonian in the BCS approximation are

$$\Delta_k^{Gogni} = \sum_{k'>0} V_{k\bar{k},\bar{k}'k'} < c_{\bar{k}'} c_{k'} > \tag{8}$$

$$\Delta_k^{Exact} = G\sqrt{\alpha - \varepsilon_k} \sum_{k'>0} \sqrt{\alpha - \varepsilon_{k'}} < c_{\bar{k}'} c_{k'} > \tag{9}$$

where the single particle energies $\varepsilon_k$, the pair interaction $V_{k\bar{k},\bar{k}'k'}$, and the pair wavefunction $< c_{\bar{k}'} c_{k'} >$ are obtained from a self-consistent Gogni calculations in the canonical basis. The integrable gaps can be expressed as $\Delta_k^{Exact} = C\sqrt{\alpha - \varepsilon_k}$, having a square root dependence on the single particle energies. In figure 4 we plot the pairing gaps of $^{154}Sm$ for protons and neutrons using $\alpha = 35MeV$ as a cutoff and fitting the parameter $C$. In spite of the significant dispersion of the Gogni gaps, it is clear that the integrable gaps follow correctly the global trend. It is interesting to note that a constant pairing interaction, extensively used in the past and also exactly solvable within the rational family of RG models, would give a non reliable constant gap (horizontal line).

These preliminary results suggest that the hyperbolic model could be extremely useful in nuclear structure calculations as a benchmark realistic exactly solvable to test approximations beyond HFB. On a more ambitious respect, it might be possible to fit the pairing strength $G$ as a function of $N$ and $Z$ to the whole table of nuclides and to set up a program of self-consistent Hartree-Fock plus exact pairing. Work along these lines is in progress.

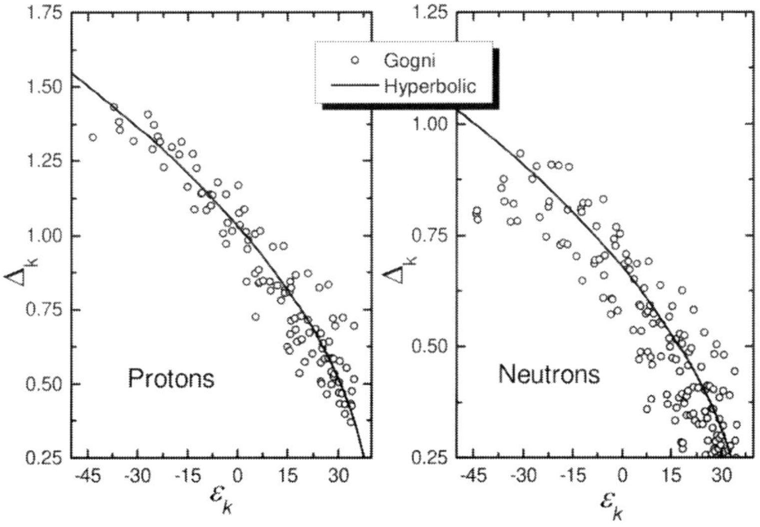


**Figure 4.** Energy dependent gaps of $^{154}Sm$ for protons and neutrons. Open circles are Gogni HFB calculations in the canonical basis while the continuous lines are the integrable pairing gaps with $\alpha = 35MeV$ and $C$ fitted to Gogni gaps.

## Acknowledgments
JD would like to thank Peter Schuck for being his teacher, collaborator and friend in the last 23 years. Furthermore, Peter has been the main collaborator at the beginning of this program on RG integrable models. We are grateful to Rayner Rodriguez-Guzmán for providing us results of the Gogni mean field calculation of $^{154}Sm$. We acknowledge support from a Marie Curie Action of the European Community Project No. 220335 and the Spanish Ministry for Science and Innovation Project No. FIS2009-07277

## 5. References
[1] Bardeen J, Cooper L N and Schrieffer J R 1957 *Phys. Rev.* **108** 1175
[2] Bohr A, Mottelson BR and Pines D 1958 *Phys. Rev.* **110** 936
[3] Dietrich K, Mang H J and Pradal J H 1964 *Phys. Rev.* **135** B22
[4] Richardson R W 1963 *Phys. Lett.* **3** 277
[5] Gaudin M 1976 *J. Physique* **37** 1087
[6] Dukelsky J and Schuck P 2001 *Phys. Rev. Lett.* **86** 4207
[7] Dukelsky J, Esebbag C and Schuck P 2001 *Phys. Rev. Lett.* **87** 066403
[8] Dukelsky J, Pittel S and Sierra G 2004 *Rev. Mod. Phys.* **76** 643
[9] Ibañez M, Links J, Sierra G and Zhao S-Y 2009 *Phys. Rev. B* **79** 180501
[10] Rombouts S M A, Dukelsky J and Ortiz G 2010 *Phys. Rev. B* **82** 224510
[11] Ortiz G, Somma R, Dukelsky J and Rombouts S M A 2005 *Nucl. Phys. B* **707** 421
[12] Gurarie V and Radzihovsky L 2007 *Ann. Phys.* **322** 2
[13] Read N and Green D 2000 *Phys. Rev. B* **61** 10267
[14] Moore G and Read N 1991 *Nucl. Phys. B* **360** 362

Journal of Physics: Conference Series **321** (2011) 012022 doi:10.1088/1742-6596/321/1/012022

# Covariant Density Functional Theory for Excited States in Nuclei.

**Ioannis Daoutidis**[1], **Elena Litvinova**[2,3], **and Peter Ring**[4]

[1] Institut d'Astronomie et d'Astrophysique, Université Libre de Bruxelles, B-1050, Belgium
[2] GSI Helmholtzzentrum für Schwerionenforschung, 64291 Darmstadt, Germany
[3] Institut für Theoretische Physik, Goethe-Universität, 60438 Frankfurt am Main, Germany
[4] Physik-Department der Technischen Universität München, D-85748 Garching, Germany

E-mail: `idaoutid@ulb.ac.be,e.litvinova@gsi.de,ring@ph.tum.de`

**Abstract.** Recent applications of covariant density functional theory (CDFT) for the description of excited states in nuclei are discussed. They are based on the Runge-Gross theorem of time-dependent density functional theory. In the adiabatic approximation, where the energy dependence of the self-energy is neglected one obtains in the small amplitude limit RPA and QRPA. In order to describe the width of giant resonances an energy dependent self energy is constructed by coupling to surface modes of the system. Recent applications with $2qp$-phonon and 2-phonon configurations are discussed.

## 1. Introduction

Since one of the underlying symmetries of QCD is Lorentz invariance covariant density functional theory [1, 2] is of particular interest in nuclear physics. This symmetry not only allows to describe in a consistent way the spin-orbit coupling which has an essential influence on the underlying shell structure, but it also puts stringent restrictions on the number of parameters in the corresponding functionals without reducing the quality of the agreement with experimental data. Density functional theory originally introduced by Kohn and Sham [3] in the sixties is based on the mean field concept and strictly valid only for a description of the ground states. In fact density functionals presently in use are adjusted to ground state properties. It is therefore an interesting question, to what extend those functionals can be used also for the description of excited states in nuclear physics.

## 2. Time dependent density functional theory.

In Coulombic system density functional theory has been extended over the years also for excited states in the framework of time-dependent density functional theory (TDDFT) [4]. Runge and Gross could derive a theorem [5] which is an extension of the Kohn-Sham method. They consider a many-body problem in a time-dependent external single potential $f_{ext}(\mathbf{r}, t)$ with the exact solution $|\Psi(t)\rangle$ for a given initial condition $|\Psi(0)\rangle$. They could show that for given initial conditions there is a one-to-one correspondence between the exact time-dependent local single particle density $\rho(\mathbf{r}, t)$ and the external potential $f_{ext}(\mathbf{r}, t)$. As in static Kohn-Sham theory there exists a fictitious system of non-interacting particles with the single particle wave functions

$\phi_i(\mathbf{r},t)$ $(i = 1 \ldots N)$ that satisfy time-dependent Kohn-Sham equations of the form

$$i\partial_t \phi_i(\mathbf{r},t) = \left[-\nabla^2/2m + v[\rho](\mathbf{r},t)\right] \phi_i(\mathbf{r},t). \tag{1}$$

where $\rho(\mathbf{r},t) = \sum_i^N |\phi_i(\mathbf{r},t)|^2$ is the exact local density. The time-dependent Kohn-Sham potential $v[\rho](\mathbf{r},t)$ is a function of $\mathbf{r}$ and $t$, but it is in addition a unique functional of the time-dependent density $\rho(\mathbf{r},t)$. The situation looks similar to static Kohn Sham theory but it is much more complicated. The mean field potential $v[\rho](\mathbf{r},t)$ contains the entire history of the system and its derivation would require the full solution of the exact Schrödinger equation for arbitrary systems with the density $\rho(\mathbf{r},t)$.

However, if the external potential $f_{ext}(\mathbf{r},t)$ is weak, the situation can be simplified considerably and linear response theory is applicable. This case applies also to nuclear physics. We do not need to know the functional $v[\rho](\mathbf{r},t)$ for arbitrarily changing densities, rather we need it only in the vicinity of the initial density, e.g. of the static ground state density $\rho_s(\mathbf{r})$ of the system and we can write $\rho(t) = \rho_s + \delta\rho(t)$. Going up to linear order in $f_{ext}$, we obtain

$$v[\rho_s + \delta\rho](\mathbf{r},t) = v_s[\rho_s](\mathbf{r}) + \int dt' \int d^3r' V[\rho_s](\mathbf{r},t,\mathbf{r}',t')\delta\rho(\mathbf{r}',t'). \tag{2}$$

Here $V$ is an effective interaction, the derivative of the mean field with respect to the density

$$V(\mathbf{r},\mathbf{r}',t-t') = \left.\frac{\delta v(\mathbf{r},t)}{\delta\rho(\mathbf{r}',t')}\right|_{\rho=\rho_s}. \tag{3}$$

It is a functional of the static density $\rho_s$ and for a time independent Hamiltonian it depends only on the time difference $t - t'$ [4]. Introducing as usual the response function $R(\mathbf{r},t,\mathbf{r}',t')$

$$\delta\rho(\mathbf{r},t) = \int d^3r' \int dt' R(\mathbf{r},\mathbf{r}',t-t') f_{ext}(\mathbf{r}',t') \tag{4}$$

we find after a Fourier transformation in time the well known linear response equation [6]

$$R(\mathbf{r},\mathbf{r}',\omega) = R_0(\mathbf{r},\mathbf{r}',\omega) + \int d^3r_1 \int d^3r_2 R_0(\mathbf{r},\mathbf{r}_1,\omega)V(\mathbf{r}_1,\mathbf{r}_2,\omega)R(\mathbf{r}_2,\mathbf{r}',\omega) \tag{5}$$

Of course, all these quantities are functionals of the exact ground state density $\rho_s(\mathbf{r})$. For a weak external field $f_{ext}$, these equations are exact, but, of course, we do not know the functional $v[\rho](\mathbf{r},t)$ in Eq. (1) nor its functional derivative at $\rho = \rho_s$. As usual in density functional theory the Runge-Gross theorem gives us no hint, how to derive these quantities. We therefore need some model for this single particle potential $v[\rho](\mathbf{r},t)$ in the Kohn-Sham equation.

## 3. The adiabatic approximation

There is a very common approximation, the so called *adiabatic approximation*. Here one neglects the memory and assumes that the density changes only very slowly, such that the potential is given at each time by the static potential $v_s$ corresponding to this density.

$$v[\rho](\mathbf{r},t) \approx v_s[\rho_s](\mathbf{r},t) \tag{6}$$

In this approximation $v[\rho]$ is no longer depending on the function $\rho(\mathbf{r},t)$ of 4 variables, but rather on the function $\rho_s(\mathbf{r}) = \rho(\mathbf{r},t)$ depending only on three variables at a given time. $t$ is here only a parameter. The effective interaction is the second derivative of the static energy functional with respect to the static density at a given time $t$. In Fourier space this interaction

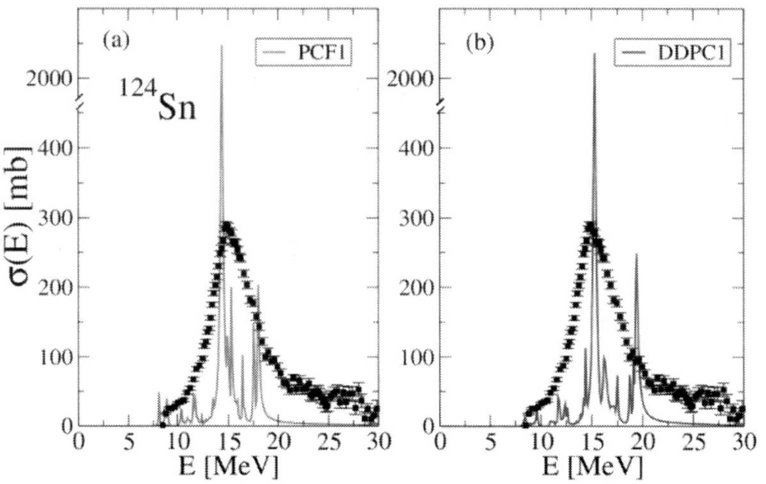


**Figure 1.** (Color online) The isovector dipole strength distribution in $^{124}$Sn for the parameter set PC-F1 of Ref. [8] and the parameter set DD-PC1 of Ref. [9]. Relativistic continuum QRPA calculations [7] are compared with the experiment (black) of Ref. [10].

does not depend on energy. This approximation corresponds to the small amplitude limit of the time-dependent mean field equations, i.e. to RPA and QRPA and it is extensively used in nuclear physics [2]. As an example we show in Fig. 1 the distribution of the isovector dipole strength in the nucleus $^{124}$Sn. In the relativistic case, so far, most of these applications have been done by diagonalizing the QRPA-matrix in a discrete basis. Here we compare recent relativistic continuum QRPA results [7] with the experimental dipole strength. As one can see, there is excellent agreement between the position of the peak of the resonance. However, it is clearly seen that the theoretical width is much smaller than the experimental value. This is typical for all heavy nuclei, where large Coulomb and centrifugal barriers prevent a coupling to the continuum. The large experimental width can only be explained by a coupling to more complicated configurations. In order to describe this coupling in a consistent microscopic way, we have to go back to Eq. 5 and take into account an energy dependence in the self-energy.

## 4. Energy dependence of the self energy $\Sigma(\omega)$.

In a relativistic many-body theory the motion of single nucleons in the nuclear medium is described by the Dyson equation

$$\big(\boldsymbol{\alpha}(\mathbf{p}-\boldsymbol{\Sigma})+\beta(m+\Sigma_s)+\Sigma_0\big)|\psi\rangle=\varepsilon|\psi\rangle. \tag{7}$$

where $\Sigma_s$ and $\Sigma_\mu$ are the self energies. They correspond to the effective single particle $v[\rho](\mathbf{r},t)$ potential of time-dependent Kohn-Sham theory in Eq. 1. We decompose it in a stationary local part $\tilde{\Sigma}(\mathbf{r})$ and an energy-dependent non-local term

$$\Sigma(\mathbf{r},\mathbf{r}';\omega)=\tilde{\Sigma}(\mathbf{r})\delta(\mathbf{r}-\mathbf{r}')+\Sigma^e(\mathbf{r},\mathbf{r}';\omega), \tag{8}$$

where $\tilde{\Sigma}$ corresponds to the scalar and vector fields of the static RMF model. $\Sigma^e(\varepsilon)$ is derived from the particle-phonon coupling model [11]. From the solution of the Dyson equation one finds a fragmentation of the single-particle states and a corresponding reduction of the single-particle strength and an increasing level density at the Fermi surface.

For the description of giant resonances we use this self-energy and derive from it the effective interaction (3). We find two parts, the static interaction of the adiabatic approximation and an

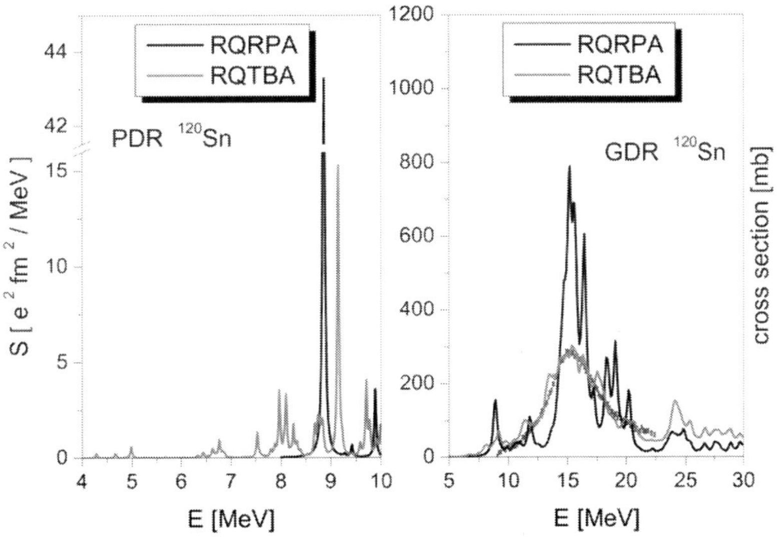


**Figure 2.** (Color online) The dipole spectra for $^{120}$Sn calculated within RQRPA (black) and RQTBA (red) are compared with data (blue) of Ref. [15]. Right panel: photoabsorption cross sections computed with the artificial width 200 keV. Left panel: the low-lying portions of the corresponding spectra in terms of the strength function, calculated with 20 keV smearing.

induced energy dependent part $\Phi(\omega)$ (for a graphical representation see the first part of Fig. 3). As, on the one-body level, the energy dependence of the self energy $\Sigma^e(\omega)$ in the Dyson equation (7) leads to a fragmentation of the single particle spectrum by including $1p\otimes phonon$ (ie. $2p-1h$) configurations one finds now, on the two-body level, in the Bethe Salpeter equation because of the energy dependence of the induced interaction a strong fragmentation of the phonon spectrum by including $1p1h\otimes phonon$ (i.e. $2p-2h$) configurations. This causes the spreading width of giant resonances and a redistribution of the pygmy strength.

Two essential approximations are used in this context: (i) the *Time-Blocking Approximation* (TBA) introduced by Tselyaev in Ref. [12] and extended to relativistic systems (RTBA) and (RQTBA) in Refs. [13, 14], (ii) a *renormalization of the effective interaction*, by replacing $\Phi(\omega)$ by the difference $\Phi(\omega) - \Phi(0)$. This avoids double counting.

In Fig. 2 we show the dipole spectrum for the tin isotope $^{120}$Sn. The right panel shows the photoabsorption cross section. The spectrum is dominated by the IVGDR. It is seen that the envelope of the calculated GDR in $^{120}$Sn between 10 and 22 MeV, where data are available, is in excellent agreement to the experimental cross section. This is particularly satisfying since these results are obtained in a fully consistent way from the parameter set NL3 without any additional fit parameter. It indicates that the mechanism of coupling to the surface phonons provides the major contribution to the width of these resonances.

The left panel of Fig. 2 shows the low-lying parts of the E1 response in terms of the strength function. The black very sharply peaked curves correspond to RQRPA and the red fragmented curves to RQTBA calculations. Experimental data are taken from the EXFOR database [15]. These figures clearly demonstrate how the two-quasiparticle states, which are responsible for the spectrum of the RQRPA excitations, are fragmented through the coupling to the collective vibrational states. It is clearly seen how the low-lying strength develops with the increase of the neutron excess. It is found that the pygmy mode show a pronounced peak on the QRPA level, which is fragmented when the phonon coupling is switched on, but it is not washed out totally.

Journal of Physics: Conference Series **321** (2011) 012022 doi:10.1088/1742-6596/321/1/012022

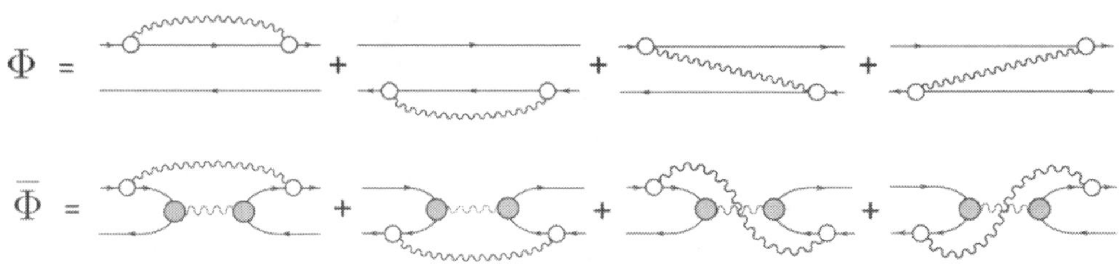


**Figure 3.** The 2q⊗phonon amplitude $\Phi$ of the conventional phonon coupling model and the two-phonon amplitude $\bar{\Phi}$ of the two-phonon model in a diagrammatic representation. The wavy curves denote phonon propagators, the empty circles represent phonon vertices, and the grey circles together with the two nucleonic lines denote RQRPA transition densities.

## 5. Coupling of two phonons

The main assumption of the relativistic quasiparticle time-blocking approximation (RQTBA) [14] discussed so far is that two types of elementary excitations – two-quasiparticle and vibrational modes – are coupled in such a way that configurations of $2qp \otimes phonon$ type with low-lying phonons strongly compete with simple $2qp$-configurations close in energy. There are however additional processes, which are not fully included in this scheme as for instance the coupling of two low-lying collective phonons, e.g. the first $2_1^+$ and the first $3_1^-$ state, to a phonon multiplet $[2_1^+ \otimes 3_1^-]$. Therefore, recently an extension of RQBTA has been introduced, which includes also the coupling to 2-phonon states [17].

In the diagrammatic representation of the amplitude $\Phi(\omega)$ in the upper line of the Fig. 3 the intermediate $1p1h$-propagator is represented by the two straight nucleonic lines between the circles denoting the amplitudes of emission and absorption of the phonon by a single particle (or hole). In RQTBA-2 one introduces the RQRPA correlations into the intermediate $1p1h$-particle propagator replacing the amplitude $\Phi(\omega)$ by the new one $\bar{\Phi}(\omega)$.

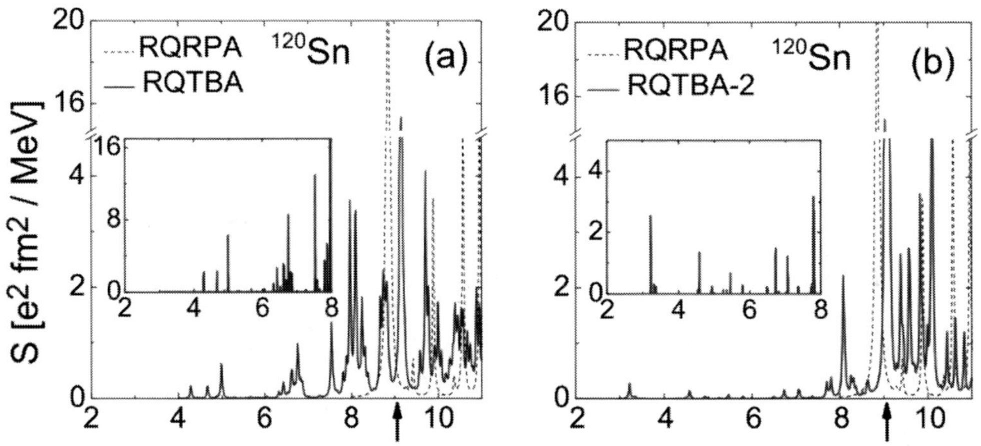


**Figure 4.** (Color online) Low-lying dipole spectrum of $^{120}$Sn calculated within the RQRPA (dashed curves), RQTBA (blue solid curve, panel (a)) and RQTBA-2 (red solid curve, panel (b)). A finite smearing parameter $\Delta = 20$ keV has been used in the calculations. The insert shows the zoomed pictures of the spectra below 8 MeV with a small value $\Delta = 2$ keV allowing to see all the states in this energy region. The arrows indicate the experimental neutron threshold.

Fig. 4 illustrates the effect of two-phonon correlations on spectra of nuclear excitations. It displays the dipole strength function of $^{120}$Sn calculated within the conventional RQTBA and the two-phonon RQTBA-2. The resulting strength functions are compared with the RQRPA strength function because both of them originate from the RQRPA by similar fragmentation mechanisms. Here the major fraction of the RQRPA pygmy mode shown by the dashed curve is pushed up above the neutron threshold by the RQTBA-2 correlations. A low lying $1^-$ state appears at 3.22 MeV with $B(E1)\uparrow = 14\times10^{-3}$ e$^2$ fm$^2$. These numbers can be compared with the corresponding data of Ref. [18] for the lowest $1^-$ state: it is observed at 3.28 MeV with a $B(E1)\uparrow = 7.60(51)\times10^{-3}$ e$^2$ fm$^2$ in $^{120}$Sn.

## 6. Summary and conclusions

Several recent applications of density functional theory for excited states have been discussed. In the adiabatic approximation, where the self-energy does not depend on energy, simple QRPA calculations in the continuum reproduce the position of giant resonance, but not the width. Energy dependent self-energies lead to induced energy-dependent interactions. On the 2*qp*-phonon level we can reproduce the width successfully, but one has to go further to include also the coupling to 2-*phonon*-configurations properly. This can be done by including higher correlations in the kernel of the Bethe-Salpeter equation.

Of course there are open problems. Some of them are of technical nature: so far pairing is treated on a very primitive level with a seniority zero force in the BCS scheme and the continuum is only on the QRPA level properly taken into account. Others are more complicated to solve. Ground state correlations for instance are, so far, only taken into account on the RPA level. One could also think about further extensions of the TBA in order to include such effects also in the kernel of the Bethe Salpeter equation. At the moment these calculations are restricted to spherical nuclei. In deformed nuclei the numerical effort is considerably larger and in addition we have the problem of symmetry violation.

## Acknowledgments

This work was supported by the FNRS (Belgium) and the Communauté française de Belgique,the Hessian LOEWE initiative through the Helmholtz International Center for FAIR, and by the excellence cluster "Origin and Structure of the Universe " (www.universe-cluster.de).

## References

[1] Ring P 1996, *Prog. Part. Nucl. Phys.* **37** 193
[2] Vretenar D, Afanasjev A V, Lalazissis G A and Ring P 2005, *Phys. Rep.* **409** 101
[3] Kohn W and Sham L J 1965, *Phys. Rev.* **137**, A1697
[4] Marques M A L, Ullrich C A, Nogueira F, Rubio, Burke A, and Gross E K U (Eds.) 2006, *Time-Dependent Density Functional Theory* (Heidelberg: Springer-Verlag)
[5] Runge E and Gross E K U 1984, *Phys. Rev. Lett.* **52**, 997
[6] Ring P and Schuck P 1980, *The Nuclear Many-Body Problem* (New York: Springer-Verlag)
[7] Daoutidis I and Ring P 2011, *Phys. Rev.* **C83**, 044303
[8] Bürvenich T, Madland D G, Maruhn J A, and Reinhard P-G, 2002, *Phys. Rev.* **C65**, 044308
[9] Nikšić T, Vretenar D, and Ring P 2008, *Phys. Rev.* **C78** 034318
[10] Fultz S C, Berman B L, Caldwell J T, Bramblett R L, Kelly M A 1969, *Phys. Rev.* **186** 1255
[11] Litvinova E and Ring P 2006, *Phys. Rev.* **C73**, 044328
[12] Tselyaev V I 1989, *Sov. J. Nucl. Phys.* **50**, 780
[13] Litvinova E, Ring P, and Tselyaev V I 2007, *Phys. Rev.* **C75**, 064308
[14] Litvinova E, Ring P, and Tselyaev V I 2008, *Phys. Rev.* **C78**, 014312
[15] Experimental Nuclear Reaction Data, http://www-nds.iaea.or.at/exfor/exfor00.htm.
[16] Litvinova E, Ring P, and Vretenar D 2007, *Phys. Lett.* **B647**, 111
[17] Litvinova E, Ring P, and Tselyaev V I 2010, *Phys. Rev. Lett.* **105**, 022502
[18] Bryssinck J *et al*, 1999, *Phys. Rev.* **C59**, 1930

Journal of Physics: Conference Series **321** (2011) 012023 doi:10.1088/1742-6596/321/1/012023

# Quasiparticle light elements and quantum condensates in nuclear matter

**G. Röpke**

Universität Rostock, Institut für Physik, 18051 Rostock, Germany

E-mail: gerd.roepke@uni-rostock.de

**Abstract.** Nuclei in dense matter are influenced by the medium. In the cluster mean field approximation, an effective Schrödinger equation for the $A$-particle cluster is obtained accounting for the effects of the surrounding medium, such as self-energy and Pauli blocking. Similar to the single-baryon states (free neutrons and protons), the light elements ($2 \leq A \leq 4$, internal quantum state $\nu$) are treated as quasiparticles with energies $E_{A,\nu}(P; T, n_n, n_p)$ that depend on the center of mass momentum $\vec{P}$, the temperature $T$, and the total densities $n_n, n_p$ of neutrons and protons, respectively. We consider the composition and thermodynamic properties of nuclear matter at low densities. At low temperatures, quartetting is expected to occur. Consequences for different physical properties of nuclear matter and finite nuclei are discussed.

## 1. Introduction

Quantum condensates are one of the amazing phenomena in many-particle physics. Isospin-triplet (proton-proton or neutron-neutron) pairing is well established [1] not only in nuclear matter at low temperatures, but also in finite nuclei. Isospin-singlet (neutron-proton) pairing may become of relevance in symmetric and asymmetric nuclear matter [2]. On the one hand, the interaction is stronger, even a bound state, the deuteron, can be formed. This would lead to higher transition temperatures and the transition from Cooper pairing to Bose-Einstein condensation. On the other hand, with increasing difference of the chemical potentials and increasing Coulomb effects, the formation of a condensate becomes more difficult, in particular in asymmetric matter. The strong interaction in the isospin-singlet channel leads to a precursor for the Bose condensate, the occurrence of a pseudogap [3] above the critical temperature. Furthermore, pairing competes with the formation of an $\alpha$-particle condensate (quartetting) [4], in particular at low densities. Therefore, the role of correlations, in particular the formation of clusters and of a quantum condensate, is an interesting item in strongly interacting quantum liquids such as nuclear matter.

Systematic quantum statistical approaches, see [5], have to be used for a treatment using appropriate concepts such as thermodynamic Green functions, spectral function, and self-energy. Within a chemical picture, a cluster decomposition of the self-energy can be performed that considers the formation of clusters in a dense medium, taking into account symmetrization (Pauli blocking), screening of the interaction, dynamical self-energy, and other many-particle effects. An interesting approximation is the cluster mean-field approximation [5, 6, 7] that considers the few-particle Schrödinger equation in a correlated medium, where the correlations in the medium have to be determined in a self-consistent way. This would allow to describe nuclear

Journal of Physics: Conference Series **321** (2011) 012023 doi:10.1088/1742-6596/321/1/012023

matter in a wide range of densities, from saturation density where a mean-field approach on the single nucleon level is possible, to the low-density region where the nuclear statistical equilibrium (NSE) applies and $\alpha$-matter can occur. We discuss the formation of clusters in nuclear matter in Sec. 2 and give some applications to finite nuclei in Sec. 3.

## 2. Nuclei in matter

The few-body problem describing $A \leq 4$ nucleons in hot and dense matter can be related to an in-medium wave equation (Bethe-Salpeter equation) that is derived from many-particle approaches, see Ref. [5]. We consider only bound states $\nu$, dropping the spin quantum number. For the light elements, $\nu = d, t, h, \alpha$ denotes the deuteron ($^2$H), the triton ($^3$H), the helion ($^3$He), and the $\alpha$ particle ($^4$He). Considering uncorrelated nucleons in the medium, the few-nucleon wave function and the corresponding eigenvalues follow from solving the in-medium Schrödinger equation

$$\begin{aligned}&[E_1^{\rm qp}(1) + \ldots + E_1^{\rm qp}(A)]\psi_{\nu P}(1\ldots A)\\&+ \sum_{1'\ldots A'}\sum_{i<j}[1-f_1(i)-f_1(j)]V(ij,i'j')\prod_{k\neq i,j}\delta_{kk'}\psi_{\nu P}(1'\ldots A') = E_\nu^{\rm qp}(P)\psi_{\nu P}(1\ldots A)\,. \quad (1)\end{aligned}$$

For brevity, the single-nucleon quasiparticle energy $E_{\tau_1}^{\rm qp}(\vec p_1)$ is denoted as $E_1^{\rm qp}(1)$. The variables $1,\ldots,i,j,\ldots,A$ denote momentum $\vec p$, spin $\sigma$, and isospin $\tau$, e.g. $1 = \{\vec p_1, \sigma_1, \tau_1\}$. The nucleon-nucleon interaction $V(ij,i'j')$ becomes medium dependent due to the Pauli blocking prefactor $[1-f_1(i)-f_1(j)]$. The phase space occupation is described by a Fermi distribution function normalized to the total density of nucleons,

$$f_1(1) = \frac{1}{\exp[E_1^{\rm qu}(1)/T - \mu_\tau/T]+1} \approx \frac{n_\tau}{2}\left(\frac{2\pi\hbar^2}{m_\tau T}\right)^{3/2} e^{-\frac{p_1^2}{2m_\tau T}} \quad (2)$$

in the low-density, non-degenerate limit ($\mu_\tau < 0$). The chemical potential $\mu_\tau$ is determined by the normalization condition $2\Omega^{-1}\sum_p f_1(p) = n_\tau$, where $\tau$ denotes isospin (neutron or proton), and has to be expressed in terms of these densities $n_n, n_p$ or the baryon density $n = n_n + n_p$ and proton fraction $Y_p = n_p/n$, and the temperature $T$.

The in-medium Schrödinger equation (1) contains the effects of the medium in the single nucleon quasiparticle shift as well as in the Pauli blocking terms. Obviously, the bound state wave functions and energy eigenvalues as well as the scattering phase shifts depend on temperature and density. In particular, we obtain the cluster quasiparticle shifts

$$E_\nu^{\rm qp}(P;T,n,Y_p) - E_\nu(P) = \Delta E_\nu^{\rm SE}(P;T,n,Y_p) + \Delta E_\nu^{\rm Pauli}(P;T,n,Y_p). \quad (3)$$

The contribution of the single nucleon energy shift to the cluster self-energy shift $\Delta E_\nu^{\rm SE}$ is easily calculated in the effective mass approximation, where the single-nucleon quasiparticle energy shift $\Delta E_1^{\rm SE}(1)$ can be represented by the energy shift $\Delta E_\tau^{\rm SE}(T,n,Y_p)$ and the effective mass $m_\tau^*(T,n,Y_p)$, see [8]. Different expressions for the single-nucleon quasiparticle shifts are available such as Skyrme, RMF, or DBHF, see [9]. Since the single-nucleon quasiparticle energy shift arises in the continuum of scattering states as well, the influence on the binding energies is small.

We consider here the Pauli blocking shift of the binding energies

$$\Delta E_\nu^{\rm Pauli}(P;T,n,Y_p) = E_\nu^{\rm qp}(P) - E_\nu(P) - (A-Z)\Delta E_n^{\rm SE} - Z\Delta E_p^{\rm SE}\,. \quad (4)$$

Different fit formula have been given [5, 9, 10] to parametrize the solution of the in-medium Schrödinger equation (1). We follow the recent work [8].

**Table 1.** Parameter values for the Pauli blocking shift $\Delta E_\nu^{\rm Pauli}(P;T,n,Y_p)$, Eq. (6), in the low-density limit

| $\nu$ | $f_{\nu,0}$ [MeV fm$^{5/2}$] | $f_{\nu,1}$ [MeV fm$^3$] | $f_{\nu,2}$ [MeV] | $f_{\nu,3}$ - | $f_{\nu,4}$ [fm$^{-1}$] |
|---|---|---|---|---|---|
| $d$ ($^2$H) | 388338 | 6792.6 | 22.52 | 0.2223 | 0.2317 |
| $t$ ($^3$H) | 159080 | 20103.4 | 11.987 | 0.85465 | 0.9772 |
| $h$ ($^3$He) | 153051 | 19505.9 | 11.748 | 0.84473 | 0.9566 |
| $\alpha$ ($^4$He) | 352965 | 36146.7 | 17.074 | 0.9865 | 1.9021 |

In the low-density limit, a linear dependence of the energy shifts on the nucleon density $n$ follows from perturbation theory. For a more general dependence on the total nucleon density we consider the expression

$$\Delta E_\nu^{\rm Pauli}(P;n,T,Y_p) = c_\nu(P;T)\left\{1-\exp\left[-\frac{f_\nu(P;T,n)}{c_\nu(P;T)}y_\nu(Y_p)n - d_\nu(P;T,n)n^2\right]\right\}, \quad (5)$$

where the dependence on the asymmetry is given by $y_d(Y_p) = y_\alpha(Y_p) = 1$, for triton $y_t(Y_p) = \left(\frac{4}{3}-\frac{2}{3}Y_p\right)$, and for helion $y_h(Y_p) = \left(\frac{2}{3}+\frac{2}{3}Y_p\right)$ that reflects the different proton and neutron content. The functions $f_\nu(P;T,0)$ can be calculated in first order perturbation theory using the unperturbed wave functions of the free nuclei. We performed model calculations with a separable potential that reproduces the empirical binding energies and rms radii. Motivated by the exact solution for $A=2$, we use the following fit for arbitrary $\nu$:

$$\begin{aligned} f_\nu(P;T,n) &= f_{\nu,1}\exp\left[-\frac{P^2/\hbar^2}{4(f_{\nu,4}^2/f_{\nu,3}^2)(1+T/f_{\nu,2})+u_\nu n}\right]\frac{1}{T^{1/2}}\frac{2f_{\nu,4}}{P/\hbar}\,\mathrm{Im}\left[e^{x^2}\mathrm{erfc}\,(x)\right], \\ x &= f_{\nu,3}(1+f_{\nu,2}/T)^{1/2}\left(1-i\frac{P/\hbar}{2f_{\nu,4}(1+T/f_{\nu,2})}\right). \end{aligned} \quad (6)$$

For zero momenta, $P=0$, the temperature dependence of $c_\nu(0;T)$ and $d_\nu(0;T,n)$ is expressed as

$$c_\nu(0;T) = c_{\nu,0} + \frac{c_{\nu,1}}{(T-c_{\nu,2})^2+c_{\nu,3}}, \qquad d_\nu(0;T,n) = \frac{d_{\nu,1}}{(T-d_{\nu,2})^2+d_{\nu,3}}. \quad (7)$$

The Pauli shift at finite momenta is fitted with $c_\nu(P;T) = c_\nu(0;T)$ not depending on $P$, but

$$d_\nu(P;T,n) = d_\nu(0;T,n)\exp\left[-\frac{P^2/\hbar^2}{v_\nu T n}\right]. \quad (8)$$

An additional dependence on $n,T$ is considered at finite values of $P$. Another additional dependence on $n,T$ for finite momenta is introduced in $f_\nu(P;T,n)$ where the dispersion relation becomes density dependent due to the parameter $u_\nu$. Parameter values are given in Tab. 1 and Tab. 2.

The in-medium Schrödinger equations (1) describe in the case $A=2$ ($d$) and $A=4$ ($\alpha$) also the onset of a quantum condensate if according to the Thouless condition the in-medium binding energy coincides with $A\mu$.

With these parameter values, it is possible to evaluate the composition of warm nuclear matter at arbitrary baryon density $n$ below the saturation density, temperature $T$, and proton fraction $Y_p$. Equations of state can be obtained for different thermodynamic properties, see also

**Table 2.** Parameter values for the Pauli blocking shift $\Delta E_\nu^{\rm Pauli}(0;T,n,Y_p)$, Eq. (5), in units of MeV and fm ($c_{\nu,0}, c_{\nu,2}, d_{\nu,2}$ - [MeV]; $c_{\nu,1}$ - [MeV$^3$]; $c_{\nu,3}, d_{\nu,3}$ - [MeV$^2$]; $d_{\nu,1}$ - [MeV$^2$ fm$^6$]; $u_\nu$ - fm; $v_\nu$ - [MeV$^{-1}$ fm])

| $\nu$ | $d$ ($^2$H) | $t$ ($^3$H) | $h$ ($^3$He) | $\alpha$ ($^4$He) |
|---|---|---|---|---|
| $c_{\nu,0}$ | 2.752 | 11.556 | 10.435 | 150.71 |
| $c_{\nu,1}$ | 32.032 | 117.24 | 176.78 | 9772 |
| $c_{\nu,2}$ | 0 | 3.7362 | 3.5926 | 2.0495 |
| $c_{\nu,3}$ | 9.733 | 4.8426 | 5.8137 | 2.1624 |
| $d_{\nu,1}$ | 523757 | 108762 | 90996 | 5391.2 |
| $d_{\nu,2}$ | 0 | 9.3312 | 10.72 | 3.5099 |
| $d_{\nu,3}$ | 15.273 | 49.678 | 47.919 | 44.126 |
| $u_\nu$ | 11.23 | 25.27 | 25.27 | 44.92 |
| $v_\nu$ | 0.145 | 0.284 | 0.27 | 0.433 |

[9]. A smooth transition from the description of nuclear matter near saturation density to the low density region where the NSE can be applied has been found. A systematic inclusion of mean-field effects is possible.

An unsolved problem is the behavior of the internal energy per nucleon $U$ at low temperatures. For symmetric matter it is shown in Fig. 1. At low densities and temperatures, $U$ approaches the binding energy per nucleon of the $\alpha$ particle. (Note that at any finite temperature, for extremely low densities the ideal gas of free nucleons will be achieved because of entropy. As described by the NSE, the $\alpha$-particle fraction decreases so that the internal energy bends upwards to $3T/2$. At $T=0$, the most strongly bound state survives in the zero density limit, i.e. $^{56}$Fe instead of $^4$He if arbitrary nuclei are admitted, so that $U \approx -8.7$ MeV.) With increasing density, the self-energy shift leads to the behavior $\partial U/\partial n < 0$ at $T=0$ that means thermodynamical instability. With the disappearance of the bound states, the internal energy approaches the values for the quasiparticle single-nucleon approximation. This occurs rather abruptly near $n = 0.02$ fm$^{-3}$ so that an intermediate region of metastability appears. However, this may be changed if better approximations for the internal energy in the intermediate region are available. Alternative approaches to the internal energy of symmetric matter at $T=0$ have been considered in connection with the formation of a quantum condensate [11, 12], see the following subsection.

## 3. $\alpha$ clustering in nuclei

Nuclear matter at low densities and temperatures is thermodynamically instable. This region of phase instability can only be reached in nonequilibrium or in inhomogeneous systems. An interesting application are nuclei, where nuclear systems at low densities can occur. In particular, excited states of symmetric $4n$ nuclei near the $n\alpha$ threshold may show an $\alpha$ cluster state such as in $^8$Be, $^{12}$C, possibly $^{16}$O, and others. As a simple ansatz to describe $\alpha$ clusters with the same c.o.m. orbits, like the pairing state, the THSR wave function was introduced [13]. This wave function is a product ansatz for the c.o.m. motion of antisymmetrized $\alpha$ clusters, similar to the BCS wave function. This approach has been proved to describe different properties of such low-density states successfully.

In particular, a relation between the density and the condensate fraction has been discussed [14] that indicates the disappearence of the condensate around $n_{\rm sat}/3 \approx 0.05$ fm$^{-3}$. Whereas $\alpha$ clusters are well developed in the Hoyle state, in the ground state of $^{12}$C no $\alpha$ clustering is present and a quasiparticle single-nucleon shell model is applicable. The reason is the antisymmetrization of the nucleonic wave functions that results in the Pauli blocking. In contrast to the calculations

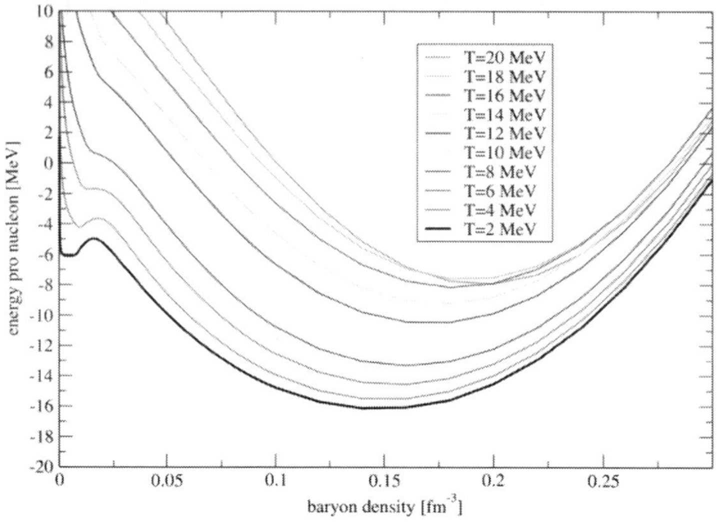


**Figure 1.** Internal energy $U$ for symmetric nuclear matter at different temperatures.

shown in Fig. 1, where the Pauli blocking is caused by uncorrelated nucleons that form a Fermi sphere, the Pauli blocking by a clustered medium is less efficient because the nucleonic wave function is more distributed in momentum space. Therefore, we expect that the crossover from $\alpha$ matter behavior, given by a cluster-virial expansion, to the quasiparticle nucleonic liquids happens not near $n = 0.02$ fm$^{-3}$, but near $n_{\rm sat}/3$ as in $^{12}$C so that a continuous decrease is expected. Then, the intermediate region of metastability where $\partial\mu/\partial n \geq 0$ would disappear.

The behavior of the quasiparticle shift $E_\alpha^{\rm qp}(P;T,n,Y_p) - E_\alpha(P)$, Eq. (3), is sensitive to the values of self-energy shift and Pauli blocking that partially compensate. At zero temperature, the self-energy shift amounts about -4000 $n$ MeV fm$^3$, whereas the Pauli blocking term is larger than 4434 $n$ MeV fm$^3$ as given for $T = 1$ MeV in [5]. If we estimate the kinetic energy of the nucleons bound in the $\alpha$ particle as $E_\alpha^{\rm kin} = 51.6$ MeV [8] and simulate the corresponding occupation in momentum space by an effective temperature $E_\alpha^{\rm kin}/6$, the Pauli blocking shift is reduced by a factor 2 so that the total quasiparticle energy shift becomes negative, around -2000 $n$ MeV fm$^3$ at $T = 0$. A decrease of the internal energy follows also from the cluster virial expansion [15] that gives about -1500 $n$ MeV fm$^3$ at $T = 1$ MeV. This value is only based on empirical $\alpha - \alpha$ scattering phase shits. The Ali-Bodmer interaction, which is also based on these phase shifts, has been considered in [12] to calculate the $\alpha$ condensate fraction at $T = 0$. Also a decrease of internal energy was found. Exploratory lattice calculations for a $T = 0$ $\alpha$ condensate state [11] gave no definite answer. Whereas for $^4$He an increase of energy with increasing density was obtained, the opposite occurs for $^{16}$O.

Another interesting region where $\alpha$ clustering is expected to occur is the surface of nuclei. Like pairing that is increasing near the surface [1] where the density drops down from the saturation value, also $\alpha$ clustering and possibly quartetting can occur. A simple way to incorporate this is a local density approach. Estimations of the role of isospin-singlet pairing and quartetting for nuclei near the $N = Z$ line have been given in [16] analysing binding energies of the ground state and of the excited state. The Wigner energy can be explained this way. This effect is seen in the light nuclei where the shift of the proton chemical potential due to the Coulomb energy is not strong compared to the condensation energy.

Cluster formation at the surface of nuclei is of interest also in the region of superheavy elements. As well known, the most dominant decay channel is $\alpha$ decay, in contrast to the emission of protons or neutrons. Therefore one expects that $\alpha$ particles are preformed in the skin of the superheavies where the density drops down. A corresponding picture is a stable core such as the double magic lead or uranium nucleus, surrounded by a skin of nucleons at reduced density where cluster formation is possible. Of course, the Pauli blocking mechanism has to be considered beyond the local density approximation, considering the density matrix due to the core orbitals. We discuss two consequences of clustering in the skin of superheavies, the binding energy and the rms radii. Further signatures can be seen in reaction processes.

The staggering of the binding energies of superheavies is well known from experimental data. It would be of interest to what extent this staggering can be explained within a quasiparticle single-nucleon approach such as shell calculations, or using a THSR-like wave function that describes the c.o.m. motion of antisymmetrized clusters.

Another effect are the rms radii of superheavy elements. Experimentally, the charge radii $\langle r^2\rangle_{\rm charge}$ are measured. The calculated values are the point radii $\langle r^2\rangle_{\rm point}$ that consider point-like nucleons. The relation between both is given by the relation [17]

$$\langle r^2\rangle_{\rm point} + \langle r^2\rangle_{\rm proton} + (N/Z)\langle r^2\rangle_{\rm neutron} = \langle r^2\rangle_{\rm charge} \tag{9}$$

that takes into account the contribution of the proton form factor due to its spatial extension, $\langle r^2\rangle_{\rm proton} = 0.743$ fm$^2$, and the neutron contribution $\langle r^2\rangle_{\rm neutron} = -0.116$ fm$^2$. If we assume that the surface is formed by $\alpha$-like clusters, we expect another relation,

$$\langle r^2\rangle_{\rm point} + \langle r^2\rangle_\alpha = \langle r^2\rangle_{\rm charge} \tag{10}$$

with the charge rms radius for the $\alpha$ particle, $\langle r^2\rangle_\alpha = 2.822$ fm$^2$. Assuming that the point rms radii are nearly the same, the effect of $\alpha$-clustering would increase the rms radius according $\langle r^2\rangle_\alpha - \langle r^2\rangle_{\rm proton} - (N/Z)\langle r^2\rangle_{\rm neutron} = 2.195$ fm$^2$. For instance, considering a heavy nucleus with $[\langle r^2\rangle_{\rm point}]^{1/2} = 6$ fm, the charge rms radius would be larger by 2.98 % if the $\alpha$ cluster determine the surface, compared to the usual nucleonic surface. The empirical values of charge rms radii of nuclei [17] have been compared with Hartree-Fock Bogoliubov shell model calculations [18], and excellent agreement has been found with exception of the superheavies like Cm where the measured values of the charge rms radii are about 2 % higher than the calculated ones. An interesting point would be whether the difference can be explained by clustering. Assuming this, $\alpha$ emitters should have an expanded charge rms radius, compared to quasiparticle single-nucleon calculations.

## References

[1] P. Ring and P. Schuck, *The Nuclear Many-Body Problem*, Springer, Berlin 1980
[2] T. Alm, G. Röpke, and M. Schmidt Z. Phys. A **337**, 355 (1990).
[3] A. Schnell, G. Röpke, and P. Schuck, Phys. Rev. Lett. **83**, 1926 (1999).
[4] G. Röpke, A. Schnell, P. Schuck, and P. Nozieres, Phys. Rev. Lett. **80**, 3177 (1998).
[5] G. Röpke, Phys. Rev. C **79**, 014002 (2009).
[6] G. Röpke, M. Schmidt, and H. Schulz, Nucl. Phys. **A424**, 594 (1984).
[7] J. Dukelsky, G. Röpke, and P. Schuck, Nucl. Phys. **A 628**, 17 (1998).
[8] G. Röpke, arXiv:1101.4685v2
[9] S. Typel *et al.*, Phys. Rev. C **81**, 015803 (2010).
[10] K. Sumiyoshi and G. Röpke, Phys. Rev. C **77**, 055804 (2008).
[11] H. Takemoto et al., Phys. Rev. C **69**, 035802 (2004).
[12] F. Carstoiu and S. Misicu, Phys. Lett. **B682**, 33 (2009).
[13] A. Tohsaki et al., Phys. Rev. Lett. **87**, 192501 (2001).
[14] Y. Funaki et al., Phys. Rev. C **77**, 064312 (2008).
[15] C. J. Horowitz and A. Schwenk, Nucl. Phys. A **776**, 55 (2006).
[16] G. Röpke, A. Schnell, P. Schuck, U. Lombardo, Phys. Rev. C **77**, 024306 (2000).
[17] I. Angeli, Atomic Data and Nuclear Data Tables **87**, (2004).
[18] L.M. Robledo, G.F. Bertsch, arXiv:1104.5453 (2011); contributions to the conference MBC2011, Paris 2011.

Journal of Physics: Conference Series **321** (2011) 012024 doi:10.1088/1742-6596/321/1/012024

# Thomas-Fermi approximation to pairing in finite Fermi systems. The weak coupling regime

**X. Viñas**[1], **P. Schuck**[2,3], **and M. Farine**[4]

[1]Departament d'Estructura i Constituents de la Matèria and Institut de Ciències del Cosmos, Facultat de Física, Universitat de Barcelona, Diagonal *647*, *E-08028* Barcelona, Spain
[2]Institut de Physique Nucléaire, IN2P3-CNRS, Université Paris-Sud, F-91406 Orsay-Cédex, France
[3] Laboratoire de Physique et Modélisation des Milieux Condensés, CNRS and Université Joseph Fourier, 25 Avenue des Martyrs, Boîte Postale 166, F-38042 Grenoble Cedex 9, France
[4] Ecole des Mines de Nantes, Université Nantes, 4, rue Alfred Kastler B.P. 20722 *44307* Nantes-Cédex 3, France

E-mail: `xavier@ecm.ub.es`

**Abstract.** We present a new semiclassical theory for describing pairing in finite Fermi systems. It is based on taking the $\hbar \to 0$, i.e. Thomas-Fermi, limit of the gap equation written in the basis of the mean field (weak coupling). In addition to the position dependence of the Fermi momentum, the size dependence of the pairing force is also taken into account in this theory. Along isotopic chains the Thomas-Fermi gaps average the well known arch structure shown by the quantal gaps. This structure can be almost recovered in our formalism if some shell fluctuations are included in the level density. We point out that at the drip line nuclear pairing is strongly reduced. This fact is illustrated with the behaviour of the gap in the inner crust of neutron stars.

## 1. Introduction

Semiclassical approaches to finite Fermi systems provide a very efficient way of extracting the average behaviour of relevant physical quantities which characterize such systems. The most well known example is the celebrated Droplet Model and its extensions developed by Myers and Swiatecki which describe nicely the average behaviour of the nuclear masses [1]. Semiclassical techniques have also been applied to study the average behaviour of other properties such as inertias [2], charge radii [3], one- and two-body matrix elements [4], etc. Our aim here is to present a new Thomas-Fermi (TF) theory, i.e. the $\hbar \to 0$ limit, for describing the average trends of the effect of pairing correlations in finite Fermi systems. Semiclassical approaches to the pairing problem can be of interest in scenarios like, e.g., cold atomic gases where the huge number of particles makes the full quantal calculation numerically very complicated. Also these calculations are useful if one is only interested in the average behaviour of the pairing gap, as is the case of the pairing term in the nuclear mass formula. The Local Density Approximation (LDA) is the standard semiclassical technique for dealing with the average behaviour of the pairing which was developed by Schuck and collaborators more than twenty years ago [10]. In LDA one considers the BCS equations in infinite homogeneous matter and replaces the Fermi momentum $k_F$ by its local version in terms of the density. The validity of

LDA applied to the pairing problem is restricted on one hand to situations where the local Fermi wavelength $2\pi/k_F(\mathbf{R})$ is small as compared with the distance where the mean field potential varies appreciably. In the case of a harmonic oscillator potential $V(\mathbf{R}) = m\omega^2 R^2/2$ this distance is the so-called oscillator length defined as $l = \sqrt{\hbar/m\omega}$. On the other hand, a second length scale introduced by pairing is the coherence length $\xi$ which measures the extension of the Cooper pairs. The validity of LDA in the pairing case also implies that the coherence length be smaller than the oscillator legth, i.e. $\xi/l < 1$, which is usually equivalent to the condition $\Delta/\hbar\omega > 1$ , where $\Delta$ is the gap in the single-particle spectrum. The condition $\xi/l < 1$ is always violated in the outer tail of the surface because the LDA coherence length behaves as $\xi \sim \Delta^{-1}$ and the gap vanishes in this region. In spite of these deficiences, integrated quantities as pairing energies may be quite accurate when considered on average [10].

In this contribution we present a novel TF theory for pairing which improves the LDA. This theory can be applied in the weak pairing regime where the chemical potential $\mu$ and the Fermi energy $\varepsilon_F$ have similar values and $\Delta/\mu << 1$. This TF theory works, for the average, in the region $\Delta < \hbar\omega$, where LDA generally fails.

An important point that will be discussed in this contribution is the possible quenching of the pairing gap when approaching the drip line. This fact is well documented in the recent nuclear physics literature [5–9] and we will see, using some examples, that this effect is a general feature when superfluid (superconducting) fermions come in a finite confining potential to an overflow situation.

The contribution is organized as follows. The basic theory is presented in the second section. The main results are discussed in the third section. Our conclusions are laid out in the last section.

## 2. Basic theory

It is well known that in the Hartree-Fock-Bogoliubov theory single-particle density matrix and the pairing tensor or anomalous density matrix are simultaneously diagonalized by the so-called canonical or natural basis, $|n_c\rangle$ [11]. As far as in this work we are only interested in the weak coupling limit where the gap is small as compared with the Fermi energy, i.e. $\Delta/\mu < 1$, one can replace with only small error the canonical basis by the basis of the normal, non-superfluid mean field (HF) Hamiltonian. In this situation the gap equation reduces to its BCS approximation and can be written as

$$\Delta_n = -\sum_{n'} V_{nn'} \frac{\Delta_{n'}}{2E_{n'}}, \tag{1}$$

where $V_{nn'} = \langle n\bar{n}|v|n'\bar{n}'\rangle$ is the matrix element of the interaction with $|\bar{n}\rangle$ the time reversed state of $|n\rangle$ and $E_n = [(\epsilon_n - \mu)^2 + \Delta_n^2]^{1/2}$ the quasi-particle energies, with $\epsilon_n$ the diagonal elements of the normal mean field Hamiltonian [11] written in the basis of the standard mean field hamiltonian, that is $H|n\rangle = \epsilon_n|n\rangle$.

At equilibrium and for time reversal invariant systems canonical conjugation and time reversal operation are related by $\langle \mathbf{r}|\bar{n}\rangle = \langle n|\mathbf{r}\rangle \Rightarrow \langle \mathbf{r}_1\mathbf{r}_2|n\bar{n}\rangle = \langle \mathbf{r}_1|\hat{\rho}_n|\mathbf{r}_2\rangle$, where $\hat{\rho}_n = |n\rangle\langle n|$ is the density matrix corresponding to the state $|n\rangle$. Therefore the pairing matrix element can be writeen as:

$$V_{nn'} = \langle n\bar{n}|v|n'\bar{n}'\rangle = \int d\mathbf{r_1} d\mathbf{r_2} d\mathbf{r'_1} d\mathbf{r'_2} \langle \mathbf{r}'_1|\hat{\rho}_n|\mathbf{r}_1\rangle\langle \mathbf{r_1}\mathbf{r_2}|v|\mathbf{r'_1}\mathbf{r'_2}\rangle\langle \mathbf{r}'_2|\hat{\rho}_n|\mathbf{r}_2\rangle. \tag{2}$$

The density matrix $\hat{\rho}_n$ fulfills the Schödinger equation

$$(H - \epsilon_n)\hat{\rho}_n = 0, \tag{3}$$

therefore we can write $\Delta_n = Tr[\hat{\Delta}\hat{\rho}_n]$ and $\epsilon_n = Tr[H\hat{\rho}_n]$ and consequently the state dependence of the gap equation (1) is fully expressed through the density matrix $\hat{\rho}_n$.

Performing the Wigner transform (WT) of Eq.(3) and taking into account that the WT of the product of two single-particle operators $\hat{A}$ and $\hat{B}$ equals, to lowest order in $\hbar$, the c-number product of the corresponding WT's, i.e. $A(\mathbf{R},\mathbf{p})B(\mathbf{R},\mathbf{p})$, one easily obtains the $\hbar \to 0$ limit of Eq.(3) [11]

$$(H_{cl.} - \epsilon) f_\epsilon(\mathbf{R},\mathbf{p}) = 0, \tag{4}$$

where $H_{cl.} = \frac{p^2}{2m^*(\mathbf{R})} + V(\mathbf{R})$ is the classical Hamiltonian which contains a local mean field potential $V(\mathbf{R})$ and a position dependent effective mass $m^*(\mathbf{R})$ and $f_\epsilon(\mathbf{R},\mathbf{p})$ is the Wigner transform of $\hat{\rho}_n$. Equation (4) has to be read in the sense of distributions. Taking into account that $x\delta(x) = 0$ one obtains the normalized distribution function

$$f_E(\mathbf{R},\mathbf{p}) = \frac{1}{g^{TF}(E)} \delta(E - H_{cl.}) + O(\hbar^2), \tag{5}$$

which corresponds to the Thomas-Fermi (TF) approximation of the normalized on-shell or spectral density matrix [4]. Its norm is equal to the level density (without spin-isospin degeneracy):

$$g^{TF}(E) = \frac{1}{(2\pi\hbar)^3} \int d\mathbf{R} d\mathbf{p} \delta(E - H_{cl.}). \tag{6}$$

The semiclassical pairing matrix element can then be written as [4]:

$$V(E,E') = \int \frac{d\mathbf{R}d\mathbf{p}}{(2\pi\hbar)^3} \int \frac{d\mathbf{R}'d\mathbf{p}'}{(2\pi\hbar)^3} f_E(\mathbf{R},\mathbf{p}) f_{E'}(\mathbf{R}',\mathbf{p}') v(\mathbf{R},\mathbf{p};\mathbf{R}',\mathbf{p}'), \tag{7}$$

where $v(\mathbf{R},\mathbf{p};\mathbf{R}',\mathbf{p}')$ is the double WT of $< \mathbf{r_1}\mathbf{r_2}|v|\mathbf{r}'_1\mathbf{r}'_2 >$. For a local translationally invariant force this matrix element reduces to $v(\mathbf{R},\mathbf{p};\mathbf{R}',\mathbf{p}') = \delta(\mathbf{R}-\mathbf{R}')v(\mathbf{p}-\mathbf{p}')$ with $v(\mathbf{p}-\mathbf{p}')$ the Fourier transform of the force $v(\mathbf{r}-\mathbf{r}')$ in coordinate space.

The gap equation in the TF approximation is obtained by replacing in (1) $\hat{\rho}_n$ and $V_{nn'}$ by their corresponding semiclassical counterparts Eqs. (5) and (7) respectively. In this way the TF gap equation reads

$$\Delta(E) = \int_0^\infty dE' g^{TF}(E') V(E,E') \frac{\Delta(E')}{2\sqrt{(E'-\mu)^2 + \Delta^2(E')}}, \tag{8}$$

Eqs.(8)-(7) can readily be solved for a given mean field and the chemical potential is fixed by the usual particle number condition.

## 3. Results

As a realistic application of our TF theory we analyze the semiclassical pairing gaps as a function of mass number along the tin isotopic chain from $^{100}$Sn to $^{132}$Sn. To this end we use the D1S Gogny force [12] for both, mean field and pairing fields. The main ingredients for solving the semiclassical pairing equation (8) are the on-shell density matrix $f_E(\mathbf{R},\mathbf{p})$ (5), which depends on the classical Hamiltonian $H_{cl}$ that is determined by the effective mass $m^*(\mathbf{R})$ and the mean field $V(\mathbf{R})$. These two quantities, namely $m^*(\mathbf{R})$ and $V(\mathbf{R})$, are obtained through the Extended Thomas-Fermi (ETF) theory for finite-range non-relativistic interactions [13, 14]. Using these quantities as input, one obtains the level density (6) and the pairing matrix element (7) which allow to solve the gap equation (8) in our TF approximation. As explained in Refs. [14, 15], the ETF energy density functional can be transformed, inspired by the Kohn-Sham scheme, into a quantal functional from where the quantal average gaps are obtained. It should be noted that within this approximation the quantal functional associated to a finite- range effective interaction becomes local [14, 15]. The quantal pairing gaps averaged with $u^2v^2$ are depicted

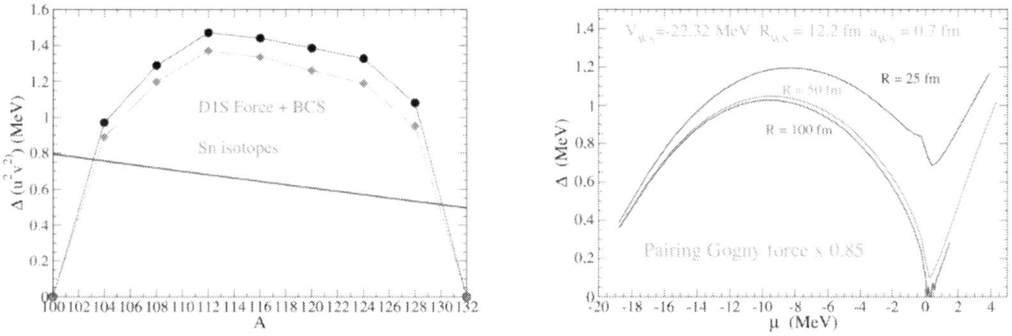


**Figure 1.** Left: Average pairing gap along the Sn isotopic chain. See text for details. Right: Average TF gaps at the Fermi energy as a function of the chemical potential in a WS potential computed in a box of radius $R$=25, 50, and 100 fm.

by circles in the left panel of Fig. 1 and show the typical arch structure. In the same panel we also display the semiclassical TF gap at the Fermi energy by a thick solid line. We see that in this case the quantal arch structure completely disappears, as expected due the absence of shell effects in this case, and that the TF gaps decrease smoothly when the neutron number increases. As it has been discussed in Ref. [16], quantal effects, i.e. the arch structure over the shell, can be almost recovered by introducing some additional quantal fluctuations in the level density and retaining the TF pairing matrix elements (7) in the gap equation (8). The average gaps obtained in these conditions are displayed by diamonds in the left panel of Figure 2. We see that the arch structure for the tin isotopic chain is recovered and that the quantal gaps are predicted quite well in this way.

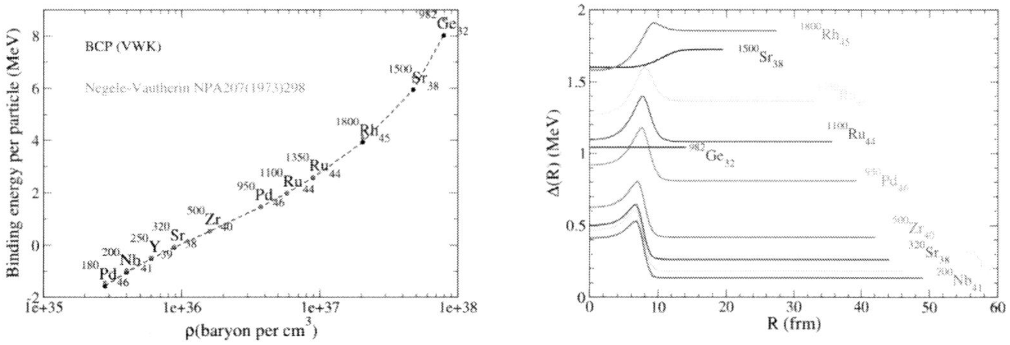


**Figure 2.** Left: Energy per baryon as a function of the average density in different WS cells corresponding to the inner crust of neutron stars. Right: Radial dependence of TF gap in the analyzed WS cells. End points indicate radius of WS cells.

An important feature of the average TF gaps is that they show a downward trend with increasing neutron number along a given isotopic chain as it can be appreciated in the left panel of Fig. 1. To investigate the behaviour of the TF gap at the Fermi energy approaching the drip line, we show in the right panel of Fig. 1 the TF gap as a function of the chemical potential employing a Woods-Saxon potential as mean field [17]. From this panel it can be seen that in the TF limit the gap vanishes just at the drip line when the chemical potential equals the depth of the single-particle potential. Actually we put the Woods Saxon potential in a large external container with radius $R$ what can simulate a Wigner-Seitz (WS) cell as relevant in the inner crust of neutron stars, see below. We see that the gap becomes more and more suppressed at the edge of the Woods-Saxon potential for increasing values of the cell radius. Continuing filling the cell with a neutron gas, the gaps raise again. This quenching of the gap when approaching the drip line has also been studied by Hamamoto [5] analyzing the effective gap in weakly bound neutron levels in spherical and deformed nuclei, finding that the presence of a $s_{1/2}$ component in the wavefunction of these levels strongly reduces the effective gap. Other gap components with $l \neq 0$ show a less decreasing tendency near the drip line probably because of the centrifugal barrier which keeps the wave functions localised. This could mask the behaviour of the state dependent gaps in real nuclei near the neutron drip line.

As mentioned already, another scenario where the quenching of the neutron gap appears is near the neutron drip line in the inner crust of neutron stars. This region is a crystal of nuclei embeded in a gas of free neutrons and electrons. The inner crust was described by Negele and Vautherin [18] at HF level by means of the energy density functional method together with a spherical WS approach to deal with the crystal structure in an approximated way. The WS cell is electrically neutral and the ground state of the system of neutrons, protons and electrons is reached when they are in $\beta$-equilibrium. We have performed a similar calculation but at TF level and using the BCP energy density functional [19]. This functional consists of a bulk part provided by a microscopic calculation complemented by a phenomenological surface term. This functional with only four adjustable parameters reproduces nuclear binding energies and charge radii of finite nuclei with the same quality as obtained with the most performant effective forces. The ground state energy per baryon obtained in this way for average densities in the WS cells ranging from $2.79 \times 10^{-4}$ to $7.89 \times 10^{-2}$ fm$^{-3}$ are displayed in the left panel of Fig. 2 (black dots) in comparison with the Negele-Vautherin results (red diamonds) finding an excellent agreement between both calculations. Although pairing correlations in the inner crust of neutron stars are mainly driven by the free neutron gas, they have, however, a noticeable influence on the composition and pairing properties of the nuclear cluster inside the WS cell [7]. On top of this TF calculation in the inner crust, we have also performed a TF pairing calculation in the studied WS cells. To this end, we have used the Gogny D1S force renormalized by a factor 0.85 to take into account that in the BCP functional the effective mass equals the physical one. A specially relevant quantity in this context is the radial dependence of the pairing gap $\Delta(\mathbf{R})$ which is obtained from the WT of the average quantal gap $\Delta_{av} = Tr[\hat{\Delta}\hat{\kappa}]/Tr[\hat{\kappa}]$ [20] which reads:

$$\Delta(\mathbf{R}) = \frac{1}{\kappa(\mathbf{R})} \int \frac{d\mathbf{p}}{(2\pi\hbar)^3} \Delta(\mathbf{R},\mathbf{p}) \kappa(\mathbf{R},\mathbf{p}) \tag{9}$$

where $\kappa(\mathbf{R},\mathbf{p}) = \int dE g^{TF}(E) \kappa(E) f_E(\mathbf{R},\mathbf{p})$ and $\Delta(\mathbf{R},\mathbf{p}) = -\int \frac{d\mathbf{p}'}{(2\pi\hbar)^3} v(\mathbf{p}-\mathbf{p}') \kappa(\mathbf{R},\mathbf{p}')$,

We see in the right panel of Fig. 2 that $\Delta(\mathbf{R})$ takes a constant value in the outer part of the WS cell which corresponds to the gap of the free neutron gas [9]. The predicted behaviour for $\Delta(\mathbf{R})$ are in qualitative agreement with previous calculations [6–9]. The nuclear cluster inside the WS cell disappears when the homogeneous phase is reached at an average density of about 0.08 fm$^{-3}$ ($^{982}Ge_{32}$). In this case the gap in the cell is practically the gap obtained in pure neutron matter at the same density. The density of the free neutron gas diminishes in approaching to drip configurations. In this situation the gap is strongly reduced not only in the gas but also inside of the nuclear cluster ($^{200}Nb_{41}$) and it even may disappear completely in the nucleus when the drip line is reached (see right panel of Fig. 1). Therefore, locally the TF $\Delta(\mathbf{R})$'s are qualitatively different from what LDA would predict.

## Conclusions and Outlook

We have presented a TF theory for pairing in finite Fermi systems for weak coupling situations where $\Delta/\varepsilon_F << 1$. This TF theory differs from the usual LDA. This essentially stems from the fact that we approximate the gap equation in configuration space and, thus, keep the size dependence of the matrix elements of the pairing force. This is not the case in LDA where the matrix elements of the force are always evaluated in plane wave basis. This semiclassical approach to pairing is only based on the usual validity criterion of Thomas-Fermi theory, namely that the Fermi wave length is smaller than the oscillator length. At no point the LDA condition that the coherence length must be smaller than the oscillator length enters the theory. Thus, the present TF approach yields for all pairing quantities the same quality as TF theory does for quantities in the normal fluid state. An interesting feature of our study is that the average gap breaks down going to the drip line. This unexpected result is confirmed by quantal calculations, though strongly masked by shell fluctuations. For systems with large numbers of particles the fluctuations should die out and, thus, the semiclassical behaviour prevail. Indeed preliminary results in a slab configuration show good agreement between quantal and TF gaps around the drip region.

## Acknowledgments

We thank Michael Urban for useful discussions and comments. This work has been partially supported by the IN2P3-CAICYT collaboration. One of us (X.V.) acknowledges grants FIS2008-01661 (Spain and FEDER), 2009SGR-1289 (Spain) and Consolider Ingenio Programme CSD2007-00042 for financial support.

## References

[1] W.D. Myers and W.J. Świątecki, Ann. of Phys. **55**, 395 (1969); **84**, 186 (1974); Nucl. Phys. **A601**.
[2] M. Farine, P. Schuck and X. Viñas, Phys. Rev. **A62**, 013608 (2000); M. Durand, P. Schuck and J. Kunz, Nucl. Phys. **439**, 263 (1985).
[3] J. Duflo, Nucl. Phys. **576**, 29 (1994); J. Piekarewicz et al, Eur. Phys. J. **A46**, 379 (2010).
[4] X. Viñas, P. Schuck, M. Farine and M. Centelles, Phys. Rev. **C67**, 054307 (2003).
[5] I. Hamamoto, Phys. Rev. **C71**, 037302 (2005).
[6] N. Sandulescu, N. Van Giai and R.J. Liotta, Phys. Rev. **C69**, 045802 (2005).
[7] M. Baldo, U. Lombardo, E.E. Saperstein and S.V. Tolokonnikov, Nucl. Phys. **A750**, 409 (2005); M. Baldo, E.E. Saperstein and S.V. Tolokonnikov, Eur. Phys. J. **A32**, 97 (2007).
[8] M. Grasso, E. Khan, J. Margueron and N. Van Giai, Nucl. Phys. **A807**, 1 (2008).
[9] N. Chamel, S. Goriely, J.M. Pearson and M. Onsi, Phys. Rev. **C81**, 045804 (2010).
[10] H. Kucharek, P. Ring, P. Schuck, R.Bengtsson and M. Girod, Phys. Lett. **B216**, 249 (1989).
[11] P. Ring and P. Schuck, *The Nuclear Many-Body Problem*, (Springer-Verlag, Berlin, 1980).
[12] J.-F. Berger, M. Giraud and D. Gogny, Comp. Phys. Comm. **63**, 365 (1991).
[13] M. Centelles, X. Viñas, M. Durand, P. Schuck and D. Von-Eiff, Ann.of Phys. **266** 207 (1998).
[14] V.B. Soubbotin and X. Viñas, Nucl. Phys. **A665**, 291 (2000).
[15] V.B. Soubbotin et al, Phys. Rev. **C67** 014324 (2003); S. Krewald et al, Phys. Rev. **C74** 064310 (2006).
[16] X. Viñas, P. Schuck and M. Farine, Int. J. Mod. Phys. **E20** 399 (2011).
[17] S. Shlomo, Nucl. Phys. **539**, 17 (1992).
[18] J.W. Negele and D. Vautherin, Nucl. Phys. **207**, 298 (1973).
[19] M. Baldo, P. Schuck and X. Viñas Phys. Lett. **B663**, 390 (2008).
[20] N. Pillet, N. Sandulescu, P. Schuck and J.-F. Berger, **C81** 034307 (2010).

Journal of Physics: Conference Series **321** (2011) 012025 doi:10.1088/1742-6596/321/1/012025

# $\alpha+\alpha+t$ cluster structures and Hoyle-analogue states in $^{11}$B

**Taiichi Yamada**[1] **and Yasuro Funaki**[2]

[1]Laboratory of Physics, Kanto Gakuin University, Yokohama 236-8501, Japan
[2]Institute of Physics, University of Tsukuba, Tsukuba 305-8571, Japan

E-mail: yamada@kanto-gakuin.ac.jp

**Abstract.** The structures of $^{11}$B are studied with the $2\alpha+t$ orthogonality condition model (OCM) for exploring Hoyle-analogue states in $^{11}$B.

Cluster picture as well as mean-field picture is important to understand the structure of light nuclei. The Hoyle state (the $0_2^+$ state at $E_x = 7.65$ MeV in $^{12}$C) is a typical cluster state with a $3\alpha$ cluster structure [1,2], which appears at only 0.38 MeV above the $3\alpha$ threshold. This Hoyle state has been reinvestigated from the viewpoint of $\alpha$ condensation [3]. The definitions and occurrences of the nuclear $\alpha$-particle condensation are discussed in detail in Ref. [4]. Many theoretical works have shown that the Hoyle state has a $3\alpha$-condensate-like structure, in which the $3\alpha$ particles occupy an identical $0S$-orbit with 70 % probability, forming a dilute $\alpha$-gas-like configuration with $(0S_\alpha)^3$, i.e. the product state of three $\alpha$'s [5,6,7].

Recently the $4\alpha$ OCM (Orthogonality Condition Model) calculation [8] succeeded for the first time in describing the structure of $^{16}$O up to $E_x \simeq 15$ MeV, and showed that the $0_6^+$ state around the $4\alpha$ threshold is a strong candidate of $4\alpha$-particle-condensate-like character, having a large $\alpha$ condensate fraction of 60 %. Similar dilute-gas states of $\alpha$ clusters have been predicted around their alpha cluster disintegrated thresholds in self-conjugate $A = 4n$ nuclei [3,9]. Besides the $4n$ nuclei, one can also expect cluster-gas states composed of alpha and triton clusters (including valence neutrons etc.) around their cluster disintegrated thresholds in $A \neq 4n$ nuclei, in which all clusters are in their respective $0S$ orbits, similar to the Hoyle state with $(0S_\alpha)^3$. The states, thus, can be called *Hoyle-analogue* in non-self-conjugated nuclei. It is an intriguing subject to investigate whether or not Hoyle-analogue states exist in $A \neq 4n$ nuclei, for example, $^{11}$B ($^{13}$C), composed of $2\alpha$ and $t$ clusters ($3\alpha$ clusters and a valence neutron).

The purposes of the present contribution are to study the structure of $^{11}$B up to around the $\alpha+\alpha+t$ and to explore the Hoyle-analogue state with the $\alpha+\alpha+t$ structure [10]. This Hoyle-analogue states are expected to appear in $1/2^+$ states because one can conjecture an appearance of the product state, $(0S_\alpha)^2(0S_t)$, in $J^\pi = 1/2^+$ of $^{11}$B (1/2 comes from the spin of triton) from the similarity of $(0S_\alpha)^3$ in the Hoyle state. The structure of the positive-parity states (isospin $T = 1/2$) in $^{11}$B so far have not been well discussed in the no-core shell model [11] and AMD [12]. In the present study, we consider the $\alpha+\alpha+t$ OCM with the Gaussian expansion method, the model space of which is large enough to cover the $\alpha+\alpha+t$ gas, the $^7$Li+$\alpha$ cluster, as well as the shell-model configurations.

The energy levels of $3/2^-$ and $1/2^+$ states (those of the positive-parity states) in $^{11}$B are shown in Fig. 1 (Fig. 2). The $3/2_1^-$ state is the ground state with a shell-model-like structure. The

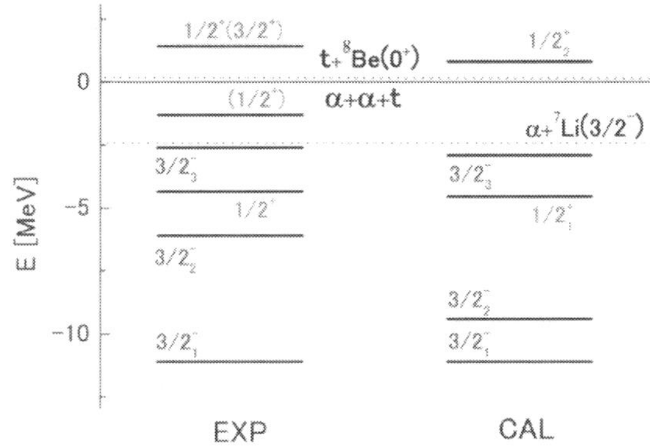


**Figure 1.** Calculated energy levels of $3/2^-$ and $1/2^+$ states in $^{11}$B with respect to the $\alpha+\alpha+t$ threshold together with the experimental data [10].

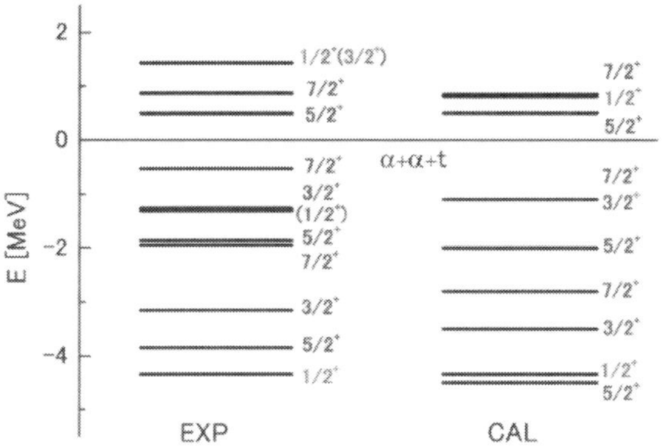


**Figure 2.** Calculated energy levels of the positive-parity states in $^{11}$B with respect to the $\alpha + \alpha + t$ threshold together with the experimental data [10].

calculated nuclear radius is $R = 2.22$ fm ($R^{\rm exp} = 2.43 \pm 0.11$ fm). The dominant configuration of this state is $\mathrm{SU}(3)[f](\lambda,\mu)_L = [443](1,3)_1$ with $Q = 7$ harmonic oscillator quanta (95 %), having the main angular momentum channel of $(L,S)_J = (1,\frac{1}{2})_{\frac{3}{2}}$. The $3/2_2^-$ state has also a shell-model-like structure with the dominant component of $(\lambda,\mu)_L = (1,3)_2$ with $Q = 7$ and $(L,S)_J = (2,\frac{1}{2})_{\frac{3}{2}}$.

The $3/2_3^-$ state appears at $E_x = 8.2$ MeV ($E = -2.9$ MeV referring to the $2\alpha + t$ threshold). The radius of $3/2_3^-$ is 3.00 fm. This value is by about 30 % larger than that of the ground state of $^{11}$B, and the rms distance between $\alpha$ and $\alpha$ (between $^8$Be($2\alpha$) and $t$) is 4.47 fm (3.49 fm). Thus, $3/2_3^-$ has a $2\alpha + t$ cluster structure. A characteristic feature of $3/2_3^-$ is that the isoscalar monopole transition rate $B$(E0:IS) is as large as $96 \pm 16$ fm$^4$ [13], comparable to that of the Hoyle state ($120 \pm 9$ fm$^4$). The present model (92 fm$^4$) reproduces well the data. It is interesting to study whether $3/2_3^-$ possesses an $\alpha$ condensate nature like the Hoyle state. In order to see it, we study the single-cluster orbits and their occupation probabilities in $3/2_3^-$ by solving the eigenvalue equation of the single-cluster density matrices [4].

Figure 3 shows the occupation probabilities of the $n$-th $L$-wave single-$\alpha$-particle (single-$t$-particle) orbits in the $3/2_1^-$ and $3/2_3^-$ states. In $3/2_1^-$, the occupation probabilities of $\alpha$ orbits spread over in several orbits, and those of $t$ orbits concentrate mainly on two orbits. These results originate from the SU(3) nature of the $3/2_1^-$ state as mentioned above. On the other hand, in $3/2_3^-$, there is no concentration of the $\alpha$ occupation probability on a single orbit. This result is in contrast with those of the Hoyle state (see Fig. 2(d)). Consequently the $3/2_3^-$ state could not be identified as the analogue of the Hoyle state. The reason why the $3/2_3^-$ state is not of the Hoyle-analogue is given as follows: The $3/2_3^-$ state is bound by 2.9 MeV with respect to the $2\alpha + t$ threshold, while the Hoyle state is located by 0.38 MeV above the $3\alpha$ threshold. This extra binding energy of $3/2_3^-$ with respect to the $2\alpha + t$ threshold suppresses strongly the growth of the gas-like $2\alpha + t$ structure.

As for the $1/2^+$ states, the $1/2_1^+$ state appears as a bound state at $E_x^{\rm exp} = 6.79$ MeV around the $^7$Li+$\alpha$ threshold. This low excitation energy indicates that $\alpha$-type correlations should play an important role in the state. In fact, we found that the $1/2_1^+$ state with the nuclear radius of 3.14 fm has a $^7$Li(g.s)+$\alpha$ structure with $P$-wave relative motion, although the $^7$Li($\alpha + t$) part is rather distorted in comparison with the ground state of $^7$Li. Since the $3/2_3^-$ state has the largest $S^2$ factor for the $^7$Li(g.s)+$\alpha$ channel with $S$-wave relative motion, the $1/2_1^+$ and $3/2_3^-$ states of

Journal of Physics: Conference Series **321** (2011) 012025 doi:10.1088/1742-6596/321/1/012025

$^{11}$B can be interpreted as the parity-doublet partners of each other.

In addition to the $1/2_1^+$ state, the $1/2_2^+$ state appears as a resonant state at $E_x = 11.95$ MeV ($\Gamma = 190$ keV) around the $2\alpha+t$ threshold with the complexed-scaling method. The large radius ($R_N = 5.98$ fm) indicates that the state has a dilute cluster structure. The analysis of the single-cluster properties showed that this state has as main configuration $(0S_\alpha)^2(0S_t)$ with about 65% probability. Thus, the $1/2_2^+$ state can be called the Hoyle-analogue with $(0S_\alpha)^2(0S_t)$. Recently, the $1/2^+$ $(3/2^+)$ state at $E_x = 12.56$ MeV with $\Gamma = 210 \pm 20$ keV (located at 1.4 MeV above the $2\alpha + t$ threshold) was observed in the $\alpha+{}^7$Li decay channel [14]. The energy and width of the 12.56-MeV state are in good correspondence to the present study. The Hoyle-analogue state in $^{11}$B, thus, may be assigned as the 12.56-MeV state. It should be reminded that $1/2_2^+$ is located by 0.75 MeV above the $\alpha+\alpha+t$ threshold, while $1/2_1^+$ is bound by 4.2 MeV with respect to the three cluster threshold. The latter binding energy leads to the suppression of the development of the gas-like $\alpha+\alpha+t$ structure in $1/2_1^+$, whereas the gas-like structure with a large nuclear radius develops in the $1/2_2^+$ state because it appears above the three-body threshold.

In $^{13}$C, the Hoyle-analogue states with the dominant configuration of $[(0S_\alpha)^3 s_{1/2}(n)]$ are expected to appear in $1/2^+$ states, because the three $\alpha$ particles (the valence neutron) are in S orbit ($s_{1/2}$). The structure of $^{13}$C has been studied with the $3\alpha$+$n$ OCM [15], the model space of which can cover the $3\alpha$+$n$ gas, the $^{12}$C+$n$ cluster, the $^9$Be+$\alpha$ cluster, the $^8$Be+$^5$He cluster, as well as the shell-model-like configurations. This model succeeded in reproducing the lowest four $1/2^-$ states and three $1/2^+$ states up to about the $3\alpha$+$n$ threshold ($E_x = 12.5$ MeV), including the shell-model-like ground state $(1/2_1^-)$ [15]. We found that the $1/2_3^+$ state around the $3\alpha+n$ threshold (see Fig. 5) has the nuclear radius of 5.40 fm with a dilute $\alpha$ condensate feature, in which the $3\alpha$ particles occupy an identical $0S$ orbit with 55 % probability and the valence neutron is in the $s_{1/2}$ state. This means that the $1/2_3^+$ state is a candidate of the Hoyle-analogue

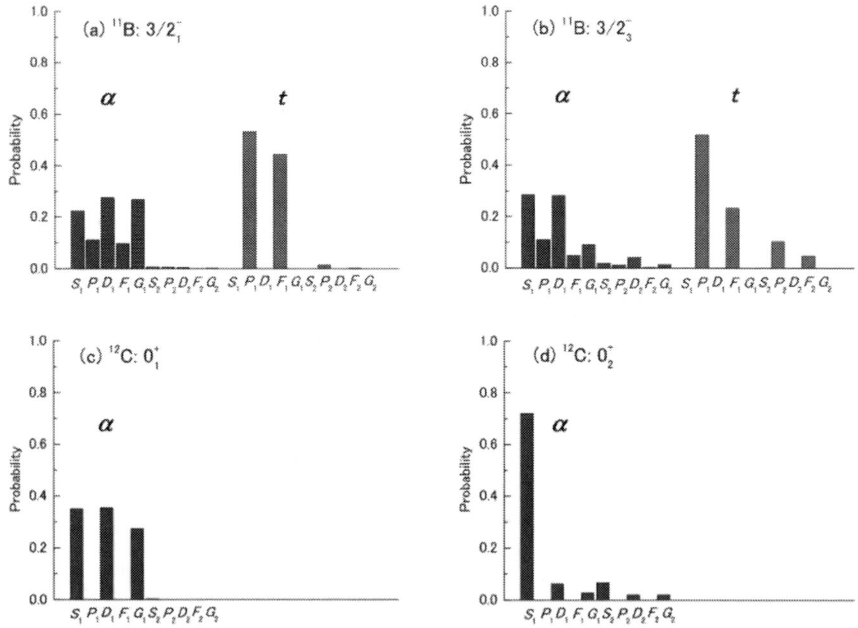


**Figure 3.** Occupation probabilities of the single-$\alpha$-particle and single-$t$-particle orbits for the (a) $3/2_1^-$ (g.s.) and (b) $3/2_3^-$ states in $^{11}$B [10]. We also show those of the single-$\alpha$-particle orbits for the (c) $0_1^+$ (g.s.) and (d) $0_2^+$ (Hoyle) states in $^{12}$C calculated by the $3\alpha$ OCM [6].

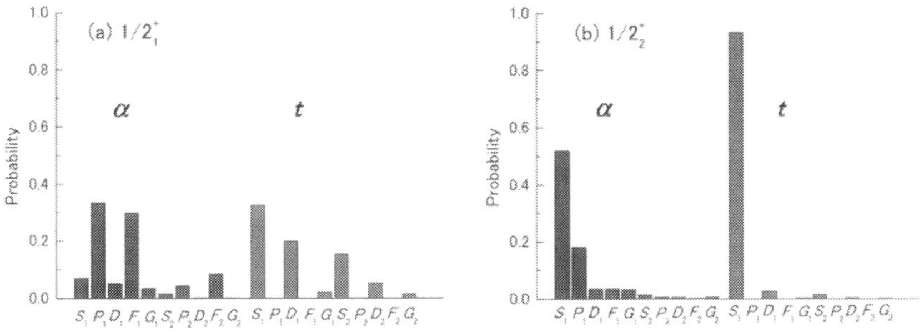

**Figure 4.** Occupation probabilities of the single-$\alpha$-particle and single-$t$-particle orbits for the (a) $1/2_1^+$ and (b) $1/2_2^+$ states in $^{11}$B [10].

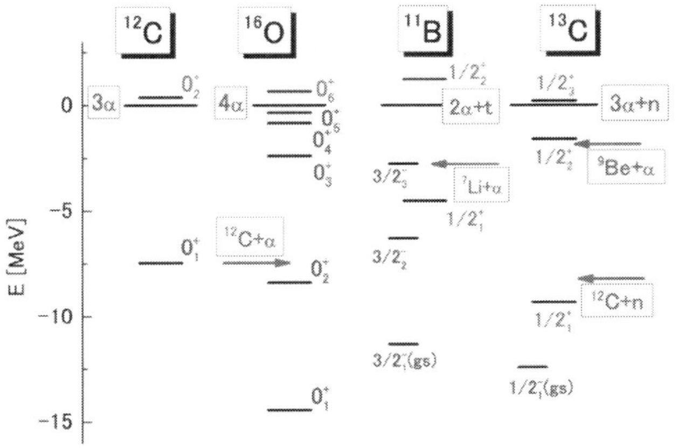

**Figure 5.** Energy levels of $0^+$ states in $^{12}$C and $^{16}$O together with those of $1/2^+$ states in $^{11}$B and $^{13}$C, with respect to their cluster disintegrated thresholds. The $0_6^+$ state of $^{16}$O, $1/2_2^+$ of $^{11}$B, and $1/2_3^+$ of $^{13}$C are the strong candidates of the Hoyle-analogue states (see text).

state in $^{13}$C. The reason why the $1/2^+_{1,2}$ states in Fig. 5 have no Hoyle-analogue character is that the two states are bound with respect to the $3\alpha+n$ threshold. This extra binding energy suppresses the development of the $3\alpha+n$ gas-like structure in the $1/2^+_{1,2}$ states. From Fig. 5, one notices as a common feature that the Hoyle-analogue states in $^{16}$O, $^{11}$B, and $^{13}$C appear around their cluster disintegrated threshold, like the Hoyle state ($0_2^+$ in $^{12}$C).

## References

[1] Uegaki E *et al.* 1977 *Prog. Theor. Phys.* **57** 1262; *ibid.* 1978 **59** 1031; 1979 **62** 1621
[2] Kamimura M 1981 *Nucl. Phys.* A **351** 456
[3] Tohsaki A, Horiuchi H, Schuck P, Röpke G 2001 *Phys. Rev. Lett.* **87** 192501
[4] Yamada T *et al.* 2011 arXiv:1103.3940
[5] Funaki T, Tohsaki A, Horiuchi H, Schuck P, Röpke G 2003 *Phys. Rev.* C **67** 051306
[6] Yamada T and Schuck P 2005 *Eur. Phys. J.* A **26** 185
[7] Matsumura H and Suzuki Y 2004 *Nucl. Phys.* A **739** 238
[8] Funaki Y, Yamada T, Horiuchi H, Röpke G, Schuck P, Tohsaki A 2008 *Phys. Rev. Lett.* **101** 082502
[9] Yamada T and Schuck P 2004 *Phys. Rev.* C **69** 024309
[10] Yamada T and Funaki Y 2010 *Phys. Rev.* C **82** 064315
[11] Navrátil P *et al.* 2009 *J. Phys.* G **36** 083101
[12] Kanada-En'yo Y 2007 *Phys. Rev.* C **75** 024302
[13] Kawabata T *et al.* 2007 *Phys. Lett.* B **646** 6
[14] Soić *et al.* 2004 *Nucl. Phys.* A **742** 271; Curtis N *et al.* 2005 *Phys. Rev.* C **72** 044320; Charity R J *et al.* 2008 *Phys. Rev.* C **78** 054307
[15] Yamada T and Funaki Y 2008 *Int. Mod. Phys.* E **17** 2101

# Damping of collective modes in ultracold trapped Fermi gases with in-medium cross section

**Michael Urban[1], Silvia Chiacchiera[2], Dany Davesne[3] and Thomas Lepers[3]**

[1] Institut de Physique Nucléaire, CNRS/IN2P3 and Univ. Paris Sud, 91406 Orsay cedex, France
[2] Centro de Física Computacional, Department of Physics, University of Coimbra, 3004-516 Coimbra, Portugal
[3] IPN Lyon, Université de Lyon, Université Lyon 1, CNRS/IN2P3, 69622 Villeurbanne cedex, France

E-mail: urban@ipno.in2p3.fr

**Abstract.** We study the damping of the quadrupole and scissors modes of ultracold trapped Fermi gases with large negative scattering length at temperatures above the superfluid critical temperature $T_c$ within the framework of the Boltzmann equation. The in-medium scattering cross-section is strongly enhanced at temperatures slightly above $T_c$. This has strong effects on the collisional damping of the collective modes and the transition from hydrodynamic to collisionless behaviour as a function of temperature. Within the method of phase-space moments of second order, it was found that the resulting damping rates are in strong disagreement with experimental data from the Innsbruck group, which seemed to favour the free cross section. However, the comparison with numerical solutions of the Boltzmann equation revealed that the method of second-order phase-space moments is insufficient. Here we show that the agreement between the calculation using the in-medium cross-section and the data can be clearly improved by including also fourth-order moments.

## 1. Introduction

The study of many-body physics with ultracold atoms has become possible in the last few years due to the enormous progress in trapping and cooling neutral atoms in magneto-optical traps combining lasers and magnetic fields. From the theorist's point of view, the trap is simply an external potential, which can be considered as harmonic near the minimum, $V_{trap}(\mathbf{r}) = \frac{1}{2}\sum_{i=x,y,z}\omega_i^2 r_i^2$. Atoms are loaded into this potential and cooled down by laser and evaporative cooling and are then ready for the actual experiment. These systems have the unique feature that almost all properties (particle number, density, interaction strength, statistics, dimensionality) can be varied in the experiments. Here, we will consider the case of fermionic atoms with two equally occupied spin states $\uparrow$ and $\downarrow$ with an attractive interaction, i.e., a negative scattering length $a < 0$. At temperatures below a certain critical temperature $T_c$, this system undergoes a phase transition to a superfluid phase, but we will concentrate on the normal phase ($T > T_c$).

A couple of experiments were done to study the collective modes of these systems. Here we are interested in interpreting the results obtained by the Innsbruck group [1, 2]. In these

Published under licence by IOP Publishing Ltd

Journal of Physics: Conference Series **321** (2011) 012026 doi:10.1088/1742-6596/321/1/012026

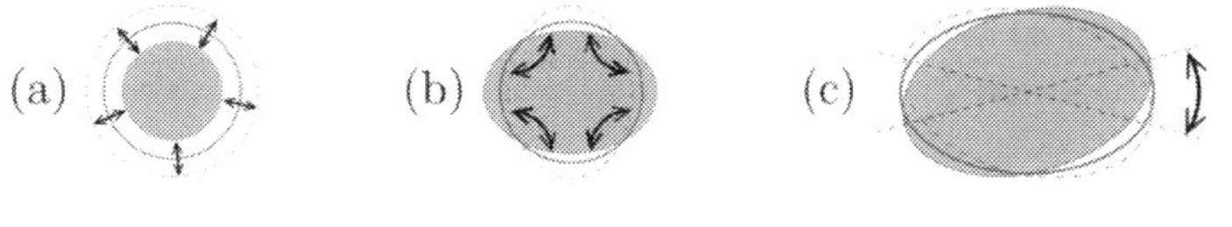


**Figure 1.** Schematic representation of different collective modes, as seen in a cut through the $z = 0$ plane: (a) radial breathing mode, (b) radial quadrupole mode, (c) scissors mode.

experiments, the trap was cigar shaped ($\omega_x \approx \omega_y \gg \omega_z$) and the following collective modes in the radial direction were studied: (a) the radial breathing mode (Fig. 1a), whose frequency depends on the equation of state of the system; (b) the radial quadrupole mode (Fig. 1b), whose frequency is different depending on whether the system behaves hydrodynamically (superfluid or normal fluid) or collisionless; (c) the scissors mode (Fig. 1c), in which the cloud rotates back and forth against the triaxial trap potential ($\omega_x \neq \omega_y$).

The question we are interested in here concerns the transition from collisionally hydrodynamic to collisionless behaviour as function of interaction strength and temperature. With increasing temperature, the system expands and gets more and more dilute, so that the collision rate decreases. In the other limit, at very low temperature, the collision rate goes down, too, due to Pauli blocking. In the regime of a strong attractive interaction, $-1 \lesssim 1/k_F a \leq 0$ ($k_F$ being the Fermi momentum at the trap center), the system behaves more or less hydrodynamically at $T \approx T_C$ (for a summary of the experimental results, see Fig. 4 of Ref. [1]).

## 2. Formalism

Our starting point is the Boltzmann equation $\dot{f} + \frac{\mathbf{p}}{m} \cdot \boldsymbol{\nabla}_r f - \boldsymbol{\nabla}_r V \cdot \boldsymbol{\nabla}_p f = -I[f]$, where $f(\mathbf{r}, \mathbf{p})$ denotes the Wigner function (phase-space distribution function), $V(\mathbf{r}) = V_{trap}(\mathbf{r}) + U$ is the sum of trap and mean-field potentials, and $I$ is the collision integral [3]. The collision integral contains the cross section $\sigma$ and Pauli-blocking factors for the final states.

In free space, the cross section is given by $\sigma_0 = 4\pi a^2/[1 + (qa)^2]$, where $q$ denotes the momentum of the atoms in the center-of-mass frame ($q = |\mathbf{p}_1 - \mathbf{p}_2|/2$). The cross section in the medium can be calculated from the in-medium T matrix in ladder approximation as shown in Fig. 2a. At $T = T_c$, the in-medium T matrix develops a pole at the Fermi surface (Thouless criterion for $T_c$). As a precursor effect, the in-medium cross section is strongly enhanced as compared to the free one at temperatures close to $T_c$ [2, 3], as shown in Fig. 2b. In the context of nuclear physics, this effect was already discussed in Ref. [4].

For the study of collective oscillations, the Boltzmann equation is linearised around equilibrium. Approximate solutions of the linearised equation can be obtained with the method of moments. One writes the distribution function as $f(\mathbf{r}, \mathbf{p}, t) = f_{eq}(\mathbf{r}, \mathbf{p}) + (df_{eq}/d\mu)\Phi(\mathbf{r}, \mathbf{p}, t)$,

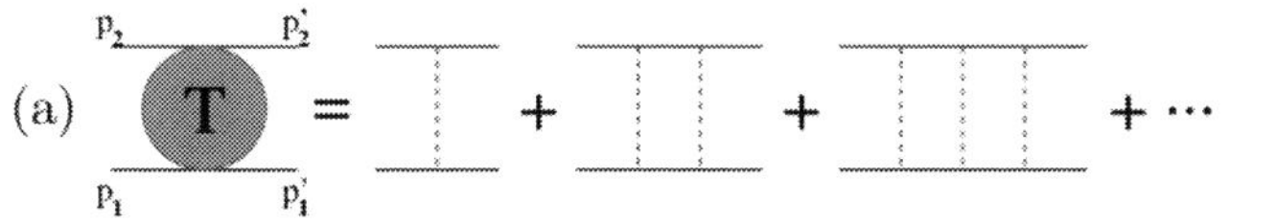


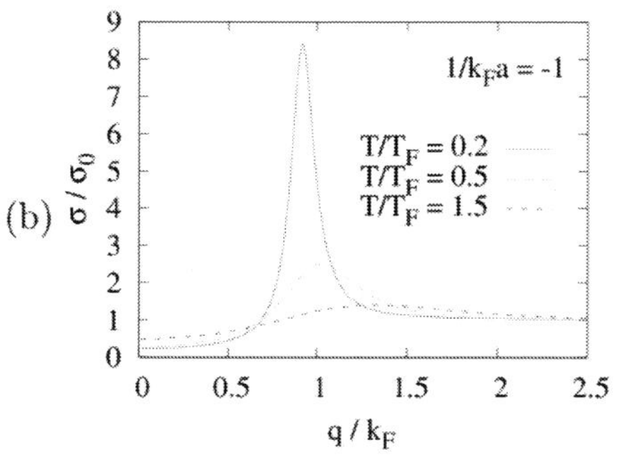


**Figure 2.** (a) Diagrams for the in-medium T matrix in ladder approximation. (b) Ratio of in-medium to free cross section for zero total momentum as function of the relative momentum $q$ for different temperatures $T$ ($T_F$ = Fermi temperature).

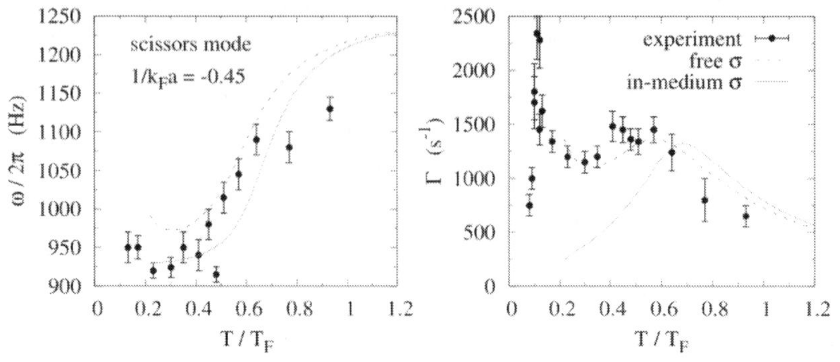


**Figure 3.** Frequency $\omega$ (left) and damping rate $\Gamma$ (right) of the scissors mode as functions of temperature. The data are from Ref. [1], the theoretical curves were obtained within the second-order moment method.

where the equilibrium distribution function $f_{eq}$ is a Fermi function with chemical potential $\mu$, and $\Phi$ is supposed to be a smooth function in phase space (contrary to $f$). Then one makes the ansatz that $\Phi$ is a polynomial in $\mathbf{r}$ and $\mathbf{p}$ with time-dependent coefficients. The time dependence of the coefficients can be determined by taking moments of the Boltzmann equation. In the previous literature [2, 3], only polynomials of second order in $\mathbf{r}$ and $\mathbf{p}$ were included. In this case, the method is equivalent to the widely used scaling ansatz. Only very recently, the method of moments has been applied to higher-order polynomials in $\mathbf{r}$ and $\mathbf{p}$ [5].

## 3. Results

Let us start by discussing results obtained within the method of second-order moments. As an example, we display in Fig. 3 the temperature dependence of the frequency and damping rate of the scissors mode for the parameters of the experiment by Wright et al. [1]. Note that the inclusion of the in-medium cross section, $\sigma$, instead of the free one, $\sigma_0$, completely spoils the agreement with the data. Identical results, except for the mean field effects, were found earlier in calculations by Bruun and Smith [2].

Having obtained such a disappointing result, one might worry that the whole Boltzmann approach is invalid, e.g., because there are no good quasiparticles in the strongly correlated pseudogap regime. But before drawing such pessimistic conclusions, one should first check whether the problem persists if one finds a more accurate solution of the Boltzmann equation.

Our first idea was to solve the Boltzmann equation numerically using the method of test particles (like in the simulation of heavy-ion collisions). So far, we were able to do so only for the case of an isotropic trap with a reduced number of atoms, with the free cross section and without mean field effects. Nevertheless, these calculations revealed that the method of second-order moments systematically overestimates the effects of collisions [5]. As a typical example, we show in Fig. 4 the strength function (excitation spectrum) of the quadrupole mode. The result obtained within the second-order moment method deviates clearly from the numerical

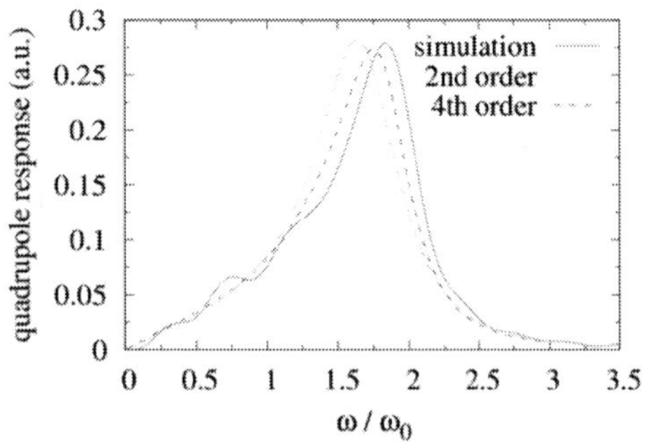


**Figure 4.** Strength function of the quadrupole mode obtained by a test-particle simulation, by the method of moments up to second order, and by the extended method of moments up to fourth-order (parameters: $N = 10000$ atoms in a spherical trap, $T/T_F = 0.4$, $1/k_Fa = -0.5$)

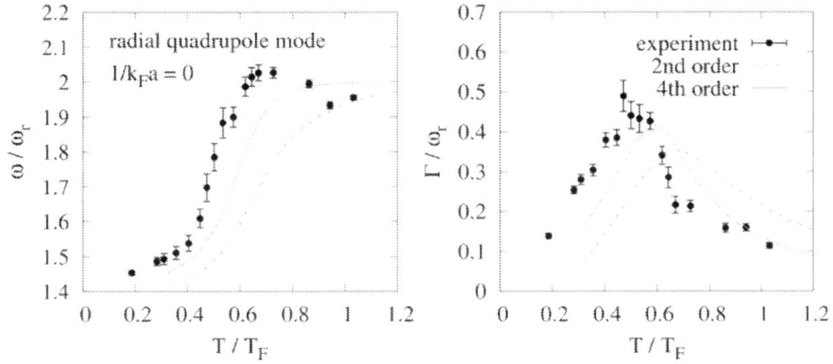


**Figure 5.** Frequency $\omega$ (left) and damping rate $\Gamma$ (right) of the radial quadrupole mode as functions of temperature. The data are from Ref. [2], the theoretical curves were obtained within the second- and fourth-order moment method, respectively.

simulation.

What is wrong with the second-order moment method? Let us look at the ansatz for the function $\Phi$ of the quadrupole mode, which reads $\Phi = c_1(x^2 - y^2) + c_2(p_x^2 - p_y^2) + c_3(xp_x - yp_y)$. The deformation of the Fermi sphere is determined by the second term ($\propto p_x^2 - p_y^2$), which is independent of $\mathbf{r}$. In reality, however, collisions are much more frequent in the centre than in the low-density regions of the cloud. Therefore, the Fermi surface deformation should be much smaller in the centre ($r = 0$) than at large $r$.

Following these arguments, we concluded that the inclusion of higher order moments should help to improve the results. For instance, in the case of the quadrupole mode, a term of the form $\propto r^2(p_x^2 - p_y^2)$ in the ansatz for $\Phi$ would account for the position dependence of the Fermi-surface deformation. We therefore decided to extend the method of moments to fourth order [5]. The corresponding result is also shown in Fig. 4, and one can clearly see that the agreement with the numerical simulation is now much better than within the second-order moment method.

Encouraged by this result, we applied the fourth-order moment method to the realistic parameters, including the in-medium cross section (but not yet the mean field). First results for the case of the radial quadrupole mode, with parameters corresponding to the conditions of the experiment by Riedl et al. [2], are displayed in Fig. 5. It can be clearly seen that the fourth order results are in much better agreement with the data than the second-order ones.

## 4. Conclusion

In the normal-fluid phase, the collective modes of trapped Fermi gases show a smooth transition from hydrodynamic to collisionless behaviour as function of temperature. Previous analyses indicated that the data concerning this transition contradict the theoretical results including the in-medium enhancement of the cross section. However, we have shown that this conclusion was at least partly an artefact of the approximate solution of the Boltzmann equation (second-order moments) and that the agreement with the data can be greatly improved by including fourth-order moments, too. In conclusion, at the present stage, the available solutions of the Boltzmann equation are not reliable enough to decide whether the dynamics of these systems can be described by the Boltzmann equation or not.

## References

[1] Wright M J, Riedl S, Altmeyer A, Kohstall C, Sánchez Guajardo E R, Hecker Denschlag J and Grimm R 2007 *Phys. Rev. Lett.* **99** 150403
[2] Riedl S, Sánchez Guajardo E R, Kohstall C, Altmeyer A, Wright M J, Hecker Denschlag J, Grimm R, Bruun G M and Smith H 2008 *Phys. Rev.* A **78** 053609
[3] Chiacchiera S, Lepers T, Davesne D and Urban M 2009 *Phys. Rev.* A **79** 033613
[4] Alm T, Röpke G and Schmidt M 1994 *Phys. Rev.* C **50** 31
[5] Lepers T, Davesne D, Chiacchiera S and Urban M 2010 *Phys. Rev.* A **82** 023609

# Bose-Fermi pair formation in the mixture of cold atomic gases

**Toru Suzuki**

Department of Physics, Tokyo Metropolitan University, Tokyo 192-0397, Japan

E-mail: suzukitr@tmu.ac.jp

**Abstract.** We study the role of the Bose-Fermi (BF) pair correlations in the cold BF-mixture of atomic gases. After summarizing the previous results on the formation of composite BF-pairs in the uniform BF mixture, we present an exactly solvable model where bosons and fermions are distributed over finite number of orbits and interact via attractive pairing-type interaction. We show that, as the interaction strength increases, BF pairs with different quantum numbers are gradually created in the ground state through a series of level crossings, until finally the system becomes a Fermi sea of BF-pairs.

## 1. Introduction

Recent development in the manipulation of cold fermionic atoms provided a means of realizing a crossover from Bardeen-Cooper-Schrieffer (BCS) superconducting phase to the Bose-Einstein Condensation (BEC) of bound fermion pairs. In these studies a full use of Feshbach resonance technique is made, whereby one adjusts magnetic field so as to bound or dissolve a number of fermion pairs. A similar situation may be expected with different types of particles as far as a similar technique is available. In fact a proposal to create composite particles in a Bose-Fermi(BF) mixture has been made[1, 2, 3, 4, 5], though no direct experimental signature has been reported up to now[6]. An obvious difference from the fermion-pair case is that because the BF-pairs are (composite) fermions one cannot expect a condensation of pairs in the latter. Instead, a new BF-pair can only be created with a different quantum number (e.g., the center-of-mass(CM) momentum), which requires an additional attraction to compensate the increase of energy (e.g., CM energy) of the pair. This suggests an extra complication in the theoretical prediction: In contrast to the fermion-pair case where Cooper-type instability signals a condensation, one has in principle to calculate the whole dispersion relation of BF-pairs in the presence of other particles and pairs. Main purpose of the present work is to make a solvable model study of the formation of BF-pairs in the BF-mixture. The model system is a schematic one which allows a calculation of exact eigenstates, so that one may exhibit a change from independent Bose-Fermi system to that of composite BF-pairs.

## 2. Boson-fermion Cooper problem

To set a physical basis we first summarize results of the previous studies on the formation of a single BF-pair in the uniform BF-mixture. In the absence of interactions the system at $T = 0$ consists of BEC and a Fermi sphere. Now one may switch on the attractive BF-interaction so that they can form a (quasi-)bound state. Since a BF pair obeys Fermi statistics, one would

Journal of Physics: Conference Series **321** (2011) 012027 doi:10.1088/1742-6596/321/1/012027

expect that, for sufficiently strong binding of the pair, they may form a new Fermi sea as far as many-body correlations are not large. As the original noninteracting BEC-Fermi sea system and the new Fermi sea of BF pairs are quite different, the nature of the transition between the two extremes against, e.g., the strength of the interaction is not self-evident.

When the BF attractive interaction is not strong, one may start with a single boson and fermion, on top of the free BF-ground state, interacting via BF-effective interaction modified by the presence of the BF-medium. We start with the Hamiltonian

$$H = \sum_k \epsilon_k (b_k^\dagger b_k + f_k^\dagger f_k) + \sum V_{q'q} (bf)^\dagger_{Pq'} (bf)_{Pq} \tag{1}$$

where $\epsilon_k$ denotes kinetic energy of a particle with momentum $k$, and $V$ is the pseudopotential matrix element between boson and fermion with total momentum $P$ and relative momentum $q, (q')$, i.e., $(bf)_{Pq} \equiv b_{\frac{P}{2}+q}, f_{\frac{P}{2}-q}$. We may calculate dispersion relation of the BF-pair by studying the pole of the two-body propagator in the BF-medium:

$$G_{\rm BF} \sim \langle \theta(t) \{ (bf)_{Pq}(t), (bf)^\dagger_{Pq'}(t') \} \rangle. \tag{2}$$

Figure 1 shows the energy $P_0$ of the pole against CM-momentum $P$ of the BF-pair for various interaction strength characterized by the scattering length $a$[4]. The latter is calculated with a suitable renormalization procedure for a short-range interaction.

We find that even before the formation of the bound state (i.e., $P_0 \leq -E_F$ at $P = 0$), a Cooper type BF-pair can be created in the ground state. This interpretation is consistent with the result of the binding energy calculation performed within a (renormalized) mean-field framework[4]. The latter shows a rapid decrease of energy below zero before the parameter $(k_F a)^{-1}$ signals a formation of a BF bound state.

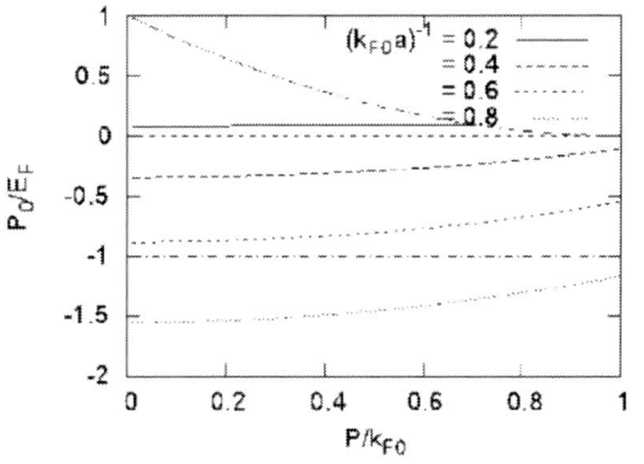


**Figure 1.** Pole energy $P_0$ against CM momentum

## 3. Solvable model study

We now consider a solvable boson-fermion pairing model which exhibits a formation of BF pairs, and study the behavior of the ground state as function of the BF-pairing strength. To construct such models we need two quantum numbers, i.e., the one which characterizes the intrinsic structure of the BF pair and the one which distinguishes the state of pairs. Let us call these quantum numbers as $k(= 1, 2, \ldots, D)$ and $p(= 1, 2, \ldots, M)$, respectively. Thus the model is a system of bosons and fermions distributed over the single-particle states which consist of $M$ orbits, each orbit having degeneracy $D$, for both bosons $(b^\dagger, b)$ and fermions $(f^\dagger, f)$. With a simple Bose-Fermi pairing type interaction, the model Hamiltonian is given by

$$H = \sum_{p,k} (\epsilon_p^b b_{pk}^\dagger b_{pk} + \epsilon_p^f f_{pk}^\dagger f_{pk}) + \sum_p G_{p'p} C_{p'}^\dagger C_{p'}, \quad C_p^\dagger \equiv \sum_{k=1}^{D} b_{pk}^\dagger f_{pk}^\dagger. \tag{3}$$

where $\epsilon^{b(f)}$ are the boson (fermion) single-particle energies, while $G_{qp}$ denotes an attractive BF interaction. The boson-fermion pair operators $C^\dagger, C$ are defined in each orbit $p$ and satisfies $(C_p)^2 = 0$, though they are not pure fermions as their anticommutator is different from unity. We assume that the interaction is diagonal in $p$, i.e., $G_{pq} = -G\delta_{pq}$ so that the pairs may be formed in each orbit with the same internal structure. Note that a presence of nondiagonal interaction in the quantum number $p$ would create a single very collective composite BF pair[7]. The model can be solved first by finding eigenstates in each orbit and then by combining these to obtain the whole system with given quantum numbers.

The problem is to find the ground state from many competing configurations with various number of BF pairs for each value of the interaction strength. For a weak attraction, favorable configurations consist of many seniority states, where 'seniority' means a number of unpaired particles. To obtain exact eigenvalues and eigenfunctions with given boson and fermion seniorities, we used a method of partial fermionization which is an extension of the partial bosonization. The latter has been developed to study quasiparticle excitations in the pairing problem of finite nuclei[8].

We first calculate the number of states. We consider a system with equal number $N$ of bosons and fermions, i.e., $N_b = N_f = N$. Particles are distributed over the $L = DM$ single-particle states, so that the total number of states is given by $W_{\rm tot} = (L+N-1)!L/(L-N)!(N!)^2$. For the present pairing type interaction the eigenstates are characterized by boson and fermion seniority numbers $\nu_p^b, \nu_p^f$ in each orbit $p$. For $N_b = N_f$ the total seniority number are given by $\nu_f = \nu_b (= \nu) = N - n_C$, where $\nu_{b,f} = \sum_p \nu_p^{b,f}$ and $n_C = \sum_p n_p^C$ is the sum of the number of BF pairs $n_p^C$ in the orbit $p$. The total number of states are then given by

$$W_{\rm tot} = \sum_{\nu=\nu_{\rm min}}^{N} W(\nu), \quad W(\nu) = \sum_{p=0}^{p_{\rm max}} \binom{M}{\lambda}\binom{M-\lambda}{p}\binom{L-M}{\nu-p}\binom{L+\nu-p-1}{\nu}, \tag{4}$$

where $\nu_{\rm min} = \max(N-M, 0), p_{\rm max} = \min(M-\lambda, \nu)$. The quantity $\nu_{\rm min}$ gives the minimum number of unpaired particles, and for $N \leq M$, where all particles can form a pair, $\nu_{\rm min} = 0$ and $p_{\rm max} = \nu$. $W(\nu)$ is the number of states with seniority $\nu$. For instance, for $D = 4, M = N = 10$, where total number of states is about $10^{20}$, there is only one $N$-pair state with $\nu = 0$, a free ground state, i.e., bosonic BEC and Fermi degenerate state, hides in the huge number of large seniority states with $W(\nu = 9) \sim 10^{18}$.

Using the method of partial fermionization we can write down the exact eigenvalues for the Hamiltonian. They are characterized by the number of BF pairs $n_p^C$ and boson and fermion seniorities $\nu_p^{b,f}$ as

$$E = \sum_p (\epsilon_p^b + \epsilon_p^f - GQ_p) n_p^C + \sum_p \epsilon_p^b n_p^b + \sum_p \epsilon_p^f n_p^f, \qquad Q_p = D + n_p^b - n_p^f, \tag{5}$$

where $n_p^{b,f} = n_p^C + \nu_p^{b,f}$ are the number of bosons/fermions in the orbit $p$. The eigenstates have large degeneracies, which in general depend on the choice of $\epsilon_k$.

In the numerical example below, we take an equal spacing single-particle orbits

$$\epsilon_p^b = \epsilon_p^f \equiv (p-1)\epsilon \qquad (p = 1, 2, \cdots, M) \tag{6}$$

and consider the case of equal number $N_b = N_f = N$ of bosons and fermions distributed over $M = N$ orbits. We study the ground state for a fixed value of the pairing strength $G$ by comparing low-lying configurations of particles for various seniority numbers $n_C$ of BF pairs, and thus for the number $\nu = \nu_b = \nu_f$ of unpaired particles.

Figure 2 shows the energies of low-lying 100 states for $D = 10, M = N = 20$ against $G$.

Journal of Physics: Conference Series **321** (2011) 012027 doi:10.1088/1742-6596/321/1/012027

It shows that the character of the ground state changes from high seniority state at small $G$ to low one at large $G$ through a series of level crossings. To understand the result we shall simplify the expression (5) for the ground state, by assuming that for a given $n_C$ (and hence $\nu$), each of the BF pairs, unpaired bosons and fermions independently takes the lowest energy configuration by neglecting blocking effect. Thus we consider that BF pairs occupy the lowest $n_C$ orbits, all unpaired bosons occupy the $p = 1$ state, while unpaired fermions are distributed over the lowest $p_F = \nu/D$ orbits, where we assume $p_F$ to be an integer.

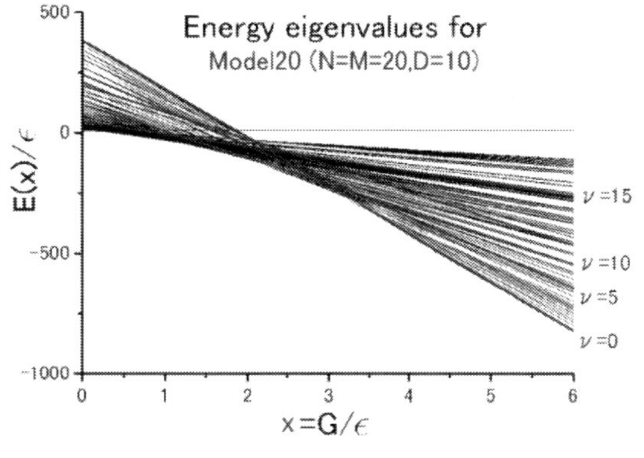


**Figure 2.**
Energies of selected eigenstates as functions of the interaction strength $x = G/\epsilon$.

The lowest energy $E_{n_C}$ for a given number $n_C$ takes the form

$$E_{n_C} = \epsilon_1(N - n_C) + 2\sum_{p=1}^{n_C} \epsilon_p + D\sum_{p=1}^{p_F} \epsilon_p - GD\ \max(p_F, n_C). \tag{7}$$

The value of $G = G_{\max}$ for which all the particles form BF pairs (i.e., the completion of the composite Fermi sea) is obtained by equating $E_N$ and $E_{N-1}$. This gives the value

$$G_{\max} = \frac{2}{D}(\epsilon_N - \epsilon_1), \quad \text{i.e.,} \quad x_{\max} = \frac{G_{\max}}{\epsilon} \sim \frac{2N}{D}. \tag{8}$$

This condition shows that the interaction energy $\sim GD$ just compensates the single particle energy of the last orbit $p = N$ of the BF pair. For the above parameters we obtain $x_{\max} \simeq 4$ in agreement with Fig.2. For intermediate values $0 \le G \le G_{\max}$ the ground state experiences a sequence of level crossings from no pair $(n_C = 0, \nu = N)$ to the fully paired $(n_C = N, \nu = 0)$ states. The crossing is studied by comparing $E_{n_C}$ and $E_{n_C - 1}$ for a general value of $n_C$. The envelope of these crossings can be calculated from eq.(7), which gives $E/\epsilon \simeq -(Dx)^2/4$ up to $x = x_{\max}$ which reproduces well the calculated results for the ground state behavior.

**Acknowledgments**
Section 2 is the collaboration work with T. Watanabe and P. Schuck to whom the author is very much grateful. He is also indebted to J. Dukelsky and H Yabu for fruitful discussions.

**References**

[1] Storozhenko A, Schuck P, Suzuki T, Yabu H and Dukelsky J. 2005 *Phys.Rev.* A **71** 063617.
[2] Yabu H, Takayama T, Suzuki T and Schuck P 2004 *Nucl.Phys.* **A738** 273.
[3] Kagan M Yu, Brodsky I V, Efremov D V and Klaptsov A V 2004 *Phys. Rev.* A**70** 023607.
[4] Watanabe T, Suzuki T and Schuck P 2008 *Phys. Rev.* **A78** 033601.
[5] Watanabe T 2010 Ph.D. thesis, Tokyo Metropolitan University.
[6] An instability of a BF-mixture was reported in Modguno G et. al. 2002 *Science* **297** 2240. This phenomenon has been interpreted as a collapse of the system due to attractive interaction.
[7] Dukelsky J and Pittel S 1990 *Phys.Rev.* C**42** 2030 and private communication.
[8] Suzuki T and Matsuyanagi K 1976 *Prog.Theor.Phys.* **56** 115.

# BCS-BEC crossover in spatially modulated fermionic condensates

**Armen Sedrakian**

Institute for Theoretical Physics, J. W. Goethe-University, Max-von-Laue-Str. 1, 60438 Frankfurt am Main, Germany

E-mail: sedrakian@th.physik.uni-frankfurt.de

**Abstract.** Several novel multi-component fermionic condensates show universal behavior under imbalance in the number of fermionic species. Here I discuss their phase structure, thermodynamics, and the transition from the weak (BCS) to strong (BEC) coupling regime. The inhomogeneous superconducting phases are illustrated on the example of the Fulde-Ferrell phase which appears in the weak coupling regime, at low temperatures and large asymmetries. The inhomogeneous phases persist through the crossover up to (and possibly beyond) the transition to the strong coupling regime.

## 1. Introduction

The last decade has seen impressive advances in the research, both theoretical and experimental, on pairing in novel fermionic systems which share a number of common features. One example is encountered in experiments with ultracold atomic vapors in magnetic traps. The atomic gases allow for a remarkable control over the parameter space, including variations in the strength of the pairing force via Feshbach resonance. The second example of interest is the nuclear/quark matter which can be either spin polarized (by strong magnetic fields) or isospin polarized, as required by the beta equilibrium in the interiors of compact stars.

The possibility of transition from Bardeen-Cooper-Schrieffer (BCS) pairing to Bose-Einstein condensation (BEC) in ensembles of attractive degenerate fermions was conjectured long ago [1]. This transition takes place when the dimensionless coupling $\lambda$, which is a product of the density of states at the Fermi surface and the dimensional coupling of the theory increases from weak coupling ($\lambda \ll 1$) to strong coupling ($\lambda \gg 1$). Experimentally, it is achieved by variations of the coupling constant via the mechanism of Feshbach resonance in atomic gases. For particles interacting via the strong force the interactions cannot be varied at will, but the range over which they are effective and the density of states at the Fermi surface will change with density. Therefore, the combined effect of density variation on the density of states and effective range of the coupling can lead to a crossover in strongly interacting systems with density gradients.

The generalization of the Nozières-Schmitt-Rink (NSR) theory to ensembles with population imbalance was carried out in the context of the strongly interacting matter in ref. [2]. Experimental realizations in the atomic vapors were achieved somewhat later [3]. The novel feature of the Hamiltonian describing these imbalanced gases or (super)fluids is the non-invariance under exchange of paired particles, i.e., there is a new scale in the problem $\delta\mu = (\mu_\alpha - \mu_\beta)/2$, where $\mu_\alpha$ and $\mu_\beta$ are the chemical potentials of the species. Furthermore,

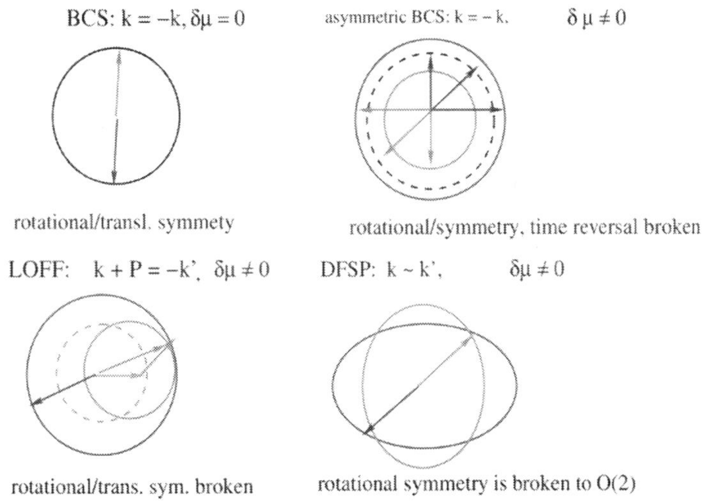


**Figure 1.** Schematic illustration of the Fermi spheres of two components (solid - majority component, dashed - minority component) and the momenta of paired fermions for four phases: BCS, asymmetric BCS, LOFF and DFSP.

for certain values of asymmetry $\delta\mu$ the system may support a current carrying state, where the counter-propagating super-current and normal current cancel each other. This phase is known as the Larkin-Ovchinnikov-Fulde-Ferrell (LOFF) phase [4, 5]. The LOFF condensate is spatially modulated, therefore, minimally, it breaks the $O(3)$ rotational symmetry of the system down to $O(2)$. However, in general, the condensate may assume a complicated structure which breaks also translational symmetries. The emergence of the LOFF phase becomes energetically favorable for the following reason: setting up the condensate and the normal excitations in motion costs kinetic energy, which is always positive and thus unfavorable; however, the current changes the way the phase-space is sampled in the vicinity of the Fermi surface, which in turn enhances the condensation energy. The interplay of these two effects tells us whether the LOFF phase is favored or not. Indeed, in the low-temperature and large asymmetry sector of the phase diagram the LOFF phase is favored over the BCS state. The way the LOFF phase evolves into the BEC regime is not fully understood. I shall provide some answers to this problem below. An alternative to the LOFF phase, is the phase which allows deformations of the Fermi surfaces of the species [6]. The emergence of deformed Fermi surface pairing (DFSP) can be explained in full analogy to the LOFF phase and I will not repeat the arguments above.

## 2. Homogeneous phases

Let us start our discussion with the homogeneous phases, i.e., the ordinary BCS phase and the asymmetrical BCS phase. Once the pairing interaction is specified, the mean-field solutions are obtained by solving the gap equation together with the equations for the densities for each fermionic component (see, e. g., refs. [6, 7]). The solutions of the coupled integral equations then provide the value of the gap parameter for fixed net density and asymmetry in the populations of fermions measured either in terms of the density asymmetry (hereafter $\alpha$) or the shift in the chemical potentials $\delta\mu$. Figures 2 and 3 show the temperature dependence of various quantities for fixed density asymmetries [8]. The pairing correlations are clearly suppressed for large asymmetries; in the high-temperature domain $T \to T_c$ the temperature dependence of the functions is analogous to the BCS theory for balanced systems. The low temperature domain

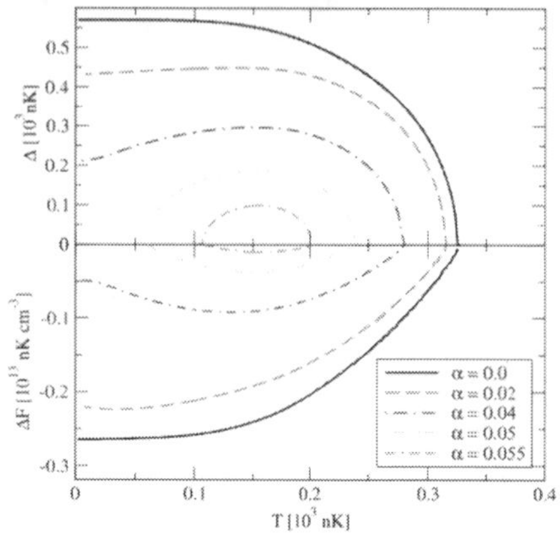


**Figure 2.** Dependence of the gap (upper panel) and the free energy (lower panel) on the temperature for an ultracold fermionic gas ($Li^6$) for several asymmetries.

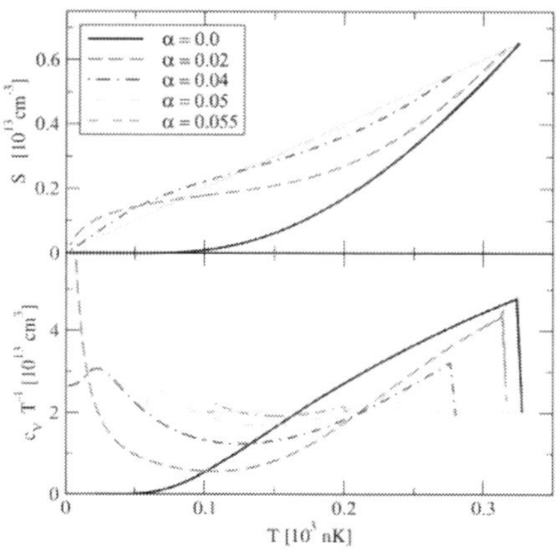


**Figure 3.** Same as in Fig. 2, but for the entropy (upper panel) and the specific heat (lower panel) for several asymmetries.

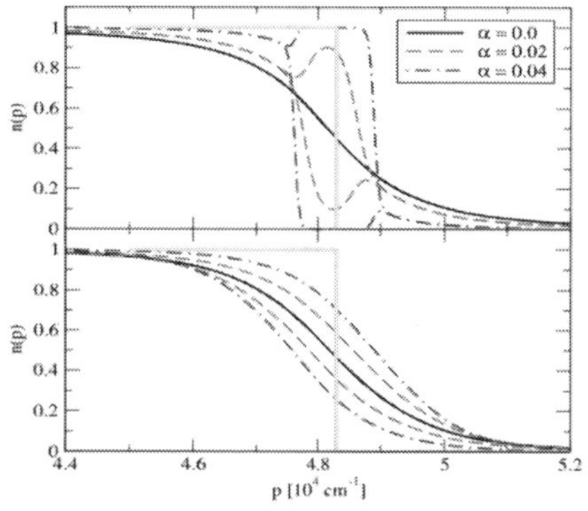


**Figure 4.** The occupation numbers of two components for various asymmetries. Upper panel corresponds to the low temperature limit, the lower panel - to the high temperature limit. The minority component is extinct close to its Fermi surface for large asymmetries and low temperatures. This effect is "smeared out" at higher temperatures (lower panel).

$T \to 0$, on the contrary, is anomalous, i.e., all the quantities show temperature dependence that is different from the BCS theory. The reduction of the gap, for example, is the result of the loss of coherence between the paired fermions (or equivalently, overlap between the thermal bands around the Fermi surfaces). For large asymmetries this leads to a complete disappearance of the gap in the limit of $T \to 0$, which manifests itself in the anomalous jump of the specific heat (Fig. 3). While such a jump could be a signature of the asymmetrical BCS state, we will see below that the temperature anomalies are removed if one allows for the LOFF (or DFSP) phase. This, of course, does not exclude the possibility of observing the second jump in the specific heat if the gas is prepared at high temperatures, where the BCS phase is stable, and is cooled down into the low temperature unstable phase [8].

Another interesting observation is the emergence of the empty shell around the Fermi surface of the minority component for large asymmetries seen in Fig. 4. This leads to the notion of "breached pairing" superconductivity [9, 10]. Clearly the topology of the minority Fermi surface changes as the asymmetry is increased. However, the topological phase transition disappears in the high temperature regime (Fig. 4 lower panel).

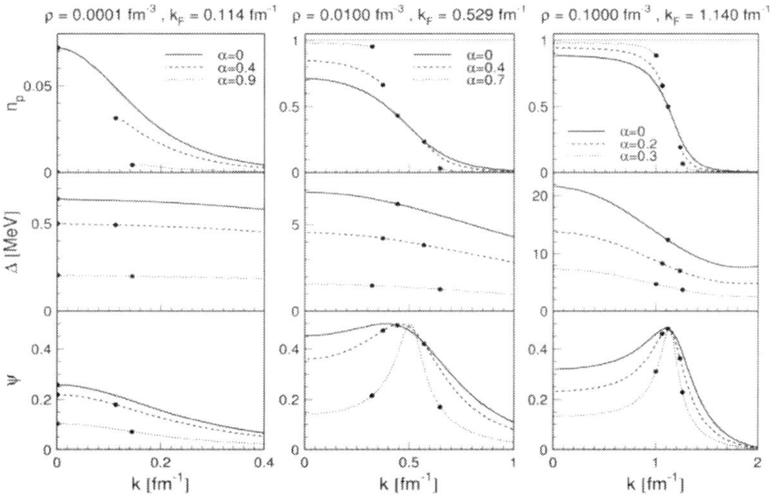


**Figure 5.** BCS (right column) to BEC (left column) crossover in an imbalanced system. Upper panel: the occupation number of minority component; middle panel: the pairing field; lower panel: the condensate wave-function. Figure taken from ref. [2].

Next we turn to the question on how the Nozières–Schmitt-Rink theory is modified in asymmetrical systems. This problem was first studied in the context of pairing in low-density nuclear matter in the attractive $^3S_1$-$^3D_1$ coupled channel, where neutron-proton Cooper pairs in the BCS limit transform to a BEC of deuterons [2]. Following the NSR conjecture one can directly extrapolate the BCS equations in the strong coupling regime to obtain the properties of the deuteron condensate. The pair wave-function, which is defined as the correlation function of two fermionic creation operators $\psi(k) = \langle a^\dagger_{\vec{k}} b^\dagger_{-\vec{k}} \rangle$, where $a^\dagger$ and $b^\dagger$ are the creation operators of neutrons and protons, obeys a Schrödinger type equation with an eigenvalue equal $2\mu$, with $\mu$ being the chemical potential in the symmetrical case. The state that emerges in the strong coupling limit corresponds to a mixture of deuterons (strongly bound two-body states) plus a fluid of unbound neutrons at arbitrary large asymmetries. At low densities the Pauli-blocking plays a minor role and its modifications to the deuteron wave-function are marginal. As a consequence large asymmetries are compatible with the existence of a deuteron condensate. Figure 5 shows the evolution of occupation numbers of the minority component (upper panel), pairing gap (intermediate panel) and condensate wave-function (lower panel) across the BCS-BEC transition (from left to right). In this particular case, the transition is enforced via the reduction of the net density. A remarkable feature of the transition is the change in the occupation numbers of the minority component: the gap (breach) is widened across the transition and in the BEC regime an empty (unoccupied) sphere emerges inside the Fermi sphere. Clearly, the topology of the Fermi sphere changes from having a strip of empty states to a empty interior region; thus, the BCS-BEC transition is associated with a topological phase transition in the shape of the minority Fermi sphere at non-zero asymmetries. The BCS vs BEC nature of the condensate is most clearly seen in the form of the condensate wave-function. In the momentum space it is strongly peaked in the BCS limit and is flat in the BEC limit (Fig. 5 lower panel). In the real space this picture translates into loosely bound state whose wave-function oscillates over many periods in the BCS limit. In the opposite BEC limit one finds tightly bound state

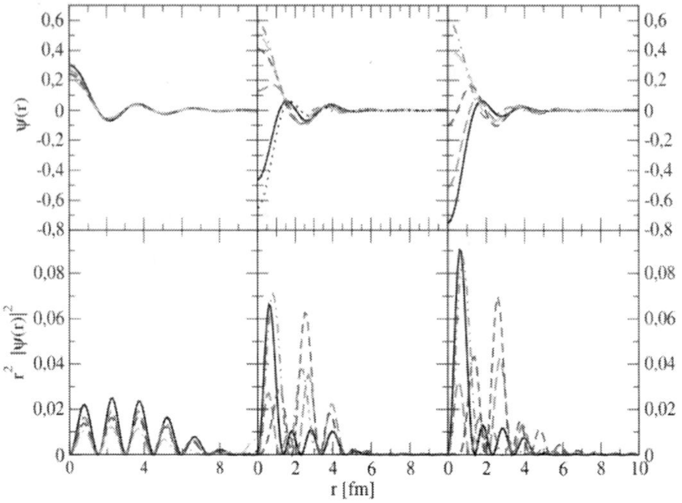


**Figure 6.** BCS-LOFF-BEC crossover in an imbalanced system. The dependence of the condensate wave-function $\psi(r)$ and the quantity $r^2|\psi(r)|^2$ on the spatial coordinate; the coupling increases from left to right. For an explanation see the text.

with a strongly localized wave-function.

## 3. Inhomogeneous phases

In this section I will explore the modifications that arise when inhomogeneous phases are allowed to exist along with the symmetrical and asymmetrical BCS phases. I will illustrate the main points on the example of color superconductivity in cold quark matter with pairing among the lightest quarks of different flavors ($u$ and $d$) in the color anti-triplet state (see e.g. refs. [11, 12]). Our Ansatz for the inhomogeneous condensate corresponds to the Fulde-Ferrell (FF) phase which postulates a single plane-wave dependence of the gap on the total momentum $\Delta(Q) \propto \Delta_0 \exp(iQr)$ [13]. Let us first concentrate on the BCS limit. One remarkable feature of the FF state is that, once it is allowed for, the anomalies seen in Figs. 2 and 3 are removed, i.e., the functional dependence of the physical quantities on the temperature are the same as in the ordinary BCS theory. The only effect of asymmetry is the reduction of the magnitude the gap [13, 14]. Thus, the $T$-dependence of quantities for $\alpha \neq 0$ are self-similar to the symmetrical BCS phase. The low-temperature topology of the Fermi spheres in the FF state is the same as in the asymmetrical BCS case, but only for selected directions, such as along in the direction orthogonal to the axis of symmetry breaking. This direction is chosen by the condensate spontaneously. At other angles the appearance of the empty strip (breach) mentioned above is suppressed. The FF phase occupies the low-temperature and large asymmetry portion of the $T$-$\alpha$ phase diagram and there exists a tri-critical Lifshitz point where the normal phase, the BCS phase, and the FF phase all meet (for more details see ref. [13]). Now let us comment on the transition from the BCS-FF state to the BEC state. Generally, as one moves towards the BEC limit, the portion of the $T$-$\alpha$ phase diagram occupied by the FF phase becomes smaller. A key question is whether the FF state exists only in the weakly coupled regime of the phase diagram or does it persist into the strongly coupled regime. The answer is best provided by the form of the condensate wave-function. Fig. 6 displays the evolution of the condensate wave-function from the weak (left column) to the strong (right column) coupling regime. In the weak coupling limit the wave-function and the quantity $r^2|\psi(r)|^2$ show oscillations with a period given by

$l \simeq 2\pi k_F^{-1}$, where $k_F$ is the Fermi wave-number. These sustained oscillations in space evidence the long-range correlations characteristic to BCS state; in this regime Cooper-pair size is much larger than the inter-particle spacing. In the strong coupling limit and for small asymmetries (solid line) a pronounced peak in these quantities corresponds to a well-localized bound state. However, for large asymmetries the picture is more complicated; there are several peaks which are not necessarily localized at the origin. These features indicate that a perfect BEC condensate has not been formed yet. Since the condensate is in the FF state sector of the $\alpha$-$T$ phase diagram (see ref. [13]) one may conjecture that the FF state has a strong-coupling counterpart, where tightly bound states carry a super-current.

**Acknowledgments**
The results discussed above are based mainly on the papers [2, 8, 13] and I would like to thank my colleagues for their contribution and insight. This work was partially supported by the Deutsche Forschungsgemeinschaft (Grant SE 1836/1-2).

## References

[1] Nozières P and Schmitt-Rink S 1985 *J. Low Temp. Phys* **59** 195
[2] Lombardo U Nozières P Schuck P Schulze H-J and Sedrakian A 2001 *Phys. Rev.* **C 64** 064314
[3] Regal C A and Jin D S 2006 *Adv. in Atom. Mol. and Opt. Phys.* **54** 1
[4] Larkin A I and Ovcihnnikov Yu N 1964 *Zh. Eksp. Teor. Fiz.* **47** 1136 [1965 *Sov. Phys. JETP* **20** 762 ]
[5] Fulde P and Ferrell P A 1964 *Phys. Rev.* **135** A550
[6] Müther H and Sedrakian A 2002 *Phys. Rev. Lett.* **88** 252503
[7] Sedrakian A Mur-Petit J Polls A and Müther H 2005 *Phys. Rev.* **A 72** 013613
[8] Sedrakian A Müther H and Polls A 2006 *Phys. Rev. Lett.* **97** 140404
[9] Liu W V and Wilczek F (2002) *Phys. Rev. Lett.* **90** 047002
[10] Forbes M M Gubankova E Liu W V and Wilczek F 2005 *Phys. Rev. Lett.* **94** 017001
[11] Bowers J A and Rajagopal K 2002 *Phys. Rev.* **D 66** 065002
[12] Casalbuoni R Gatto R Mannarelli M and Nardulli G 2002 *Phys. Rev.* **D 66** 014006
[13] Sedrakian A and Rischke D H 2009 *Phys. Rev.* **D 80** 074022
[14] He L Jin M and Zhuang P 2007 *Phys. Rev.* **D 75** 036003

Journal of Physics: Conference Series **321** (2011) 012029 doi:10.1088/1742-6596/321/1/012029

# Cluster formation and dissolution in a generalized relativistic density functional approach for dense matter

**Stefan Typel**

Excellence Cluster *Origin and Structure of the Universe*, Technische Universität München, Boltzmannstraße 2, D-85748 Garching, Germany

GSI Helmholtzzentrum für Schwerionenforschung mbH, Theorie, Planckstraße 1, D-64291 Darmstadt, Germany

E-mail: s.typel@gsi.de

**Abstract.** The formation and dissolution of clusters in dense matter is described in a generalized relativistic mean-field model with density-dependent nucleon-meson couplings. Nucleons, light clusters and two-body correlations in the continuum are treated as effective degrees of freedom with medium dependent properties. The resulting generalized relativistic density functional successfully describes the transition from clusterized matter at low densities to uniform neutron-proton matter at high densities with the correct low-density limits. The formation of heavy clusters is modeled by employing inhomogeneous calculations in the Thomas-Fermi approximation within spherical Wigner-Seitz cells.

## 1. Introduction

Correlations in nuclear matter modify the thermodynamical properties of the equation of state (EoS) and affect the chemical composition. At low densities, two-, three-, . . . , many-body correlations due to the short-range nuclear interaction lead to the formation of bound states that appear as new particle species. At high densities around and above nuclear saturation density, homogeneous and isotropic matter is expected that is composed of neutron and proton quasiparticles and mean-field effects dominate. In between at low temperatures, there is a transition from inhomogenous to homogeneous matter passing through various "pasta" phases where the interplay of surface effects and the long-range Coulomb interaction strongly affects the spatial distribution of the nucleons. For the application in astrophysical models, e.g. simulations of core-collapse supernovae or neutron stars, an EoS of dense matter is required that takes correlations into account and that consistently interpolates between the low-density and high-density limits. The effects strongly depend on the temperature and the neutron-proton asymmetry of the system.

## 2. Theoretical approaches

There are a number of models for the dense matter EoS with different points of view in the approach. In a chemical picture, the system is described as a mixture of different inert nuclei and nucleons in chemical equilibrium. A typical example are nuclear statistical equilibrium

(NSE) approaches, e.g. [1]. Interactions between the particles can be included in a virial EoS that encodes two-body correlations in the second virial coefficient. In the quantum mechanical formulation, the two-body density of states enters. It depends on experimentally known quantities such as the binding energies of the nuclei and the scattering phase shifts [2]. However, the properties of the constituents are independent of the medium and the Mott effect, i.e. the dissolution of the nuclei at high densities, is not described. It has to be modeled by simple concepts like the excluded volume mechanism [1]. In consequence, these models are limited to low densities.

In a physical picture, nucleons can be regarded as the basic constituents and the formation of bound states and resonances in the continuum is explicitly considered. Here, the main problems are the treatment of three-, ..., many-body correlations and the choice of an appropriate interaction. Considering the different paths to describe dense nuclear matter, it is tempting to formulate a model that combines the various approaches.

A step beyond the standard virial EoS is the generalized Beth-Uhlenbeck approach that is able to describe the Mott effect. It is a quantum statistical method employing a finite-temperature Green's function formalism and an expansion of the spectral function beyond the quasiparticle approximation [3]. The main features are medium-dependent shifts of the nucleon self-energies, mainly due to the Pauli principle, and the use of generalized scattering phase shifts from the in-medium T-matrix.

Around nuclear saturation density, phenomenological approaches with effective nuclear interactions depending on parameters that are fitted to properties of finite nuclei are very successful. Typical examples are nonrelativistic Hartree-Fock or relativistic mean-field (RMF) models that give information on the nucleon self-energies. Another class of models are *ab initio* approaches, e.g. (Dirac) Brueckner-Hartree-Fock calculations, that use a given realistic nucleon-nucleon interaction. Unfortunately, these models are limited to the description of homogenous and isotropic nuclear matter.

## 3. Generalized relativistic density functional

The approach of the present contribution is an extension of an RMF model with density-dependent meson-nucleon couplings with parameters that are fitted to properties of finite nuclei [4]. In addition to nucleons, light clusters (deuterons, tritons, helions, $\alpha$ particles) are introduced as additional degrees of freedom [5]. Continuum contributions of two-body correlations are represented by effective, temperature-dependent resonances. The binding and resonance energies are shifted depending on density, temperature and asymmetry. Theses shifts are taken in parametrized form from generalized Beth-Uhlenbeck calculations. A density functional is constructed that depends on temperature, neutron and proton chemical potentials, the meson fields and their spatial derivatives. The field equations can be derived and contain additional rearrangement contributions due to the medium dependence of the self-energy shifts, reflecting the interaction of the clusters inside the medium. All thermodynamical quantities can be derived consistently. In addition to the bound states of light clusters in Ref. [5] also two-body scattering correlations are included. The formation of heavy clusters is described by applying the Thomas-Fermi (TF) approximation in spherical Wigner-Seitz cells.

## 4. Results

The generalized RMF (gRMF) model has to approach the correct limits at low densities. For temperature $T = 0$ MeV, only a single heavy cluster appears, possibly surrounded by a gas of nucleons for large isospin asymmetries beyond the drip lines. Since the original RMF parametrisation DD2 [5] was based on a Hartree approximation for the description of nuclei and the gRMF model for the EoS uses the TF approximation, different results are obtained. Experimental binding energies can be reproduced by a rescaling of the $\sigma$ meson mass and of the

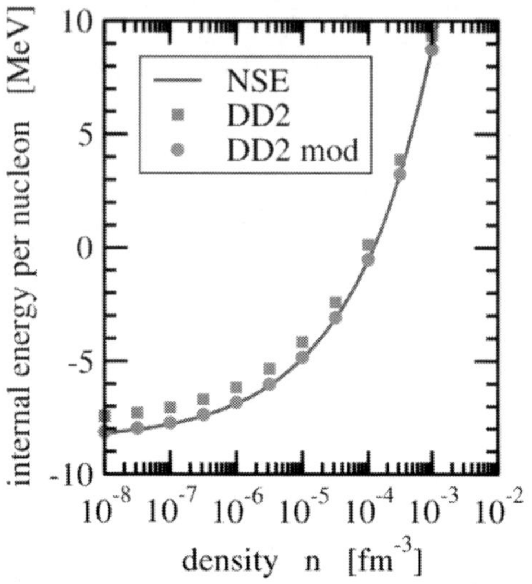


**Figure 1.** Internal energy per nucleon in charge neutral matter as a function of the total baryon number density $n$ for $T = 0$ MeV and proton fraction $Y_p = 0.5$.

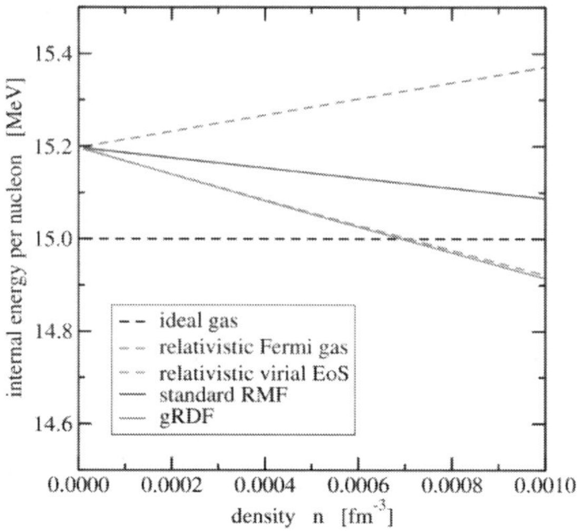


**Figure 2.** Internal energy per nucleon for neutron matter at temperature $T = 10$ MeV as a function of the total neutron number density $n$ in different models.

coupling in a modified parametrization DD2 mod such that the properties of uniform nuclear matter are not affected. As depicted in figure 1, an excellent agreement with NSE calculations at low density is obtained.

At finite temperatures, only nucleons and light clusters survive at low densities and the standard virial EoS results are reproduced. This is achieved by exploiting new relations that connect the temperature dependent resonance energies in the continuum contributions and the scattering lengths in the various nucleon-nucleon scattering channels. In addition, a relativistic correction is found that becomes more important than the virial two-body correlation correction at low densities. In figure 2 the various effects that modify the low-density behavior of the internal energy per nucleon in neutron matter are compared. It can be seen that the generalized relativistic density functional (gRDF) nicely reproduces the relativistically corrected virial EoS.

The nucleon-nucleon correlations in the continuum, that are explicitly included in the present model, also affect the particle fractions of the nucleons and light clusters, in particular for the deuteron as depicted in figure 3. At low-densities, two-body correlations, that are not considered in standard EoS for astrophysical applications, are most important. With increasing densities, the dissolution of the clusters, i.e. the Mott effect, is observed in the gRDF approach. Thus both low- and high-density limits are correctly reproduced in this model.

At low densities and not too high temperatures, nuclear matter becomes inhomogeneous. Heavy nuclei are formed that are surrounded by a low-density background of nucleons and light clusters. The charge of the hadronic system is compensated by an almost constant distribution of electrons, hence total charge neutrality is achieved. In the present approach, the density distribution of nucleons, clusters and electrons inside a spherical Wigner-Seitz cell is calculated using a Thomas-Fermi approximation beyond the local density approximation since gradient terms in the functional are considered. The formation of clusters reduces the size of the heavy nucleus. A specific feature of the model is the enhanced probability of finding clusters at the surface of the heavy nucleus.

Another consequence of the cluster formation at low densities and temperatures is the increase

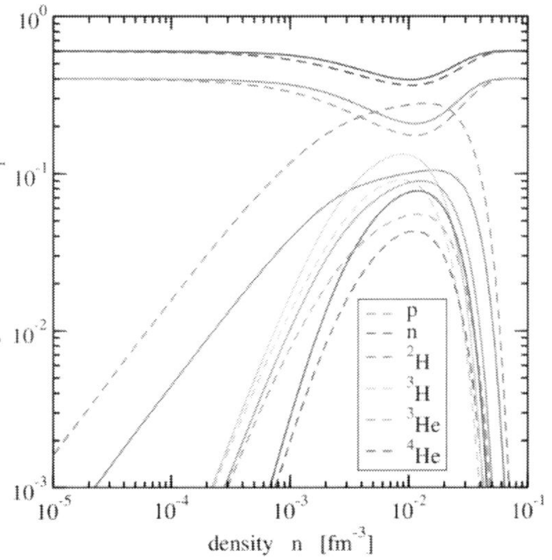


**Figure 3.** Particle fractions $X_i = A_i n_i / n$ in nuclear matter with proton fraction $Y_p = 0.4$ and at temperature $T = 10$ MeV in the generalized RMF approach without (- - - -) and with (——) continuum correlations.

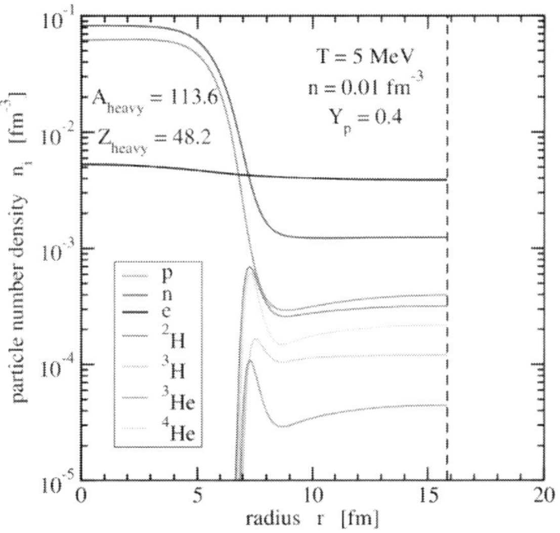


**Figure 4.** Particle densities $n_i$ as a function of the radius $r$ inside the spherical Wigner-Seitz cell for given temperature $T = 5$ MeV, total baryon number density $n = 0.01$ fm$^{-3}$ and proton fraction $Y_p = 0.4$.

of the symmetry energy as compared to pure mean-field models. See Ref. [6] for the experimental evidence of this effect and a comparison with models that consider cluster formation.

## 5. Conclusions

The relativistic mean-field model with density-dependent nucleon-meson couplings was extended to a relativistic density functional that includes light clusters as additional degrees of freedom and considers two-body correlations in the continuum. The model describes the formation and dissolution of nuclei on a more microscopic level than NSE models with excluded volume mechanism. Using a suitable parametrization of the effective interaction, the correct limits at low and high densities are obtained. In the future, the new approach will be used in the construction of an EoS table for astrophysical applications.

## Acknowledgments

This work was supported by the DFG Cluster of Excellence "Origin and Structure of the Universe". The author acknowledges the collaboration with G Röpke (U Rostock), T Klähn (U Wrocław), D Blaschke (U Wrocław), H H Wolter (LMU München) and M Voskresenskaya (GSI Darmstadt).

## References

[1] Hempel M and Schaffner-Bielich J 2010 *Nucl. Phys.* A **837** 210
[2] Horowitz C J and Schwenk A 2006 *Nucl. Phys.* A **776** 55
[3] Schmidt M, Röpke G and Schulz H 1990 *Ann. Phys. (NY)* **202** 57
[4] Typel S and Wolter H H 1999 *Nucl. Phys.* A **656** 331
[5] Typel S, Röpke G, Klähn T, Blaschke D and Wolter H H 2010 *Phys. Rev.* C **81** 015803
[6] Natowitz J B, Röpke G, Typel S, Blaschke D, Bonasera A, Hagel K, Klähn T, Kowalski S, Qin L, Shlomo S, Wada R and Wolter H H 2010 *Phys. Rev. Lett.* **104** 202501

Journal of Physics: Conference Series **321** (2011) 012030 doi:10.1088/1742-6596/321/1/012030

# Excitations in superfluid Neutron Star matter

**M. Baldo**
INFN, Sez. Catania, Via S. Sofia 64, 95123 Catania, Italy
E-mail: marcello.baldo@ct.infn.it

**C. Ducoin**
CFC, Department of Physics, University of Coimbra, P3004-516, Coimbra, Portugal
E-mail: camille@teor.fis.uc.pt

**Abstract.** We present a study of the excitation spectra and of the spectral functions for the elementary excitations in superfluid neutron star matter below the crust. We restrict the analysis to the proton and electron components within the Random Phase Approximation generalized to the superfluid case. The excitation spectrum shows the interplay of these two components and their relevance for the excitation modes. The results are relevant for neutrino physics and thermodynamical processes in neutron stars.

## 1. Introduction

Homogeneous matter below neutron star crust is expected to have neutron and proton superfluid components [1]. The elementary excitations of the matter affect the whole thermodynamics and long term evolution of the star. Collective modes in asymmetric nuclear matter have been studied previously, e.g. in Refs. [2, 3, 4]. In the astrophysical context, a study of the collective excitations in normal neutron star matter on the basis of the relativistic mean field method has been presented in Ref. [5]. The elementary excitations in superfluid neutron star matter have been studied by several authors, often with controversial results [6, 7, 8, 9, 10, 11]. As it is well known a neutral superfluid must present a Goldstone mode at low momentum, and this can have a strong influence on e.g. neutrino emission [6, 7, 8, 9, 10, 11, 12] or mean free path. For a single charged superfluid the Coulomb interaction suppresses the mode, which is replaced by the plasmon mode. In neutron star matter the physical situation is complicated by the multi-component structure, since the plasmon mode is in fact mainly an electron excitation, and furthermore the nuclear interaction couples neutron and proton excitation modes. We present the general theoretical scheme within the conserving approximations [13, 14], which guarantee current conservation and the fulfilment of the related Generalized Ward Identities (GWI) [15]. However, we will focus the study only on the proton component, leaving the complete treatment to future works. The reason of this choice is that the role of Coulomb interaction is quite crucial in this case, while the coupling with neutron is expected to be weak [16]. The aim of the paper is to have a qualitative picture of the overall excitation spectrum and of the corresponding strength functions. Therefore, for simplicity, in the numerical applications we will consider the effective mass equal to the bare one.

## 2. Formalism

According to the conserving approximation scheme [14] the density-density correlation function $\Lambda$, defined by ($\rho$ is the density)

$$\Lambda(12;3) \; = < T(\rho(3)\psi^\dagger(1)\psi(2) > \tag{1}$$

satisfies the integral equation

$$\Lambda(12;3) \;\; = \;\; \Lambda_0(12;3) + \Lambda_0(12;\overline{1'2'})\mathcal{V}(\overline{1'2'};\overline{45})\Lambda(\overline{45};3) \tag{2}$$

In these equations $i \equiv (\mathbf{r}_i, t_i, \sigma_i, \tau_i)$ stands for coordinate, time, spin and isospin variables for single particle states, and

$$\Lambda_0(12;3) \;\; = \;\; \frac{1}{i}\mathbf{G}(13)\mathbf{G}(32) \tag{3}$$

The quantity $G$ is the single particle Green' function,

$$\mathbf{G}(12) \; = \; -i < T\{\psi(1)\psi^\dagger(2)\} > \tag{4}$$

where $\psi^\dagger, \psi$ are the creation and annihilation operators for the considered particles. Finally in Eq. (2) a bar over a symbol $i$ indicates integration and summation over the corresponding set of variables. The effective particle-hole interaction $\mathcal{V}$ is the key quantity and if the approximation has to be conserving it must be expressed by the functional derivative

$$\mathcal{V}(12;45) \;\; = \;\; i\frac{\delta\Sigma(12)}{\delta\mathbf{G}(45)}$$

where $\Sigma$ is the single particle self-energy, defined according the Dyson's Equation

$$\mathbf{G}^{-1}(12) \;\; = \;\; \mathbf{G}_0^{-1}(12) - U(12) - \Sigma(12) \tag{5}$$

with $\mathbf{G}_0$ the Green's function for non-interacting particles and $U$ a possible external single particle potential, eventually local in space and time. In Eq. (5) the functional derivative is meant performed considering the Green's function as a functional of the external potential. If the derivative is taken at $U = 0$, then $\Lambda$ describes the linear response of the system. In this framework, if we express the self-energy in terms of the Green's function according to some approximate scheme, the coupled Equations (2) and (5) for $\Lambda$ and $\mathbf{G}$ ( or $\Sigma$ ) define the corresponding conserving approximations. It has to be stressed that $\mathbf{G}$ and $\Sigma$ must be calculated self-consistently. The simplest approximations for $\Sigma$ are the Hartree or Hartree-Fock ones, where the self-energy is a linear functional of the Green's function, with the bare or effective interaction. In particular the bare Coulomb interaction for the protons and electrons can be treated in this way. In this case Eq. (1) for $\Lambda$ is the Random Phase Approximation (RPA). The excitations of the proton-electron system in Neutron Star conditions via Coulomb interaction only have been extensively studied in ref. [17]. It was shown that the electron screening suppresses the proton plasma excitation, which is then replaced by a sound-like mode, where the proton and electron components move in phase and are both strongly excited. On the contrary, the electron plasma excitation is almost unaffected, with a slightly reduced strength, and remains almost a pure electron mode. The inclusion of superfluidity can be formally achieved by including in the above scheme a further discrete variable $\alpha$ labeling the destruction ($\alpha = 1$) and creation ($\alpha = -1$) operators, so that now the collective index $i \equiv (\mathbf{r}_i, t_i, \sigma_i, \tau_i, \alpha_i) \equiv (x_i, \alpha_i)$. Special care must be taken in the generalization of the basic equations, e.g. the one defining the inverse single particle Green's function.

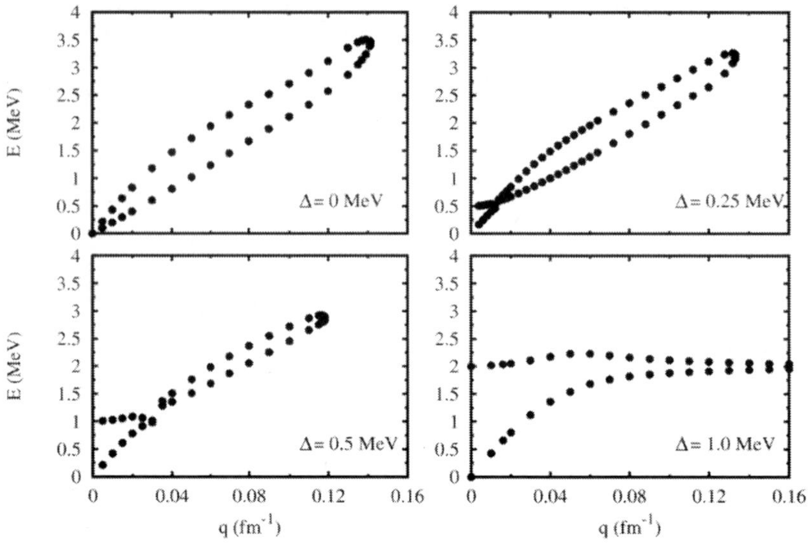


**Figure 1.** Proton excitation spectrum at saturation baryon density in Neutron Star matter, corresponding to proton Fermi momentum $k_F = 0.5596$ fm $^{-1}$, as the pairing gap $\Delta$ is varied. The four panels correspond to the gap values indicated by the corresponding labels. The electron component is included.

## 3. Results

The calculation of the proton pairing gap is a difficult task, because of the several competing many-body effects that must be taken into account. We will consider the value of the proton pairing gap as a parameter to be varied in a reasonable interval that can be considered typical of nuclear matter. The pairing interaction will be therefore adjusted accordingly within the BCS scheme. It is interesting to follow the evolution of the spectrum as the pairing gap $\Delta$ is changed. In Fig. 1 are reported the branches of the spectrum corresponding to excitations where the proton component is large. The left upper panel corresponds to the normal system. One identifies in the upper branch the sound mode, while the lower branch is strongly over-damped, and it does not correspond to a real excitation. The spectrum terminates at a certain momentum and the excitations are strongly damped beyond this momentum. As the gap increases the spectrum develop in a well defined manner. Below $2\Delta$ only one branch is present. It corresponds to a pseudo-Goldstone mode. This mode is present for superfluid systems where only the pairing interaction is present, and it corresponds to the Goldstone mode associated with the breaking of gauge invariance. When other interactions are active, the mode is mixed with the sound mode, but the branch is still present below $2\Delta$ with a different velocity. In any case, the energy of the mode is linear in the momentum at low enough momentum. Besides the pseudo-Goldstone, a mode above $2\Delta$ appears, that can be described as a pair-breaking mode. The two branches meet at an energy close to $2\Delta$, where they undergo a quasi-crossing phenomenon. The pseudo-Goldstone mode merges then into the ordinary sound mode, while the pair-breaking mode merges into the over-damped branch of the normal system. As the pairing gap increases, the part of the spectrum above $2\Delta$ shrinks and finally disappears. It has to be stressed that besides these branches, other branches are present, not shown in the figure, where the electron component is dominant. Again a lower branch is over-damped, while the upper branch is mainly the electron plasmon mode. However, as the pairing gap increases, this plasmon mode starts to be damped because of the coupling with the proton excitations, and it disappears even at moderately low

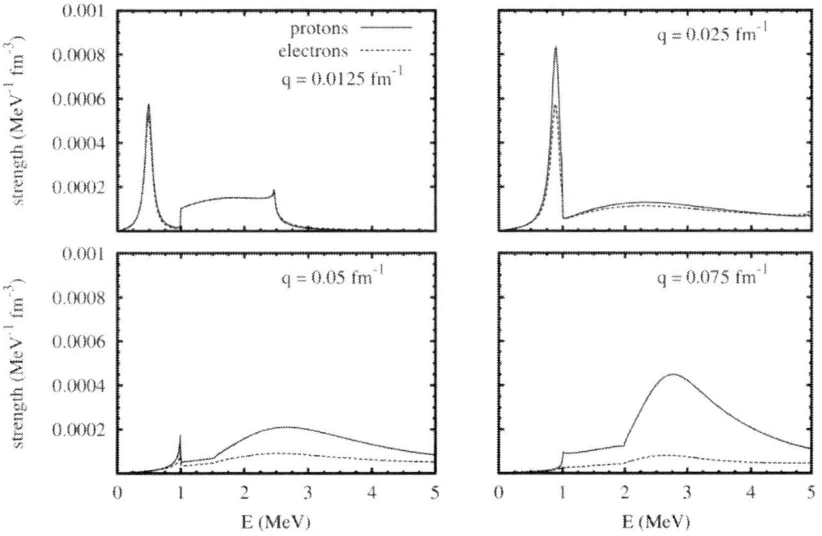


**Figure 2.** Proton and electron spectral function at saturation baryon density in Neutron Star matter, corresponding to Fermi momentum $k_F = 0.5596$ fm $^{-1}$. The pairing gap is $\Delta = 0.5$ MeV. The four panels correspond to the momentum values indicated by the corresponding labels.

values of the gap, typically 100-200 KeV. In Fig. 2 are reported the spectral functions of protons and electrons at density 0.16 fm$^{-3}$ of the neutron star matter as the momentum of the excitation is varied. The pairing gap is $\Delta = 0.5$ MeV. At low momenta the peak of the pseudo-Goldstone below $2\Delta$ is apparent. Above $2\Delta$ some strength is present, corresponding to a pair-breaking mode. As the momentum increases, the pseudo-Goldstone merges into the sound mode of the normal system. All that is consistent with the structure of the spectrum discussed above.

This work was supported by CompStar, a Research Networking Program of the European Science Foundation.

## References

[1] S.L. Shapiro and S.A. Teukolsky, *Black Holes, White Dwarfs and Neutron Stars* (Jhon Wiley and Sons, New York, 1983).
[2] P. Haensel, Nucl. Phys. A627, 53 (1978).
[3] F. Matera and V. Yu. Denisov, Phys. Rev. C 49, 2816 (1994).
[4] V. Greco, M. Colonna, M. Di Toro, and F. Matera, Phys. Rev. C 67, 015203 (2003).
[5] C Providência, L. Brito, A. M. S. Santos, D. P. Menezes and S.S. Avancini, Phys. Rev. C **74**, 045802 (2006).
[6] S.Reddy, M. Prakash, J. Lattimer and J. Pons, Phys. Rev. C 59, 2888 (1999).
[7] J. Kundu and S. Reddy, Phys. Rev. C 70, 055803 (2004).
[8] L.B. Leinson and A. Perez, Phys. Lett. B 638, 114 (2006).
[9] A. Sedrakian, H. Müther and P. Schuck, Phys. Rev. C 76, 055805 (2007).
[10] L.B. Leinson, Phys. Rev. C 78, 015502 (2008)
[11] E. E. Kolomeitsev and D. N. Voskresensky, Phys. Rev. C77, 065808 (2008) ; arXiv:1003.2741 [nucl-th].
[12] D. G. Yakovlev, A. D. Kaminker and K. P. Levenfish, A&A 343, 650 (1999).
[13] G. Baym and L.P. Kadanoff, Phys. Rev. 124, 287 (1961).
[14] G. Baym, Phys. Rev. 127, 1391 (1962).
[15] J.R. Schrieffer, *Theory of Superconductivity*, W.A. Benjamin, Inc., N.Y., 1964.
[16] M. Baldo and C. Ducoin, Phys. Rev. C79, 035801 (2009).
[17] M. Baldo and C. Ducoin, Phys. of Atomic Nuclei 72, 1188 (2009).

Journal of Physics: Conference Series **321** (2011) 012031 doi:10.1088/1742-6596/321/1/012031

# Thermalization time of hot neutron star crust

**J. Margueron[1], M. Fortin[2], F. Grill[3], D. Page[4] and N. Sandulescu[5]**

[1] Institut de Physique Nucléaire, IN2P3-CNRS and Universié Paris-Sud, F-91406 Orsay CEDEX, France

[2] École Normale Supérieure, Departement de Physique, 24 rue Lhomond, 75005 Paris, France

[3] Dipartimento di Fisica, Universitá degli Studi di Milano, Via Celoria 16, 20133 Milan, Italy

[4] Departamento de Astrofísica Teórica, Instituto de Astronomía, Universidad Nacional Autónoma de México, 04360 Mexico D.F., Mexico

[5] National Institute of Physics and Nuclear Engineering, 76900, Bucharest, Romania

E-mail: `jerome.margueron@ipno.in2p3.fr`

**Abstract.** We discuss the thermalization process of the neutron stars crust described by solving the heat transport equation with a microscopic input for the specific heat of baryonic matter. The heat equation is solved with initial conditions specific to a rapid cooling of the core. To calculate the specific heat of inner crust baryonic matter, i.e., nuclear clusters and unbound neutrons, we use the quasiparticle spectrum provided by the Hartree-Fock-Bogoliubov approach at finite temperature. In this framework we analyze the dependence of the crust thermalization on pairing properties and on cluster structure of inner crust matter. It is shown that the pairing correlations reduce the crust thermalization time by a large fraction. The calculations show also that the nuclear clusters have a non-negligible influence on the time evolution of the surface temperature of the neutron star.

The thermalization process of the neutron stars crust can give important informations concerning the properties of the crust matter. This is indeed the case in the cooling of isolated neutron stars [1] and in the thermal after-burst relaxation of neutron stars from X-ray transients [2, 3]. In the cooling process an important quantity is the cooling or thermalization time of the crust, defined as the time needed for the crust matter to equilibrate its temperature to the temperature of the colder core. The outer crust, due to its thinness and the very small amount of matter it contains, has a very short thermal time-scale. Hence, the cooling time is essentially determined by the inner crust thickness, as well as by the composition of the inner crust matter formed by nuclear clusters, unbound neutrons and ultra-relativistic electrons [1]. Moreover several studies have shown that the thermalization time depends significantly on the superfluid properties of the inner crust baryonic matter [1, 4, 5, 6, 7]. This dependence is induced through the specific heat of unbound neutrons, strongly affected by the pairing energy gap. Since the neutron pairing gap is influenced by the presence of the nuclear clusters [8, 5, 9], a reliable calculation of neutron specific heat should take the clusters into account. How the intensity of pairing correlations affects the specific heat of the neutrons in the presence of nuclear clusters was analyzed in Ref. [10]. Using the framework of the Hartree-Fock-Bogoliubov (HFB) approach at finite temperature it was shown that the specific heat can change by several orders of magnitude if the pairing gap is adjusted to describe two possible scenarios for neutron matter superfluidity, i.e., one corresponding to BCS approximation and the other to calculations schemes which take into account in-medium effects [11]. The impact which these changes in the specific

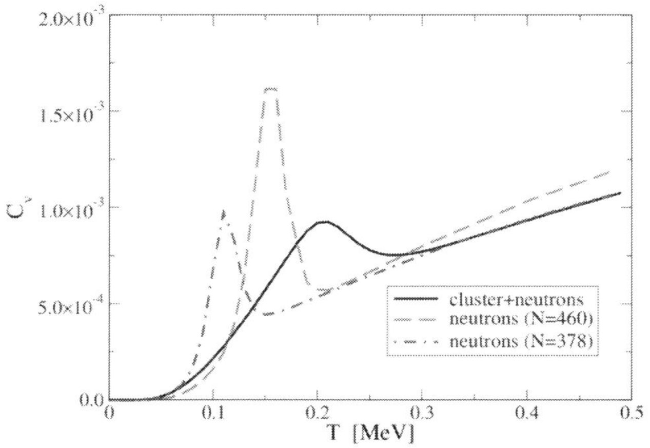


**Figure 1.** Neutron-specific heat in the the Wigner-Seitz cell N=460 neutrons and Z=40 protons [14] for strong pairing. The results corresponds to various approximations discussed in the paper. The specific heat is given in units of Boltzman constant $k_B$.

heat could have on the thermalization time was discussed in Ref. [6]. Employing a simple random walk model for the cooling [5, 12], in which the diffusion of the heat towards the core was calculated without taking into account the dynamical change of the temperature through the whole crust, it was shown that the thermalization times corresponding to the two pairing scenarios mentioned above differ by a large fraction. The scope of this work is to perform more accurate estimations of the thermalization time obtained by employing a more realistic cooling model based on dynamical solutions of the heat equations [13] and on a state-of-the-art description of the specific heat for baryonic matter.

## 1. Specific heat of the inner crust matter

The specific heat of the inner crust has contributions from electrons, the lattice and unbound neutrons. They are calculated for a given set of densities in the Wigner-Seitz approach. In the inner crust the unbound neutrons are superfluid in the $^1S_0$ channel. To take into account the superfluid properties of the neutrons as well as the non-uniform cluster structure of the inner crust, we perform HFB calculations at finite-temperature for each Wigner-Seitz cell [14]. The details of the HFB calculations in a Wigner-Seitz cell are given in Ref. [10]. In the HFB calculations the mean field is described with the same interaction used in the Douchin and Haensel EoS [17], i.e., the Skyrme force SLy4.

We present now in more detail the results for the cell which contains N=460 neutrons and Z=40 protons [14]. In this cell the HFB calculations predict 378 unbound neutrons. It is interesting that in spite of the large number of neutrons in the cell, the number of the bound neutrons in the cluster with Z=40 protons is equal to the magic number 82, as for the dripline nucleus $^{122}$Zr (see, e.g., Ref. [15]). The specific heat given by the HFB spectrum, in which the contribution of the cluster is included, is shown in Fig. 1 by the full line. In the same figure are shown the specific heat corresponding to two approximations employed in some studies [1, 5]: the uniform distribution of neutrons is calculated in the WS cell removing all the protons. The number of neutrons is fixed to be equal either to the total number of neutrons in the cell (dashed line) or only to the number of the unbound neutrons (dash-dotted line). The latter case is considered as an effective way of taking into account the influence of the cluster [5]. How these approximations work is seen in Fig. 1. To make the comparison meaningful, the calculations for the uniform neutron gas are done solving the HFB equations with the same boundary conditions as for the non uniform system, i.e., neutrons+cluster. As seen in Fig. 1, the transition from the superfluid to the normal phase, localized at the peak of the curves, takes place at a lower temperature in the case of uniform neutron gas, especially when are considered

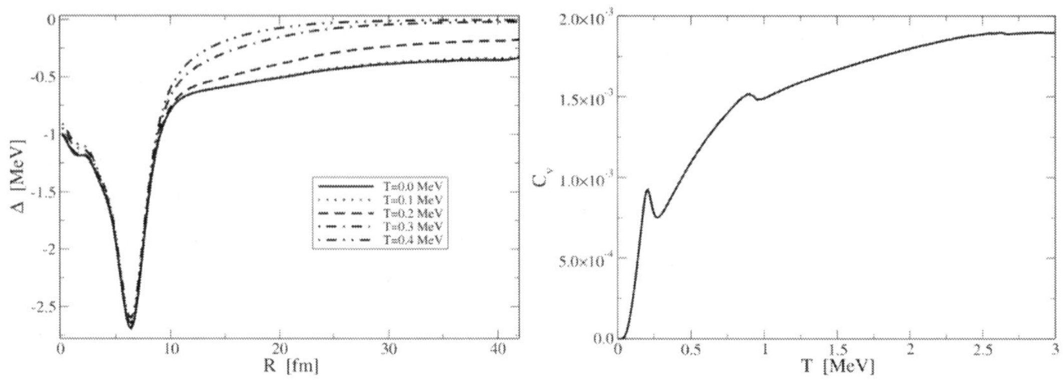


**Figure 2.** On the left is shown the radial distribution of the pairing field for the Wigner-Seitz cell N=460 neutrons and Z=40 protons [14] and for various temperatures. The results correspond to a strong pairing as in Fig. 1. On the right is represented the neutron-specific heat for the same Wigner-Seitz cell. The results are shown up to high temperatures in order to illustrate the second transition region around T= 900 keV. The specific heat is given in units of $k_B$.

only the unbound neutrons. We can also notice that, in contrast to the uniform system, in the non uniform system the transition from the superfluid to the normal phase is smooth.

To see better what happens in a non uniform system, in Fig. 2 (left) is shown the evolution with the temperature of the pairing field in the same cell as in Fig. 1. The pairing field is defined by

$$\Delta(r) = V_0 \left[1 - \eta \left(\frac{\rho(r)}{\rho_0}\right)^{\alpha}\right] \kappa(r), \tag{1}$$

where $\kappa(r)$ is the local pairing tensor. Notice that the pairing field is taken with negative values. From Fig. 2 (left) we can see that at zero temperature the pairing field is much larger at the surface of the cluster than in the bulk region. Due to this fact by increasing the temperature the pairs corresponding to the neutrons localized preferentially at the surface region of the cluster are destroyed gradually and much slower compared with the pairs formed by neutrons localized far from the surface of the cluster. In fact, as shown in Fig. 2 (right), the non uniform system shows two transition regions, one around T=200 keV, corresponding mainly to the neutrons located far from the surface of the cluster, and another one, much less pronounced, around T=900 keV, which corresponds to the neutrons localized in the surface region of the cluster.

## 2. Crust thermalization

The cooling process is described by solving the heat equation in the whole volume of the neutron star (more details and references can be found in Ref. [16]). The initial temperature distribution in the star is chosen to be constant, $T(r, t = 0) = T_i$. The cooling calculations presented here are for a neutron star of a mass equal to 1.6 $M_\odot$. For this mass and for the Douchin and Haensel EoS [17] the TOV equations predict a total star radius of 11.49 km and a central density of 4.06 $\rho_0$, where $\rho_0 = 2.9 \times 10^{14}$ g cm$^{-3}$ is the saturation density of symmetric nuclear matter. The three needed microphysical ingredients, the conductivity $K$, the specific heat $C_V$, and the neutrino emission rate$Q_\nu$, as well as pairing of neutron and protons in the core, are described in Ref. [16].

The overall dependence of $K$ on density is shown in Fig. 3 (left) for three temperatures, i.e., $T = \{10^7, 10^8, 10^9\}$ K, typical for the cooling calculations performed in this study. In the

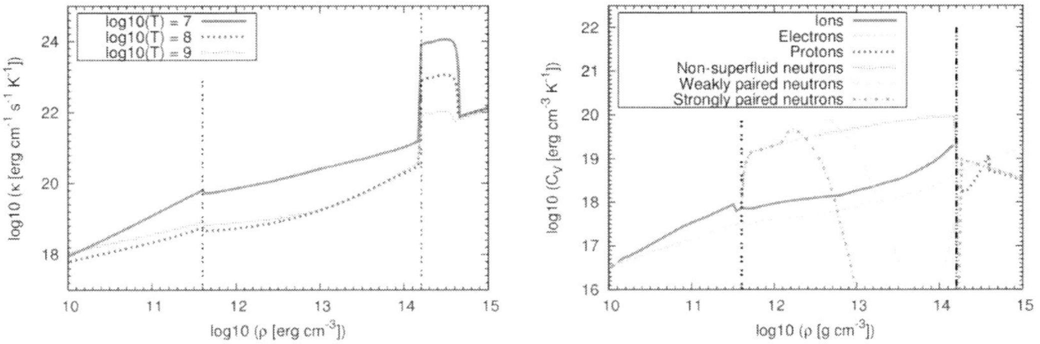


**Figure 3.** On the left the total conductivity in the neutron star is shown as a function of the density and for the temperatures $T = \{10^7, 10^8, 10^9\}$ K (see the text for more details). The bounds of the inner crust are indicated by the vertical lines. On the right the specific heat in the crust of the neutron star is represented for the different components of the star matter and for a temperature of $10^9$ K (see the text for more details).

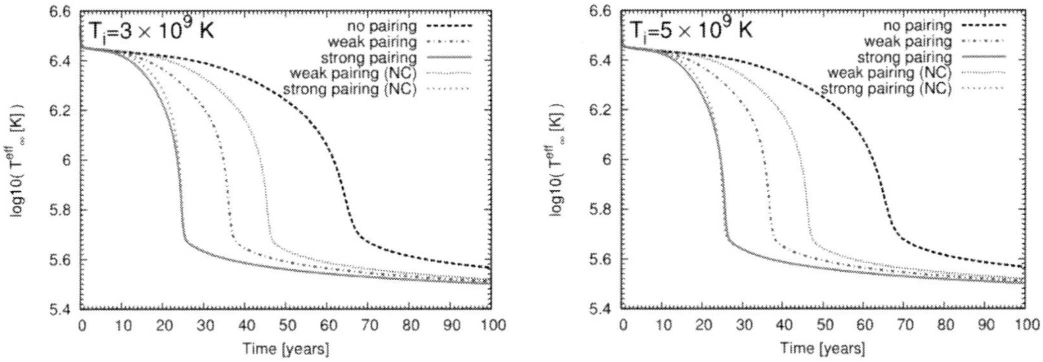


**Figure 4.** Time evolution of the apparent surface temperature for the initial temperatures $T_i = \{300, 500\}$ keV. NC indicates the results of the calculations obtained by neglecting the effect of the clusters.

crust the conductivity shown in Fig. 3 (left) is similar to that of Ref. [4]. In the core Fig. 3 (left) clearly exhibits the $\propto T^{-1}$ behavior at densities $\rho < 5 \times 10^{14}$ g cm$^{-3}$ where protons are superconductors, contrasted by the T-independent values at higher densities. This high density behavior of $K^e$ was not included in the work of Ref. [4] which was previous to Ref. [18].

The dependence of the specific heat on the density is illustrated in Fig. 3 (right) for a temperature equal to $10^9$ K. In the outer crust the specific heat has contributions from electrons and ions while in the core from electrons, neutrons and protons. These specific heats are evaluated using standard approximations (see, e.g., [19]). The specific heat is strongly dependent on the physical state of the neutrons in the inner crust. Indeed, the pairing strongly supress the contribution of the neutrons to the specific heat which otherwise would give a dominant contribution, see Fig. 3 (right). In the region where neutrons are paired the ions provide the major contribution to the specific heat, eventually being supplanted by the electrons since $C_V^{(\text{ion})} \propto T^3$ while $C_V^{(\text{e})} \propto T$.

The time evolution of the effective surface temperature $T_\infty^{\rm eff}$ is displayed in Fig. 4 for two initial temperatures of the crust, $T_i = \{300, 500\}$ keV. The effective surface temperature shown in Fig. 4 is obtained from the temperature at the bottom of the crust, $T_b = T(\rho_b)$, where $\rho_b = 10^{10}$ g cm$^{-3}$, using the relationship given in Ref. [20] for a non-accreted envelope. It can be seen that the time evolution of the surface temperature does not depend much on the initial temperature of the crust. In Fig. 4 the apparent surface temperatures obtained by neglecting the effect of the clusters is shown, i.e., supposing that the neutron specific heat is given solely by that of the neutron gas. In this case the specific heat of the neutrons is calculated from the quasiparticle spectrum of BCS equations solved for infinite neutron matter at a density corresponding to that of the external neutrons in the WS cell [21]. It can be seen that for the weak pairing scenario the clusters have a non-negligible effect on the time evolution of the surface temperature. For the strong pairing the initial temperature being already lower than the critical temperature there is almost no impact of the cluster on the thermal relaxation.

## 3. Conclusion

In this work we have studied how the thermalization of neutron stars crust depends on pairing properties and the cluster structure of the inner crust matter. The thermal evolution was obtained by solving the relativistic heat equation with initial conditions specific to a rapid cooling process. The specific heat of neutrons was calculated from the HFB spectrum, taking into account consistently the effects of nuclear clusters, pairing correlations and temperature. The thermal evolution of the inner crust was analyzed using for the neutrons two sets of specific heats obtained with a strong and a weak pairing force which simulate two possible scenarios for the intensity of pairing correlations in neutron matter. The results show that the crust thermalization is strongly influenced by the intensity of pairing correlation. We have also shown that the cluster structure of the inner crust affects significantly the time evolution of the surface temperature.

**Acknowledgment.** This work was supported by ESF through the project " The New Physics of Compact Stars", by the ANR NExEN, and by CNCSIS through the grant IDEI no. 270.

[1] J. M. Lattimer, K. A. Van Riper, M. Prakash, and M. Prakash, Astrophys. J. **425**, 802 (1994)
[2] P. S. Shternin, D. G. Yakovlev, P. Haensel, and A. Y. Potekhin, Mont. Not. R. Astron. Soc. **382**, L43 (2007)
[3] E. F. Brown, and A. Cumming, Astrophys. J. **698**, 1020 (2009)
[4] O. Y. Gnedin, D. G. Yakovlev, A. Y. Potekhin, Mont. Not. R. Astron. Soc. **324**, 725 (2001)
[5] P.M. Pizzochero, F. Barranco, E. Vigezzi, and R. A. Broglia, Astrophys. J. **569**, 381 (2002)
[6] C. Monrozeau, J. Margueron, and N. Sandulescu, Phys. Rev. C **75**, 065807 (2007)
[7] N. Sandulescu, Eur. Phys. J. (Special Topics) **156**, 265 (2008)
[8] F. Barranco, R. A. Broglia, H. Esbensen, and E. Vigezzi, Phys. Rev. C **58**, 1257 (1998)
[9] N. Sandulescu, N. Van Giai, and R. J. Liotta, Phys. Rev. C **69**, 045802 (2004)
[10] N. Sandulescu, Phys. Rev. C **70**, 025801 (2004)
[11] U. Lombardo, in *Nuclear Methods and the Nuclear Equation of State*, edited by M. Baldo (World Scientific, Singapore, 1999), pp. 458-510
[12] G. E. Brown, K. Kubodera, D. Page, and P. M. Pizzochero, Phys. Rev. D **37**, 2042 (1988)
[13] D. Page, U. Geppert, and F. Weber, Nucl. Phys. **A 777**, p. 497-530 (2006); D. Page and S. Reddy, Annu. Rev. Nucl. Part. Sci. **56**, 327 (2006)
[14] J. W. Negele and D. Vautherin, Nucl. Phys. **A 207**, 298 (1973)
[15] N. Sandulescu, L. S. Geng, H. Toki, and G. C. Hillhouse, Phys. Rev. C **68**, 054323 (2003)
[16] M. Fortin, F. Grill, J. Margueron, D. Page, and N. Sandulescu, Phys. Rev. C **82**, 065804 (2010)
[17] F. Douchin and P. Haensel, Astron. Astrophys. **380**, 151167 (2001)
[18] P. S. Shternin, and D. G. Yakovlev, Phys. Rev. D **75**, 103004 (2007)
[19] D. Page, J. M. Lattimer, M. Prakash, and A.W. Steiner, Astrophys. J. Suppl. **155**, 623 (2004)
[20] A. Y. Potekhin, G. Chabrier, and D. G. Yakovlev, Astron. Astrophys. **323**, 415 (1997)
[21] K. P. Levenfish, D. G. Yakovlev, Astron. Rep. **38**, 247 (1994)

# New microscopic model for dilute alpha-gas states and its extension

**Akihiro Tohsaki**
Research Center for Nuclear Physics, Osaka University, 567-0047, Japan
E-mail: tohsaki@rcnp.osaka-u.ac.jp

**Abstract.** We show a road map of the microscopic $\alpha$-condensation model, which fully takes into account the Pauli principle between all the constituent nucleons and effective inter-nucleon forces simultaneously, and which can play an important role in reproducing an $\alpha$-gas-like nature thanks to $\alpha$-condensation.

It is twelve years, which are just one cycle of years in the Oriental tradition, since four old nuclear theorists started in thinking about $\alpha$-condensation in nuclear system. A decade ago, at first, they proposed a toy model with a complete consideration of the Pauli principle for all the nucleons unlike a precise microscopic $\alpha$-cluster model [1]. Nevertheless, the toy has been powerful for accounting for an excited state of $n\alpha$-nuclei around $n\alpha$ breakup energy including the Hoyle state of $^{12}C$ until now. An idea of the toy wave function is to change variational parameters in the Hill-Wheeler equation from the relative distances between $\alpha$-clusters, as the traditional $\alpha$-cluster model employs, to their density. In other words, we consider only one-parameter contribution for $\alpha$-condensation in spite of entangled parameters coming from relative motion of $\alpha$-clusters. The toy wave function has been successfully applied to $\alpha$ nuclei from 2 to 5 $\alpha$-clusters of which an excited state around $n\alpha$ break-up energy is a dilute $\alpha$-gas state being regarded as the $\alpha$-condensation [2]. We think that the Pauli principle, the effective inter-nucleon forces and the Coulomb force are key terms in the toy model. This is because it has absolutely a long range tail, which could not be artificially cut off, and which could play an important role in guaranteeing the stability of the $\alpha$-gas. It is necessary condition to employ the effective inter-nucleon force, which reproduces the saturation property of nuclei for a wide range of mass numbers and the experimental phase shifts of the $\alpha$-$\alpha$ elastic scattering simultaneously. It is truly new phenomena in physics that, unlike atomic He condensation, the $\alpha$-condensation is in self-stability.

The original toy wave function can correctly reproduce the $0^+$ phase shift of $\alpha$-$\alpha$ RGM calculation, and can conjecture the Hoyle state as a dilute $\alpha$-gas state for $^{12}C$. In this meaning, it should be regarded as a spherical $\alpha$-condensation, and its extension enables us to estimate the deformed dilute $\alpha$-gas. As an attempt in the early stage, after a young scientist Funaki has joined them, an elliptical deformation has been successfully introduced in $^{12}C$ nucleus [3, 4]. In the norm kernel, the variable to be considered is reduced only the ratio of the box size concerning the $\alpha$-condensation strength and the size of $\alpha$-cluster. A new development has come to a general deformation with odd states [5]. Its expansion coefficients with respect to the ratio correspond to the eigenvalue of $n\alpha$-system. For example, the eigenvalues of the norm kernel for $3\alpha$-system are shown in Table I, along the three eigenvectors. The $p_{k_t}$ stands for the case of spherical

**Table 1.** The expansion coefficients $p_{k_1 k_2 k_3}$ for $n\alpha$-condensation, where $k_t = k_1 + k_2 + k_3$. The numerical values are symmetric for the exchange of suffices. The former corresponds to the case of the spherical $\alpha$ condensation and the latter means that of the deformed condensation. The superfices in the numerical values are the power of 10.

| $k_t$ | $p_{k_t}$ | $k_1$ | $k_2$ | $k_3$ | $p_{k_1 k_2 k_3}$ |
|---|---|---|---|---|---|
| 0 | 0 | 0 | 0 | 0 | 0 |
| 1 | 0 | 0 | 0 | 1 | 0 |
| 2 | 0 | | 0 | 2 | 0 |
| | | | 1 | 1 | 0 |
| 3 | 0 | 0 | 0 | 3 | 0 |
| | | 0 | 1 | 2 | 0 |
| | | 1 | 1 | 1 | 0 |
| 4 | $0.3955^{-1}$ | 0 | 0 | 4 | 0 |
| | | 0 | 1 | 3 | 0 |
| | | 0 | 2 | 2 | 0.1187 |
| | | 1 | 1 | 2 | $0.7910^{-1}$ |
| 5 | 0.1112 | 0 | 0 | 5 | 0 |
| | | 0 | 1 | 4 | $0.1978^{-1}$ |
| | | 0 | 2 | 3 | 0.1854 |
| | | 1 | 1 | 3 | 0.1384 |
| | | 1 | 2 | 2 | 0.2299 |
| 6 | 0.1981 | 0 | 0 | 6 | $0.1159^{-1}$ |
| | | 0 | 1 | 5 | $0.5407^{-1}$ |
| | | 0 | 2 | 4 | 0.2372 |
| | | 0 | 3 | 3 | 0.2905 |
| | | 1 | 1 | 4 | 0.1931 |
| | | 1 | 2 | 3 | 0.3210 |
| | | 2 | 2 | 2 | 0.3889 |
| 7 | 0.2863 | 0 | 0 | 7 | $0.3549^{-1}$ |
| | | 0 | 1 | 6 | $0.9603^{-1}$ |
| | | 0 | 2 | 5 | 0.2841 |
| | | 0 | 3 | 4 | 0.3595 |
| | | 1 | 1 | 5 | 0.2450 |
| | | 1 | 2 | 4 | 0.3852 |
| | | 1 | 3 | 3 | 0.4247 |
| | | 2 | 2 | 3 | 0.4801 |
| 8 | 0.3678 | 0 | 0 | 8 | $0.6742^{-1}$ |
| | | 0 | 1 | 7 | 0.1401 |
| | | 0 | 2 | 6 | 0.3276 |
| | | 0 | 3 | 5 | 0.4120 |
| | | 0 | 4 | 4 | 0.4356 |
| | | 1 | 1 | 6 | 0.2928 |
| | | 1 | 2 | 5 | 0.4356 |
| | | 1 | 3 | 4 | 0.4911 |
| | | 2 | 2 | 4 | 0.5392 |
| | | 2 | 3 | 3 | 0.5683 |
| 9 | 0.4396 | 0 | 0 | 9 | 0.1029 |
| | | 0 | 1 | 8 | 0.1829 |
| | | 0 | 2 | 7 | 0.3672 |
| | | 0 | 3 | 6 | 0.4549 |
| | | 0 | 4 | 5 | 0.4894 |
| | | 1 | 1 | 7 | 0.3358 |
| | | 1 | 2 | 6 | 0.4771 |
| | | 1 | 3 | 5 | 0.5391 |
| | | 1 | 4 | 4 | 0.5564 |
| | | 2 | 2 | 5 | 0.5821 |
| | | 2 | 3 | 4 | 0.6230 |
| | | 3 | 3 | 3 | 0.6467 |

condensation, and the states $(0, 0, k_3)$ correspond to the linear condensation. The $(0, k_2, k_3)$ states explain the plane $\alpha$-condensation. We think that those two types of condensations are the next interest.

Usually the $\alpha$-condensation assumes to take place around the origin, but it is easily possible to shift to an arbitrary position in the parameter space. In this sense, we can naturally treat several $\alpha$-condensation systems separated each other. In other words, the interaction between the $\alpha$-condensations or between the $\alpha$-condesation and another nucleus should be also the next interest. However we will be sincerely careful to remove the spurious c.m. component. This serious difficulty is common in microscopically treating $\alpha$-cluster model with reasonable size parameters. It is also fascinating to know a mathematical structure of the $\alpha$-condensation wave function. This is because the mathematical information is inevitably important to investigate the multi $\alpha$-consideration more than 7 $\alpha$-clusters. As an easier subject than the $\alpha$-condensation, we have to address ourselves to the di-nucleon condensation. We think that it has a secret key to see the $\alpha$-condensation.

As mentioned before, we have the following road map to the future:

(i) The linear and the plane $\alpha$-condensations are easier to calculate than the other subjects. The former one is related to the $\alpha$-chain structure, and the latter is the disc structure.

(ii) The odd states of $\alpha$-condensation for $^{12}C$ are also interesting. We have two approaches to carry out it. The first one depends upon the deformed description, another is in treating the interaction of $2\alpha$ and $\alpha$.

(iii) To clarify the interaction of $3\alpha$ and nucleon is amusing as well as to see that of $3\alpha$ and $\alpha$. The levels of $^{16}O$ with $\alpha$-cluster structure are mostly elucidated in studying the coupling of $3\alpha$-$\alpha$ and $2\alpha$-$2\alpha$ systems. However, the selection of the effective inter-nucleon force is absolutely important.

(iv) The mathematical formulation for the $\alpha$-condensation is also expecting throughout that for the di-nucleon condensation. It is also funny to see whether the infinite numbers of $\alpha$-clusters make the $\alpha$-condensation or not.

We are now studing an essence of the $\alpha$-cluster model in nuclei which is the most important one for their excited state.

This study is carried out under collaboration with Y. Funaki, H. Horiuchi, P. Schuck, G. Röpke and T. Yamada. We expect young scientists world over to join us!

## References

[1] A. Tohsaki, H. Horiuchi, P. Schuck, and G. Röpke, Phys. Rev. Lett. **87**, 192501 (2001).
[2] A. Tohsaki, H. Horiuchi, P. Schuck, and G. Röpke, Nucl. Phys. **A738**, 259 (2004).
[3] Y. Funaki, H. Horiuchi, A. Tohsaki, P. Schuck, and G. Röpke, Prog. Theor. Phys. **108**, 297 (2002).
[4] Y. Funaki, A. Tohsaki, H. Horiuchi, P. Schuck, and G. Röpke, Phys. Rev. C **67**, 051306(R) (2003).
[5] A. Tohsaki, to be published in Frontier of Physics (China).

# Alpha clustering and condensation in nuclei

**Y. Funaki[1], T. Yamada[2], H. Horiuchi[3], G. Röpke[4], P. Schuck[5] and A. Tohsaki[3]**

[1] Nishina Center for Accelerator-Based Science, The institute of Physical and Chemical Research (RIKEN), Wako 351-0198, Japan
[2]Laboratory of Physics, Kanto Gakuin University, Yokohama 236-8501, Japan
[3]Research Center for Nuclear Physics (RCNP), Osaka University, Osaka 567-0047, Japan
[4]Institut für Physik, Universität Rostock, D-18051 Rostock, Germany
[5]Institut de Physique Nucléaire, 91406 Orsay Cedex, France

E-mail: funaki@riken.jp

**Abstract.** Low density states near the $3\alpha$ and $4\alpha$ breakup thresholds in $^{12}$C and $^{16}$O, respectively, are discussed in terms of the $\alpha$-particle condensation. Calculations are performed in OCM (Orthogonality Condition Model) and THSR (Tohsaki-Horiuchi-Schuck-Röpke) approaches. The $0_2^+$ state in $^{12}$C and the $0_6^+$ state in $^{16}$O are shown to have dilute density structures and give strong enhancement of the occupation of the $S$-state c.o.m. orbital of the $\alpha$-particles. The possibility of the existence of $\alpha$-particle condensed states in heavier $n\alpha$ nuclei is also discussed.

## 1. Introduction

It is well established that $\alpha$-clustering plays a very important role for the structure of lighter nuclei [1, 2]. The importance of $\alpha$-cluster formation has also been discussed in infinite nuclear matter, where $\alpha$-particle type condensation is expected at low density [3]. In nuclei the bosonic constituents always are only very few in number, nevertheless possibly giving rise to clear condensation characteristics, as is well known from nuclear pairing [4]. Concerning $\alpha$-particle condensation, the Hoyle state, i.e. the $0_2^+$ state in $^{12}$C has clearly been established. Several papers of the past [5, 6, 7, 8] and also more recently [9, 10, 11, 12, 13] have by now established beyond doubt that the Hoyle state, only having about one third of saturation density, can be described, to good approximation, as a product state of three $\alpha$-particles, condensed, with their c.o.m. motion, into the lowest mean field $0S$-orbit [14, 15]. This shall be the definition of an $\alpha$-particle condensed state in finite nuclei, clearly reflecting the situation found in infinite matter in Ref. [3]. The establishment of this novel aspect of the Hoyle state naturally leads us to the speculation about $4\alpha$-particle condensation in $^{16}$O.

In the present contribution, a brief review of the status of work on the Hoyle state is given, and the recent progress on the investigation of $4\alpha$-particle condensate state in $^{16}$O is shown.

## 2. Description of the Hoyle state as the $\alpha$ condensate

Based on the investigations of the possibility of $\alpha$-particle condensation in low-density nuclear matter [3], the present authors proposed a conjecture that near the $n\alpha$ threshold in self-conjugate $4n$ nuclei there exist excited states of dilute density which are composed of a weekly interacting gas of self-bound $\alpha$ particles and can be considered as an $n\alpha$ condensed state [9]. This conjecture was backed by examining the structure of $^{12}$C and $^{16}$O using a new $\alpha$-cluster wave function of the $\alpha$-cluster condensate type, which has been denoted as THSR (Tohsaki-Horiuchi-Schuck-Röpke) wave function, as follows: $\Phi_{n\alpha} \propto \mathcal{A}\,[\prod_{i=1}^{n} \exp(-2(\boldsymbol{X}_i - \boldsymbol{X}_G)^2/B^2)\phi(\alpha_i)]$. Here $\phi(\alpha_i)$, $\boldsymbol{X}_i$, $\boldsymbol{X}_G$, and $B$ are the internal wave functions, c.o.m. coordinates, of the $i$-th alpha cluster, the total c.o.m. coordinate, and a variational parameter, respectively. It should be noted that this wave function contains two limits exactly: the one of a shell-model relevant at higher densities and the one of an ideal $\alpha$-particle condensate in the dilute limit [9]. An interesting

The detailed structure analyses of $^{12}$C [10] showed that the $0_2^+$ wave function of $^{12}$C which was obtained in the past by solving the full three-body problem of the microscopic $3\alpha$ cluster model is almost completely equivalent to the wave function of the $3\alpha$ THSR state. This result gives us strong support to our opinion that the $0_2^+$ state of $^{12}$C has a gas-like structure of $3\alpha$ clusters with "Bose-condensation". The rms radius for this THSR state was calculated as $R(0_2^+)_{\rm THSR} = 4.3$ fm which fits well with experimental data for the form factor of the Hoyle state, see Ref. [12, 16]. It confirms the assumption of low density as a prerequisite for the formation of an $\alpha$-cluster structure for which the Bose-like enhancement of the occupation of the $S$ orbit is possible.

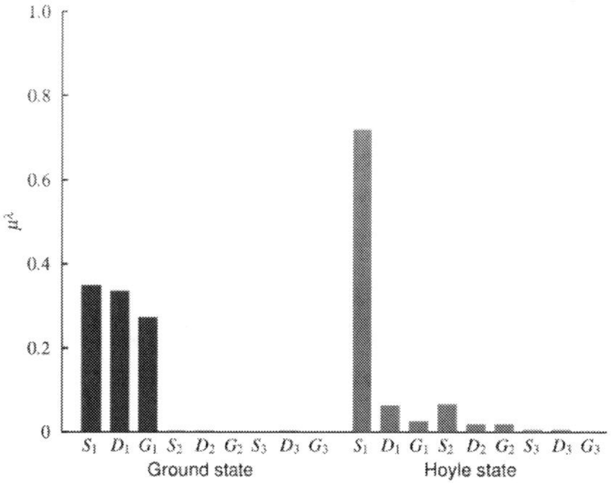


**Figure 1.** Occupation of the single-$\alpha$ orbitals of the ground state of $^{12}$C compared with the Hoyle state [15]. For explanation see the text.

analysis of the applicability of the THSR wave function can be performed by comparing with stochastic variational calculations [14] and OCM calculations [15]. The $\alpha$ density matrix $\rho(\boldsymbol{r}, \boldsymbol{r}')$ is diagonalized to study the single-$\alpha$ orbits and occupation probabilities in $^{12}$C states. FIG. 1 shows the occupation probabilities of the $L$-orbits with $S$, $D$ and $G$ waves belonging to the $k$-th largest occupation number (denoted by $L_k$), for the ground and Hoyle state of $^{12}$C obtained by diagonalizing the density matrix $\rho(\boldsymbol{r}, \boldsymbol{r}')$. We found that in the Hoyle state the $\alpha$-particle $S$ orbit with zero node is occupied to more than 70 % by the three $\alpha$-particles (see also Ref. [14] and Fig. 1). Taking into account the finite size of the nucleus, a reduction of the condensate fraction from 100 % to about 70 % is not surprising, and the remaining fraction (about 30 %) is due to higher orbits originating from antisymmetrization among nucleons. This huge percentage means that an almost ideal $\alpha$-particle condensate is realized in the Hoyle state.

## 3. $4\alpha$ condensation in $^{16}$O

The situation in $^{16}$O with respect to cluster states is already much more complicated than in $^{12}$C. In $^{12}$C, exciting one $\alpha$-particle out of the ground state necessarily leads to a dilute $\alpha$-gas state, since the remaining nucleus, $^8$Be, is itself a loosely bound two $\alpha$-object. On the other hand exciting an $\alpha$-particle out of the $^{16}$O ground state can lead to multiple $\alpha +^{12}$ C configurations. It is well documented [17] that the $0_2^+$ state at 6.06 MeV is an $\alpha$-particle orbiting in a $0S$-wave around the ground state of $^{12}$C. However, there are many more possibilities. The $\alpha$ particle can be in higher nodal $S$-wave, the $\alpha$ can orbit in a $0D$-wave around the $2_1^+$ of $^{12}$C and couple to a $0^+$ state. It also can orbit in an odd parity wave around the $1^-$ and $3^-$ states in $^{12}$C, etc. The $^{12}$C can also be in the Hoyle state which then leads us to the four $\alpha$ gas state in $^{16}$O, which is the state of our interest. Experimentally, there are six $0^+$ states in $^{16}$O up to around the four $\alpha$

disintegration threshold at 14.4 MeV: the ground state $0_1^+$, $0_2^+$ at 6.06 MeV, $0_3^+$ at 12.05 MeV, $0_4^+$ at 13.6 MeV [18], $0_5^+$ at 14.01 MeV, and $0_6^+$ at 15.1 MeV.

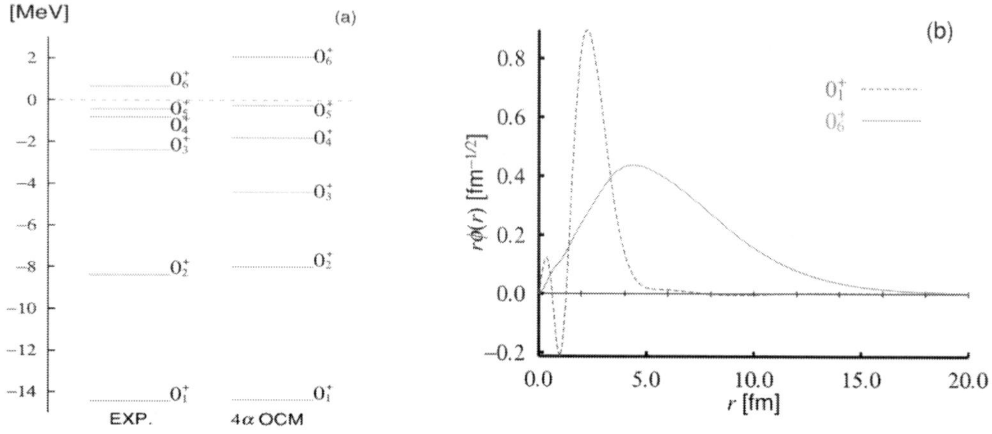


**Figure 2.** (a): Comparison of energy spectra between experiments [23] and $4\alpha$ OCM calculation [19]. The $0_4^+$ state in experiment is given in Ref. [18]. (b): Single-$\alpha$ orbits with $L = 0$ belonging to the largest occupation number, for the ground and $0_6^+$ states. The radial part of the orbits are shown.

In analogy to the above mentioned OCM calculation for $^{12}C$ [15], we recently performed a quite complete OCM calculation also for $^{16}O$, including many of the cluster configurations [19]. We were able to reproduce the full spectrum of the observed $0^+$ states. As shown in FIG. 2(a), we tentatively make a one to one correspondence of our states with the six lowest $0^+$ states of the experimental spectrum. In view of the complexity of the situation, the agreement can be considered as very satisfactory. Considering the properties of the various states in more detail, a certain number of rather big surprises came up. The analysis of the diagonalisation of the $\alpha$-particle density matrix $\rho(\boldsymbol{r}, \boldsymbol{r}')$ (as was done in [14, 15, 20, 21]) showed that the newly discovered $0^+$ state at 13.6 MeV [18], as well as the well known $0^+$ state at 14.01 MeV, corresponding to our states at 12.6 MeV and 14.1 MeV, respectively, have, contrary to what we assumed previously [22], very little condensate occupancy of the $0S$-orbit (about 20 %). On the other hand, the sixth $0^+$ state at 16.5 MeV calculated energy, to be identified with the experimental state at 15.1 MeV, has 61 % of the $\alpha$-particles being in the $0S$-orbit. The corresponding single-$\alpha$ $0S$ orbit is shown in FIG. 2(b). It has a strong spatially extended behaviour, and in the interior no node ($0S$), showing almost no effect of the Pauli principle acting on nucleons belonging to different alpha particles. This indicates that $\alpha$ particles are condensed into the very dilute $0S$ single-$\alpha$ orbit, see also Ref. [24]. Thus, the $0_6^+$ state clearly has a $4\alpha$ condensate character. We should note that the orbit is very similar to the single-$\alpha$ orbit of the Hoyle state [14, 15]. We also show in FIG. 2(b) the single-$\alpha$ orbit for the ground state. It has maximum amplitude at around 3 fm and oscillations in the interior with two nodal ($2S$) behaviour, due to the Pauli principle and reflecting the shell-model configuration. The $0_6^+$ state also has a very large radius of 5.6 fm, though this value may be somewhat over estimated because of the 10 % too high energy of the $0_6^+$ state.

## 4. Conclusion and remarks

Multiple successful theoretical investigations, concerning the Hoyle state in $^{12}C$, have established, beyond doubt, that it is a dilute gas-like state of three $\alpha$-particles, held together only by the Coulomb barrier, and describable very well by a wave function of the THSR type, where the three bosons are condensed into the $0S$-orbital. There is no objective reason, why in $^{16}O$,$^{20}Ne$,$\cdots$ there should not exist similar "Hoyle"-like states. At least the calculations with THSR and OCM approaches show this to be the case, systematically. In this contribution, we give results of a complete OCM calculation which reproduces the six first $0^+$ states of $^{16}O$ to rather good accuracy. In that calculation the $0_6^+$ state at 16.5 MeV, to be identified with the experimental $0^+$ state at 15.1 MeV, shows the characteristics typical for the Hoyle state.

An interesting question is how many $\alpha$'s can maximally be in a self bound $\alpha$-gas state. In this respect, a schematic investigation in an $\alpha$-gas mean field calculation of the Gross-Pitaevskii type was performed [25]. Because of the increasing Coulomb repulsion, the Coulomb barrier fades away and our estimate yields a maximum of about eight $\alpha$-particles that can be held together in a condensate. However, a few extra neutrons can have a strong additional binding effect (see $^9$Be and $^{10}$Be [26, 27]) and may stabilize larger condensates.

We should remark that an interesting idea concerning $\alpha$-particle condensates was put forward by W. von Oertzen and collaborators [28, 29]. $\alpha$-particles outside a strongly bound core (e.g. $^{40}$Ca) can form a condensate at the multi-$\alpha$-particle threshold [28]. For the condensate with a fixed particle number, the emission of two $\alpha$'s and three $\alpha$'s must be enhanced. In fact the observation of the emission of $^{12}$C in the state from the compound nucleus $^{52}$Fe has been observed [30] and a very strong deviation from statistical model predictions is observed. Also interesting recent experimental works on loosely bound $\alpha$-structures in light nuclei have been performed [31, 32, 33].

## References

[1] K. Wildermuth and Y. C. Tang, *A Unified Theory of the Nucleus* (Vieweg, Braunschweig, 1977).
[2] K. Ikeda *et al.*, Prog. Theor. Phys. Suppl. No. 68, 1 (1980).
[3] G. Röpke *et al.*, Phys. Rev. Lett. **80**, 3177 (1998).
[4] P. Ring, and P. Schuck, *The Nuclear Many-Body Problem* (Springer-Verlag, Berlin, 1980).
[5] H. Horiuchi, Prog. Theor. Phys. **51**, 1266 (1974); **53**, 447 (1975).
[6] M. Kamimura, Nucl. Phys. A **351**, 456 (1981).
[7] E. Uegaki *et al.*, Prog. Theor. Phys. **57**, 1262 (1977); **59**, 1031 (1978); **62**, 1621 (1979).
[8] P. Descouvemont *et al.*, Phys. Rev. C **36**, 54 (1987).
[9] A. Tohsaki *et al.*, Phys. Rev. Let. **87**, 192501 (2001).
[10] Y. Funaki *et al.*, Phys. Rev. C **67**, 051306(R) (2003).
[11] Y. Funaki *et al.*, Eur. Phys. J. A **24**, 321 (2005).
[12] M. Chernykh *et al.*, Phys. Rev. Lett. **98**, 032501(2007); **105**, 022501(2010).
[13] Y. Funaki *et al.*, Phys. Rev. C **80**, 064326 (2009).
[14] H. Matsumura *et al.*, Nucl. Phys. A **739**, 238 (2004).
[15] T. Yamada *et al.*, Euro. Phys. J. A **26**, 185 (2005).
[16] Y. Funaki *et al.*, Eur. Phys. J. A **28**, 259 (2006).
[17] H. Horiuchi and K. Ikeda, Prog. Theor. Phys. **40**, 277 (1968); Y. Suzuki, Prog. Theor. Phys. **55**, 1751 (1976).
[18] T. Wakasa *et al.*, Phys Lett B **653**, 173 (2007).
[19] Y. Funaki *et al.*, Phys. Rev. Lett. **101**, 082502 (2008).
[20] Y. Suzuki *et al.*, Phys. Rev. C **65**, 064318, (2002).
[21] T. Yamada *et al.*, Phys. Rev. A **78**, 035603 (2008).
[22] Y. Funaki *et al.*, Nucl. Phys. News Lett., **17**(04), 11 (2007).
[23] F. Ajzenberg-Selove, Nucl. Phys. **A 460**, 1 (1986).
[24] Y. Funaki *et al.*, Phys. Rev. C **77**, 064312 (2008).
[25] T. Yamada, P. Schuck, Phys. Rev. C **69**, 024309 (2004).
[26] S. Okabe and Y. Abe, Prog. Theor. Phys. **61**, 1049 (1979).
[27] W. von Oertzen *et al.*, Phys Rep. **432**, 43 (2006).
[28] W. von Oertzen et al, Eur. Phys. J. A **29**, 133 (2006).
[29] Tz. Kokalova *et al.*, Phys. Rev. Lett. **96**, 192502 (2006).
[30] T. Kokalova *et al.*, Eur. Phys. J. A **23**, 19 (2005).
[31] T. Kawabata et al., Phys. Lett. B **646**, 6 (2007).
[32] M. Freer *et al.*, Phys. Rev. C **71**, 047305 (2005); **76**, 034320 (2007).
[33] Ad. R. Raduta *et al.*, arXiv:1004.3234; Int. J. Mod. Phys. E **20**, 902 (2011).

Journal of Physics: Conference Series **321** (2011) 012034 doi:10.1088/1742-6596/321/1/012034

# Experimental evidence for alpha-particle condensation from the Hoyle state deexcitation

**B. Borderie[1], Ad. R. Raduta[1,2], E. Geraci[3], N. Le Neindre[1,4], P. Napolitani[1], M. F. Rivet[1], the ISOSPIN Collaboration**

[1]Institut de Physique Nucléaire, CNRS/IN2P3, Université Paris-Sud 11, Orsay, France
[2]National Institute for Physics and Nuclear Engineering, Bucharest-Magurele, Romania
[3]INFN, Sezione di Catania and Dipartimento di Fisica e Astronomia, Università di Catania, Italy
[4]LPC, CNRS/IN2P3, Ensicaen, Université de Caen, Caen, France

E-mail: borderie@ipno.in2p3.fr

**Abstract.** Fragmentation of quasi-projectiles from the nuclear reaction $^{40}$Ca+$^{12}$C was used to produce excited states candidates to alpha-particle condensation. The methodology relies on high granularity $4\pi$ detection coupled to multi-particle correlations. Complete kinematic characterization reveals that 7.5±4.0% of the Hoyle state particle decays correspond to direct decay into three equal-energy alpha-particles. This is the first direct evidence for alpha-particle condensation in nuclei.

## 1. Introduction

Bose-Einstein condensation is known to occur in weakly and strongly interacting systems such as dilute atomic gases and liquid $^4$He. Since more than ten years it was also theoretically shown that for symmetric nuclear matter, below a critical density, $\alpha$-particle condensation is favored [1]. It is expected to occur at densities smaller than a fifth of the nuclear saturation density. This new possible phase of nuclear matter may have its counterpart in low-density states of self conjugate lighter nuclei, in the same way as superfluid nuclei are the finite-size counterpart of superfluid nuclear and neutron matter. This means that under some circumstances, the alpha condensation, i.e. bosonic properties, might dominate over the nucleon properties even in finite nuclei. Thus, by showing that the Hoyle state (i.e. the first excited state $0^+$ at 7.654 MeV of $^{12}$C) and the sixth $0^+$ state at 15.097 MeV of $^{16}$O are described by $\alpha$-particle condensate type functions, [2, 3] advance the idea that these states are candidates to observe $\alpha$ condensation. A common feature of these states is their diluteness. For instance, calculations [2] and recent data analysis [4] show that the rms radius of the Hoyle state exceeds by 45% the radius of $^{12}$C in its ground-state. According to the present understanding of the Hoyle state: a gas-like structure of 3 $\alpha$-particles which occupy dominantly the lowest S orbit, to obtain a direct evidence of $\alpha$-particle condensation we must search for a simultaneous emission of 3 $\alpha$-particles with very low kinetic energy dispersion. As it is experimentally difficult to identify and measure $\alpha$-particles with low kinetic energy, probably the most appropriate strategy should involve high velocity reaction products in the laboratory (to take advantage of velocity boosts) detected by a high granularity-high solid angle particle array (to precisely reconstruct the directions of velocity vectors).

## 2. Experimental selection of the Hoyle state

The data reported here have been obtained with the nuclear reaction $^{40}Ca+^{12}C$ at 25 MeV per nucleon incident energy performed at INFN, Laboratori Nazionali del Sud in Catania, Italy. The beam impinging on a thin carbon target (320 $\mu g/cm^2$) was delivered by the Superconducting Cyclotron and the charged reaction products were detected by the CHIMERA $4\pi$ multi-detector [5]. The beam intensity was kept around $10^7$ ions/s to avoid pile-up events. CHIMERA consists of 1192 silicon-CsI(Tl) telescopes mounted on 35 rings covering 94% of the solid angle, with very high detection granularity at forward angles. Details on A and Z identifications can be found in [5]. The energy resolution for alpha particles varies between 1.0 and 2.5%.

Invariant $v_{par} - v_{per}$ plots have shown the dominant binary character of the collisions with the formation of two emitting sources, a quasi-target (QT) and a quasi-projectile (QP). These latter may be easily separated according to the smaller or larger velocity of reaction products with respect to $v_{proj}/2$. After a first event filtering according to the total $\alpha$ multiplicity ($m_\alpha \geq 3$) and the total detected charge ($Z_{tot} \leq Z_P + Z_T = 26$), we have focused exclusively on QP decay products with $m_\alpha = 3$ to study in particular the Hoyle state deexcitation. Then, we have used

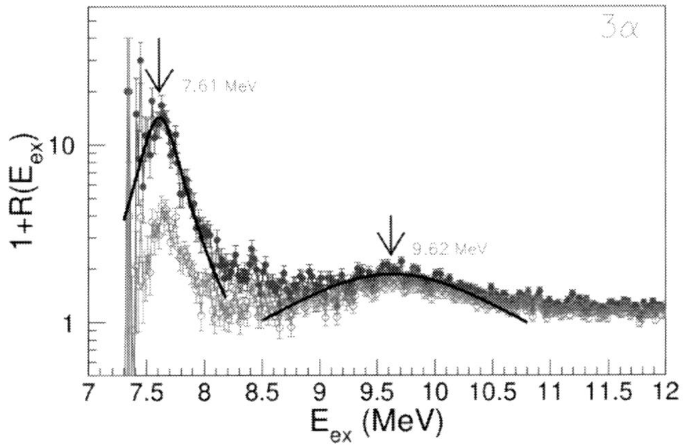


**Figure 1.** Correlation functions in excitation energy for 3-$\alpha$ emission. The distribution plotted with solid (open) circles corresponds to the case in which the uncorrelated yield is built by considering that each particle comes from a different event (that two particles come from the same event). The arrows correspond to centroids of Breit-Wigner distributions (solid lines).

multi-particle correlation functions (CF) [6, 7] to judge the quality of energy calibrations and select the Hoyle state. Multi-particle CF are defined as:

$$1 + R(E) = \frac{Y_{corr}(E)}{Y_{uncorr}(E)}, \tag{1}$$

The correlated yield $Y_{corr}$ corresponds to the physical spectrum, while the uncorrelated yield $Y_{uncorr}$ is built by mixing particles belonging to different physical events. The role of the generic variable $E$ is usually played by the total kinetic energy of the particles of interest in their center-of-mass frame $E_{tot}$ or by the excitation energy of their emitting source/state, $E_{ex} = E_{tot} - Q$. The CF plotted in figure 1 corresponds to the whole lot of events with $m_\alpha = 3$. The distribution plotted with full symbols correspond to the case in which the uncorrelated spectrum was built by taking each particle from a different event (full event mixing). It shows two peaks centered at $E_{ex}$=7.61 MeV ($\Gamma$=0.33 MeV) and at $E_{ex}$=9.62 MeV ($\Gamma$=1.14 MeV). The first peak corresponds to the Hoyle state ($E_{ex}^{exp}$=7.654 MeV, $\Gamma^{exp}$=8.5 eV), while the second one is due to the complex excited region characterized by the strong $E_{ex}^{exp}$=9.64 MeV, $3^-$ state and by the broad $E_{ex}^{exp}$=10.3 MeV, $0^+$ state submerging a possible $2^+$ state at 9.7 MeV [8, 9]. The broadening of correlation peaks is the genuine consequence of detector finite granularity and energy resolution. Numerical simulations indicate that the experimental distributions are compatible with an average energy resolution $R_E$ of 2%. The CF stability in the resonance regions was checked against the way in which the background is built. The open symbol curve corresponds to the case in which partial event mixing (PEM) [10] is employed in order to mimic sequential decays proceeding via $^8$Be. As one may notice, neither the centroids of the two peaks nor their widths are sizably modified. The CF peak height dependence on the recipe according to which the background is built is a first qualitative indication of a decay mechanism competition for the Hoyle state.

## 3. Evidence for alpha-particle condensation

We demonstrated so far that the $^{40}$Ca+$^{12}$C nuclear reaction at 25 MeV/nucleon populates excited states of $^{12}$C nuclei which decay by 3-$\alpha$ emission. We have shown in [11] that the complex excited region around 9.6 MeV deexcites via $^8$Be. Let us focus on the Hoyle state. One has to identify the direct decays and estimate their branching ratio. This may be done by comparing the experimental distribution with numerical simulations filtered by the multi-detector replica. We have considered $^{12}$C nuclei excited in the Hoyle state with the experimental velocity distribution and make them to decay by considering the following pure cases: direct alpha decay with equal energies (DDE), sequential decay proceeding via the g.s. of $^8$Be with isotropic emission out of $^8$Be (SD) and direct alpha decay from a linear chain (DDL) [12, 13]. In each situation and, for SD at each decay step, the available energy ($E_{ex}$+Q) is shared among the decay products such as to conserve the linear and angular momenta. Detection efficiencies are obtained and vary from 41 to 49% depending on the different decays. Analyses of filtered simulations show, as expected, very little sensitivity of $Y_{corr}(E_{ex})$ or, equivalently, $Y_{corr}(< E_\alpha >)$ on the decay mechanism even if the energy resolution is not perfect. Here $< E_\alpha >$ stands for the average alpha kinetic energy in the $^{12}$C center-of-mass. By contrast, the kinetic energy dispersion

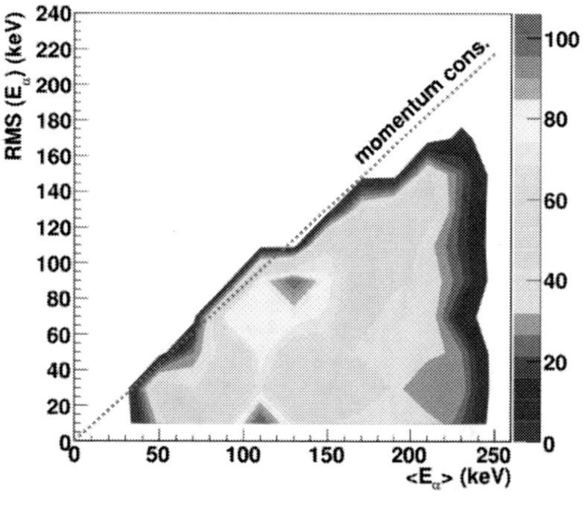


**Figure 2.** Three-$\alpha$ correlation function expressed as a function of average kinetic energy - RMS of $\alpha$ particles. The uncorrelated yield is built such as to allow for decay through $^8$Be.

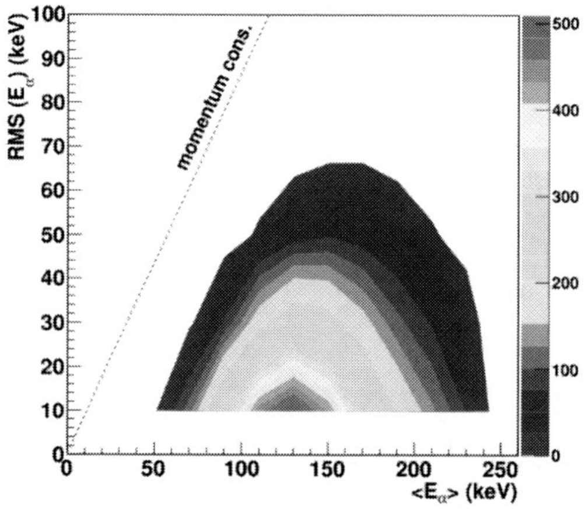


**Figure 3.** Same as figure 2 for simulated three-$\alpha$ (with equal energies) correlation function.

(RMS) in the emitter center-of-mass allows one to discriminate among the different decays. For the CHIMERA granularity and $R_E$=2%, $Y_{corr}$(RMS) are peaked at 30, 70 and 90 keV for DDE, SD and, respectively, DDL. For this reason, we adopt the best agreement between experimental and simulated $Y_{corr}(< E_\alpha >$,RMS) as the criterion to quantify each decay channel. For a better reproduction of the experimental data we add to the pure simulated events the same proportion of background events that exists in the data. These are produced by PEM of experimental events. The comparison is restricted to the excitation energy domain, $7.40 \leq E_{ex} \leq 7.80$ MeV to reduce the importance of background events to 40%. Applying a $\chi^2$ minimization procedure and correcting for efficiencies, the following contributions are derived: DDE = 7.5±4.0%, DDL = 9.5±4.0% and SD = 83.0±5.0%. Error bars are estimated by taking into account statistical, $\chi^2$ and background errors. Note that a limitation to DDE and SD as possible decays leads to a significant increase of the $\chi^2_{min}$ without changing the DDE percentage.

To better visualize the results, the CF expressed as function of $< E_\alpha >$ and RMS, for data, is plotted in figure 2. The relative separation in terms of RMS among DDE, SD and DDL allows one to identify, with the help of the simulation, the three competing mechanisms. The peak localized at very low RMS $\leq 25$ keV corresponds to an equal sharing of the available energy of the Hoyle state among the three $\alpha$ particles as demonstrated by the simulation (figure 3). The broad peak around $< E_\alpha > =$ 100-150 keV and RMS $\approx 70$ keV corresponds to the sharing of the available energy between the two $\alpha$s of $^8$Be and the remaining $\alpha$. Finally the peak at $< E_\alpha > =$ 120-140 keV and RMS = 90 keV corresponds to the direct decay of a linear chain with an $\alpha$ at rest and an equal sharing of the available energy of the Hoyle state between the two other $\alpha$ particles.

## 4. Conclusion

The Hoyle state is known to play a decisive role in stellar nucleosynthesis of $^{12}$C [14, 15, 16] with the assumption that the only contribution to the alpha-particle width comes from the $^8\mathrm{Be}_{g.s.}$ + $\alpha$ reaction. Up to now an upper limit of 4%, obtained from a low incident energy experiment, was proposed for the contribution from the direct $3\alpha$ channel to the alpha width [16]. Supposing that equal values of kinetic energy of the direct emitted $\alpha$-particles represent a sufficient criterion for deciding in favor of $\alpha$-particle condensation, we found that 7.5±4.0% of events corresponding to the Hoyle state decay fulfil this criterion. To our knowledge, this is the first direct proof of $\alpha$-particle condensation in nuclei. This work also demonstrated the presence of direct emission from a linear $\alpha$-chain.

## Acknowledgments

The authors are indebted to P. Schuck for numerous discussions and one of the authors Ad. R. R. acknowledges the partial financial support from ANCS, Romania, under grant Idei nr. 267/2007.

## References

[1] Röpke G, Schnell A, Schuck P and Nozieres P 1998 *Phys. Rev. Lett.* **80** 3177
[2] Tohsaki A, Horiuchi H, Schuck P and Röpke G 2001 *Phys. Rev. Lett.* **87** 192501
[3] Funaki Y, Yamada T *et al.* 2008 *Phys. Rev. Lett.* **101** 082502
[4] Ohkubo S and Hirabayashi Y 2004 *Phys. Rev.* C **70** 041602(R)
[5] Pagano A *et al* 2004 *Nucl. Phys.* A **734** 504
[6] Charity R J *et al* 1995 *Phys. Rev.* C **52** 3126
[7] Tăbăcaru G *et al* 2003 *Eur. Phys. J.* A **18**, 103
[8] Itoh M *et al* 2004 *Nucl. Phys.* A **738** 268
[9] Freer M *et al* 2010 *Nucl. Phys.* A **834** 621c
[10] Grenier F *et al* 2008 *Nucl. Phys.* A **811** 233
[11] Raduta Ad R *et al* 2011 *Int. J. Mod. Phys.* E **20** 902
[12] Matsumura H and Suzuki Y 2004 *Nucl. Phys.* A **739** 238
[13] Epelbaum E, Krebs H, Lee D and Meiβner U G 2011 *Phys. Rev. Lett.* in press (arXiv:1101.2547)
[14] Hoyle F, Dunbar D N F, Wenzel W A and Whaling W 1953 *Phys. Rev.* **92** 1095
[15] Cook C W, Fowler W A, Lauritsen C C and Lauritsen T 1957 *Phys. Rev.* **107** 508
[16] Freer M *et al* 1994 *Phys. Rev. C* **49** R1751

Journal of Physics: Conference Series **321** (2011) 012035 doi:10.1088/1742-6596/321/1/012035

# Signatures of $\alpha$-particle condensation in $N$=$Z$ nuclei

**W. von Oertzen**
Helmholtz Zentrum Berlin, Hahn-Meitner Platz 1, 14109 Berlin, Germany
E-mail: oertzen@helmholtz-berlin.de

**Abstract.** At the thresholds for multi-$\alpha$ particle decay a phase transition of second order appears in $N$=$Z$ nuclei, because the chemical potentials for nucleons in nuclear matter and in $\alpha$-particles become equal. The Bose gas created by the $\alpha$'s strongly interacting via $0^+$ states in $^8$Be and $^{12}$C$^*$, has coherent properties because of the fact that the de-Broglie wavelength is larger than the nuclear radius. Experimental observabels are discussed, which show these coherent properties, like large radial extensions and the coherent emission of several $\alpha$'s.

## 1. Alpha-particles in nuclei

The $\alpha$-particle is the most strongly bound cluster in nuclei, it's binding energy in nuclei follows the curve of binding energy per nucleon. The binding energy of nucleons in alphas is 7.03 MeV, a value which is somewhat smaller than in medium size nuclei (8.3 MeV/nucleon). Nuclei can have a certain probability of $\alpha$-clustering in their medium, as a mixed phase of nucleons and alphas [1, 2]. The nucleon and $\alpha$-binding energies depend on the excitation energy. At the thresholds for cluster decay (for multiple clusters of different size) a cluster-sub-structure [3] can appear, as predicted with the Ikeda diagram. The excitation energies, where the binding energy for only $\alpha$'s approaches zero, is at $E_x$ = 7.03 MeV/nucleon, at these values an $\alpha$-particle condensates will be formed in N=Z nuclei. Although for heavier ($N$=$Z$) nuclei these $\alpha$-condensed states are at rather high excitation energies in the continuum of nucleonic states, they may have collective properties [5, 6], which can give them a smaller observable width. The $\alpha$-particle condensed states have a larger radial extension due to the properties of the $\alpha$-$\alpha$ potential, a fact which can be observed in inelastic scattering to the excited $0^+$-states. For example in $^{16}$O we expect 5-6 such $0^+$ states, the state at 15.1 MeV having the most pronounced properties of a condensed state.

The basic equation for the formation of $\alpha$'s is the "reaction" of four "free" nucleons (two protons and two neutrons coupled to total values of spin and isospin of zero) :
$(N_1 + N_2 + N_3 + N_4) \longleftrightarrow \alpha$-particle + 28.3 MeV. The free nucleons, $N_i$, should have a definite volume and pressure, in order to define thermodynamic quantities and where the density allows the occurrence of the mentioned reaction. We can assume that the particles interact in a well defined volume created by a self consistent mean field for the $\alpha$-clusters (the Gross-Piatevski approach for bosons). For the case that the difference in chemical potentials vanishes a second order phase transition will occur.

This process occurs also in models like the anti symmetrized molecular dynamics, AMD [7]. A certain number of nucleons are confined in a volume with a positive kinetic energy, as suggested in Fig. 1. In this model a cooling method is applied to find the states of the lowest energy. In the AMD approach during the cooling process a certain $\alpha$-cluster phase is observed, before the formation of states with bound fermions, normal nuclei, with a binding energy per nucleon of 8.2 MeV or more, a value which is higher than in the $\alpha$-cluster (7.073 MeV). These two values define the difference in the chemical potentials in the two phases (Fig. 1). Starting from the ground states of normal nuclei the nucleons will form an $\alpha$-cluster phase with increasing temperature of the nucleus, *e.g.* with increasing excitation energy, see Fig. 1, ("heating"). We expect the decay of these excited states into many $\alpha$-particles, a decay not described by the Hauser-Fehsbach formalism for statistical compound nucleus decay (see below).

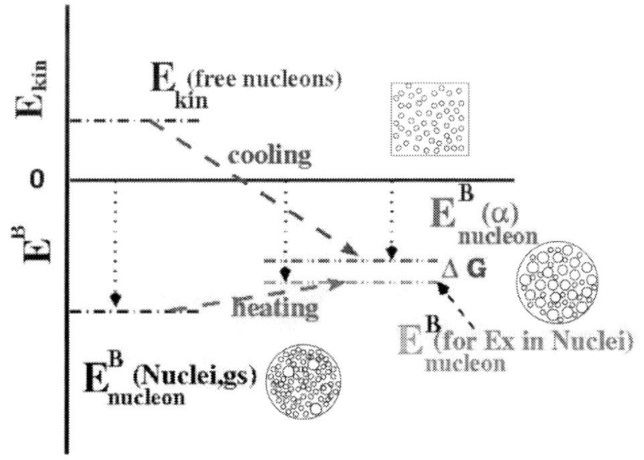


**Figure 1.** Schematic illustration of the relative values of the energies of free nucleons, their binding energies in nuclei (8.2 MeV), and in $\alpha$-clusters (7.07 MeV). The difference $\Delta$G between these two binding energies decreases in normal (N=Z) nuclei with increasing excitation energy, by ("heating"). At a critical value the two binding energies become equal, $\Delta G = 0$, a collective state of bosons (potentially mixed with fermions), the coherent $\alpha$-particle gas can be formed. In the AMD this state is approached by "cooling" from a fermion gas. From ref. [2]

## 2. Properties of $\alpha$-particle condensed states

In medium size nuclei (Z<20) the $\alpha$-condensates, calculated using the known alpha-alpha-potential with a self-consistent approach (based on the Gross-Piatevski equation [4]), will have Coulomb barriers for the decay into multiple $\alpha$'s. With these barriers the states will have sufficiently small width for potential studies by inelastic scattering, they can be excited via monopole transitions. However, the heaviest nucleus for which this barrier can create a quasi-bound state [4], is estimated to be around $^{40}$Ca. In heavier nuclei these states will be embedded high in the continuum of the fermionic states, their decay is expected to be non statistical, as the most characteristic property to study. In heavier nuclei we further expect the formation of condensates, with the $\alpha$-gas outside strongly bound cores, like $^{16}$O or $^{40}$Ca.

The coherent $\alpha$-particle emission should occur into the same (identical) angle. This will lead to the situation that the observation of unbound resonances becomes very enhanced, such as $^8$Be$(0^+, 2^+)$ and the excited states of $^{12}$C, the $^{12}$C$^\star(0_2^+, 0_3^+)$-clusters. This feature in fact has been observed in the recent data [2, 8, 9, 10], where the multi- $\alpha$-decay does not follow the independent decay steps predicted in compound decay by the Hauser-Fehsbach statistical theory.

Most important for the properties of the $\alpha$-particle gas is, that they do not represent the "ideal" gas, they interact via an interaction which has similarities with a van der Waals interaction, with a strongly repulsive core due to the Pauli principle. Two $\alpha$-particles form, as the lowest state, the ground state of $^8$Be, a resonance at $E_x^* = 92$ keV. We can calculate the *de Broglie wave length*, $\lambda = h/\sqrt{(2\mu E_x^*)}$, for this case we have $\lambda = 67$ fm (relative motion between the two $\alpha$-particles). If for higher excitation we incorporate the $2^+$ at 3.04 MeV the value of $\lambda$ is still 12.4 fm. Similarly for three $\alpha$-particles which form the Hoyle-state just above the three $\alpha$-particle threshold in $^{12}$C, the $0^+$ at 7.654 MeV (288 keV above the threshold), we again get a similarly large de Broglie wave length of relative motion. Overall we have values for $\lambda$ in the condensed states larger (by factors 2-5) then the radial extension of the nucleus. The condensed states at the binding energy threshold consisting of $\alpha$-particles will form coherent super-fluid states. The calculations of THSR-wave function based on a local $\alpha$-$\alpha$ potential reproduce the states of $^8$Be, and the threshold states in other light nuclei. The local potentials for the system of $^{16}$O+$\alpha$-particle, supports the formation of $\alpha$-condensates with a $^{16}$O-core, but also with a $^{40}$Ca-core (e.g. $^{52}$Fe $=$ $^{40}$Ca+$6\alpha$) [10, 11]. The coherent decay of an $\alpha$-gas will be identified by the fact that the $\alpha$-particles are emitted preferentially into the same angle, and by the enhanced formation of $^8$Be and the $^{12}$C$^*(0_2^+)$ states.

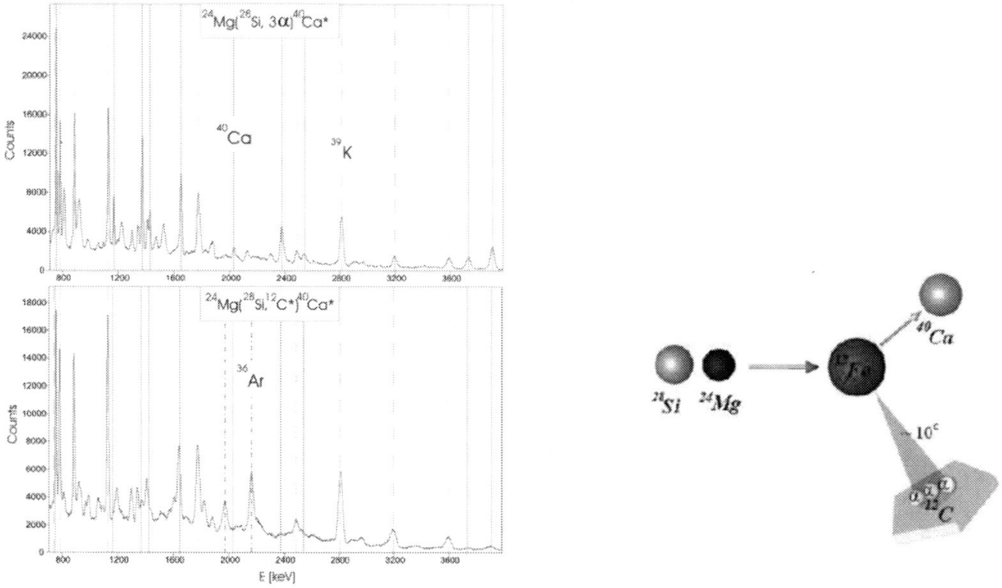


**Figure 2.** Coincident $\gamma$-spectra gated with the $\alpha$-particles from $\Delta E$-$E$-telescopes with the emission of three random $\alpha$'s at different angles in different detectors upper part, in comparison with that obtained by the triple $\alpha$ emission with $^{12}$C$^*_{0+}$-gate (lower panel). The reaction is $^{28}$Si $+^{24}$ Mg $\rightarrow^{52}$ Fe $\rightarrow^{40}$ Ca $+ 3\alpha$ at 130 MeV. Note the additional lines for $^{36}$Ar in the lower panel, which indicates that a 4-rth $\alpha$ has been emitted (courtesy of Tz. Kokalova). Right side: The kinematics for the triple $\alpha$, the $^{12}$C$*_{0+}$-observation.

## 3. Coherent multiple emission of $\alpha$-particles

The best way to study such decays is the combination of multi-detector arrays for particle detection with $\Delta E$-$E$ detectors and a "calorimeter" to observe the remaining compound nucleus residue via its $\gamma$-decay [2]. Such experiments have been performed with the large $\gamma$-detector array GASP at the Legnaro National Laboratory LNL at Padua (Italy), combined with the charged particle detector ball ISIS (details are given in ref. [9]) consisting of 42 $\Delta E$-$E$- telescopes. These experiments were performed in a study of $\gamma$-decays of compound nuclei selected with a

Journal of Physics: Conference Series **321** (2011) 012035 doi:10.1088/1742-6596/321/1/012035

particular choice particle decays [8]. The large opening angle of the individual ISIS-$\Delta$E-E telescopes, which was 27°, allows to select the spontaneous decay of the weakly unbound states, namely of $^{8}$Be into two $\alpha$'s and the $^{12}$C$^{*}$($0_2^+$ decaying into three $\alpha$-particles. With the rather modest kinetic energy of these fragments and the small decay energies of a few 100 KeV, the opening angles between the $\alpha$'s are in the range of 10° - 25°, which fit into these solid angles of individual telescopes. Therefore these prompt multiple $\alpha$-decays are observed by the pile-up of the signals produced by individual three alpha-particles. In Fig. 2 the corresponding coincident (particle gated) $\gamma$-decays are compared with the spectra obtained from statistical emission (with the same $\alpha$-multiplicity), but into different angles of the $\Delta$E-E telescopes [9].

The observation of the emission of a 4-th $\alpha$, contradicts strongly to the prediction of the statistical compound decay (3 $\alpha$'s, Hauser-Fehsbach prediction), the effect is due to a strongly lowered barrier for the $\alpha$-particle emission, which is due to the larger diffusivity and the larger radial extension of the $\alpha$-condensed state. This fact has been reproduced in calculations by Kokalova, Itagaki et al. in ref. [10], where the $\alpha$-folding model for a $\alpha$-condensed state with a $^{40}$Ca core has been performed. The result is strongly lowered emission barrier (by 10 MeV) for $^{12}$C$^{*}$($0_2^+$ emission.

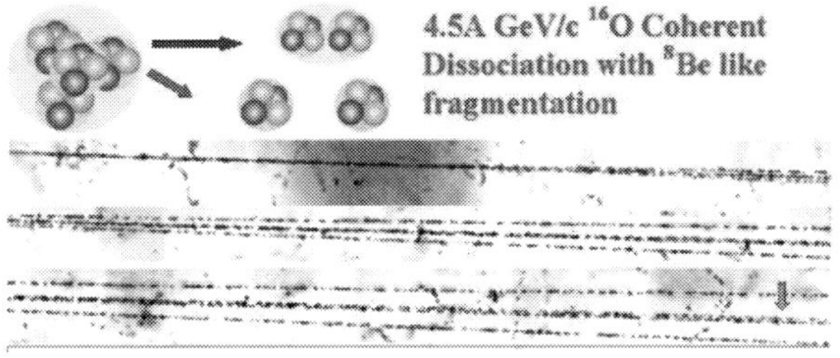


**Figure 3.** Break up of $^{16}$O into $\alpha$'s and $^{8}$Be at relativistic energies, courtesy of P.Zarubin [12]

Another spectacular effect is observed in the collective decay into $\alpha$'s and $^{8}$Be after monopole excitation in high energy scattering of $^{16}$O. This measurement shown in Fig. 3 has been obtained at 4.5 GeV/N at Dubna [12]. using emulsions. With a clear visibility of tracks at small relative angles. The break-up occurs via Coulomb excitation at the smallest angles on a Ag-nucleus in the emulsion A sequence of three panels shows the total track.

## 4. References

[1] W. von Oertzen, Eur. Phys. J. A **29**, 133 (2006).
[2] W. von Oertzen, in Lecture Notes in Physics, Vol.**818**, ed. C Beck, p.109 (2011).
[3] H. Horiuchi, K. Ikeda, Prog. Theor. Phys. (Japan) A **40**, 277 (1968).
[4] T. Yamada and P. Schuck, Phys. Rev. C **69**, 024309 (2004).
[5] P. Schuck, Y. Funaki, H. Horiuchi *et al.*, Nucl. Phys. A **738**, 94 (2004).
[6] Y. Funaki, H. Horiuchi, W von Oertzen *et al.*, Phys. Rev. C **80**, 064326 (2009).
[7] H. Horiuchi and Y. Kanada-En'yo, Nucl. Phys. A **616**, 394c (1997).
[8] S. Torilov *et al.*, Eur. Phys. J. A **19**, 307 (2004).
[9] Tz. Kokalova *et al.*, Eur. Phys. J. A **23**, 19 (2005).
[10] Tz. Kokalova, N. Itagaki *et al.*, Phys. Rev. Lett., **96**, 192502 (2006).
[11] N. Itagaki, Tz. Kokalova, W. von Oertzen, Phys. Rev. C **82**, 014312 (2010).
[12] P. Zarubin *et al.*, private communication (2009)

# Excited states of $^{12}C$ above the alpha-decay threshold

**M. Freer,[1] S. Almaraz-Calderon,[2] A. Aprahamian,[2] N.I. Ashwood,[1] M. Barr,[1] B. Bucher,[2] P. Copp,[3] M. Couder,[2] N. Curtis,[1] X. Fang,[2] F. Jung,[2] S. Lesher,[3] W. Lu,[2] J.D. Malcolm,[1] A. Roberts,[2] W.P. Tan,[2] C. Wheldon,[1] and V.A. Ziman[1]**
[1]School of Physics and Astronomy, University of Birmingham, Birmingham, B15 2TT, UK.
[2]Institute for Structure and Nuclear Astrophysics, Department of Physics, University of Notre Dame, Notre Dame, IN 46556, USA.
[3]Department of Physics, University of Wisconsin - La Crosse, La Crosse, WI 54601, USA.

E-mail: M.Freer@bham.ac.uk

**Abstract**. The excitation energy spectrum of $^{12}C$ is important for both structural and astrophysical reasons; here we present evidence for a new state in $^{12}C$. The two reactions $^{12}C(^4He,^4He+^4He+^4He)^4He$ and $^9Be(^4He,^4He+^4He+^4He)n$ were measured using an array of four double sided strip detectors. Excited states in $^{12}C$ were reconstructed filtered by the condition that the alpha-decay proceeded via the $^8Be$ ground-state. In both measurements evidence was found for a new state at 13.3(0.2) MeV with a width 1.7(0.2) MeV. Angular correlation measurements from the $^{12}C(^4He,^4He+^4He+^4He)^4He$ reaction indicates that the state may have $J^\pi = 4^+$.

## 1. Introduction

Pinning down the structure of light nuclei is a rather important task as it provides the anchor-points for our understanding of the nucleon-nucleon interaction which governs their detailed properties. Such measurements constrain ab initio calculations whereby nuclear properties are calculated from the starting point of the nucleon-nucleon interaction. Correlations between the nucleons have important contributions to the nuclear binding and in light systems can effects which dominate structurally. One manifestation of such correlations is the appearance of clustering. The $^4He$ nucleus is a maximally correlated system, possessing two protons and two neutrons in the $0s_{1/2}$ orbit, which is responsible for the very high binding energy and a first excited state in excess of 20 MeV. The intriguing possibility is that such correlations persist across the spectrum of light nuclei. The nucleus $^8Be$ is well within the reach of ab initio calculations and its properties have been calculated within a number of different frameworks. The Antisymmetrized Molecular Dynamics (AMD) approach[1] which simulates the interaction of nucleons from the starting point of 8 Gaussian wave-packets and an effective interaction clearly indicates that the nuclear density separates out into one which reflects the presence of two alpha-clusters in the ground-state. Here the clusters are not explicitly entered into the calculations, but rather grow from the details of the interaction used. The Greens Function Monte Carlo (GFMC) calculations have a starting point of a parameterized two body nucleon-nucleon interaction and a fitted 3-body component [2]. In this way the interaction is not an effective interaction as appears in the AMD approach. Remarkably, the GFMC approach also reveals a $2\alpha$ cluster structure in the ground-state. This result is an impressive demonstration of what was already known for many years experimentally – that the $^8Be$ ground-state is highly clustered. The ground-state is unbound to alpha-decay and has a width which indicates strong preformation of the alpha-particles. Further, the rotational band associated with the ground-state has a moment of inertia commensurate with such a cluster structure.

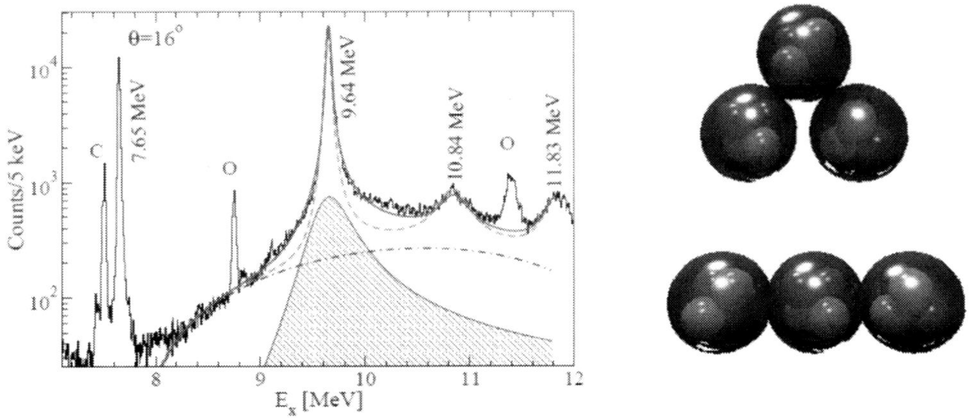


**Fig. 1**. Left-hand-side. $^{12}C$ excitation energy spectrum showing the broad $2^+$ state required to reproduce the shape of the experimental spectrum, from Ref. 8. Right-hand-side. Possible arrangements of the alpha-clusters in either a triangle (top) or chain (bottom). The experimental data indicate that the triangular structure is preferred.

## 2. Structure of $^{12}C$

The complexity of the GFMC approach prohibits its extension currently to nuclei with A>12, and $^{12}C$ currently the focus of the calculations. In this nucleus it is excited states, as opposed to the ground-state, which is the focus in terms of alpha-cluster calculations. In fact, the state of interest is the 7.65 MeV, $0^+$, Hoyle-state. The Hoyle-state because its existence was proposed by Sir Fred Hoyle some 50 years ago [3]. This state is the gateway through which carbon is synthesized via what is known as the triple-alpha process. Without this state there would be no organic life. Calculations of the properties of the Hoyle-state using the GFMC approach are still on-going. However, calculations using the AMD and its cousin the FMD approach (Fermionic Nuclear Dynamics) [4] reveal a highly clustered state. Its' very dilute nature, together with the 3alpha cluster structure has been interpreted in terms of an alpha-condensate by Peter Schuck and co-workers [5], this would be a new bosonic state of nuclear matter. Very recent calculations using the framework of chiral effective field theory have been able to closely reproduce the energy of the Hoyle-state [6]. Although they offer little detailed structural information they do suggest a larger charge radius than the ground-state, as observed experimentally. As with the $^{8}Be$ ground-state, the width of the 7.65 MeV state indicates a significant cluster structure and in agreement with theoretical predictions a large charge radius (see Ref. [4]) indicating a volume which is 3-4 times that of the $^{12}C$ ground state.

Revealing the collective excitations of such a state could have significant implications for understanding its structure. In the alpha-condensate model this would correspond to a $2^+$ state which is associated with the conversion of one of the bosons from s-wave to d-wave. In a more traditional cluster model it would correspond to a collective rotation and the moment of inertia would then reveal the details of the cluster arrangement.

## 3. Experimental Measurements

Recent studies of the $^{12}C(\alpha,\alpha')$ [7] and $^{12}C(p,p')$ [8,9] (see Fig. 1) reactions indicate the presence of a $2^+$ state close to 9.6-9.7 MeV with a width of 0.5 to 1 MeV. The state is only weakly populated in these reactions, presumably due to its underlying cluster structure, and is broad. Consequently, its distinction from other broad-states and dominant collective excitations (e.g. the 9.6 MeV, $3^-$) makes its unambiguous identification challenging. Further evidence for such an excitation comes from measurements of the $^{12}C(\gamma,3\alpha)$ reaction performed at the HIGS facility, TUNL. Here a measurable cross section for this process was observed in the region of 9-10 MeV which cannot be attributed to known states in this region. Furthermore, the angular distributions of the alpha-particles are consistent with an L=2 pattern, indicating a dominant $2^+$ component. Based on a rather simple description of this state in terms of three alpha-particles with radii given by the experimental charge radius, it is possible to use the 2 MeV separation between the Hoyle-state and the proposed $2^+$ excitation to draw some conclusions as to the arrangements of the clusters (see Fig. 1). This would indicate that rather than a linear arrangement of the three clusters that a more appropriate description would be a loose arrangement of the alpha-particles in something approaching a triangular arrangement – i.e. the Hoyle-state would be a monopole excitation of the ground-state.

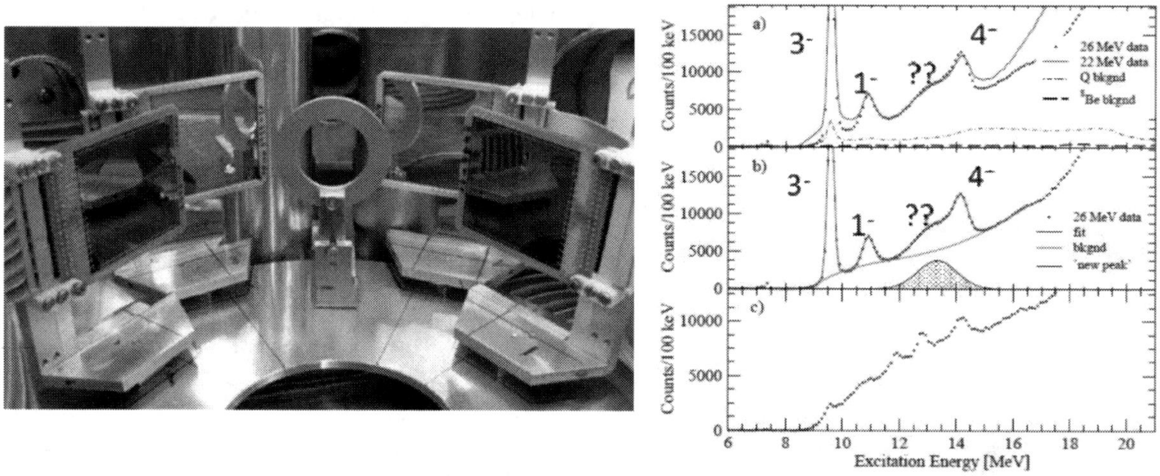


**Fig. 2**. Left-hand-side. Arrangement of the 4 silicon strip detectors used in the measurements in Ref. [11]. The beam passes from bottom to top in this picture. Right-hand-side. Carbon-12 excitation energy spectra. a) The blue line shows the measurement at 22 MeV. The spectrum corresponding to measurements at 26 MeV is shown by the dots. The backgrounds obtained by gating above the $^8$Be peak (bold dashed line) and both above and below the Q-value peak (dot-dashed line) are both illustrated. b) Fit to the 26 MeV data is given by the blue solid line. The polynomial background (red-line) and line-shape for the new peak (shaded area) is shown. c) Excitation energy spectrum for events not proceeding via the decay to the $^8$Be ground-state. The proposed new state is indicated by "??".

A natural extension of such a conclusion is that there should also be a collective $4^+$ state. Using the simple $j(j+1)$ scaling, a $4^+$excitation close to $E_x(^{12}C)$ = 14 MeV would be expected. We have performed recent measurements of the two reactions $^9Be(\alpha,3\alpha)n$ and $^{12}C(\alpha,3\alpha)^4He$ [11]. In these measurements three alpha-particles were detected in an array of four silicon strip detectors (shown in Fig 2). The analysis required that two of the three alpha-particles came from the decay of the ground-state of $^8$Be. For the decay of $^{12}$C to $^8$Be+$\alpha$ this ensures that the decay process can proceed through only natural parity states (i.e. $0^+$,$1^-$, $2^+$ ....). This restricts the complexity of the excitation energy spectrum. The measurements for both the $^9$Be and $^{12}$C targets reveal the known $3^-$, $1^-$ and $4^+$ states at

9.64, 10.84 and 14.08 MeV, respectively. However, there is an additional component to the spectrum close to 13.3 MeV with a width estimated to be 1.7 MeV (Fig. 2). It is believed that this is not a contaminant and is observed with similar properties in all spectra. Angular correlation measurements made using the $^{12}C$ target are not definitive, but indicate a $4^+$ assignment.

Such measurements are challenging and it is difficult to be absolutely certain that the feature does not correspond to an experimental artifact. Nevertheless, the appearance of a $4^+$ state close to the energy predicted based on the extension of the Hoyle-state and proposed $2^+$ excitation. Such a feature is also observed in measurements of the neutron energy produced in the $^9Be(\alpha,3\alpha)n$ reaction [12]. However, it is clear that further measurements are required to confirm, or otherwise, this new state in $^{12}C$. Moreover, analysis of beta-decay measurements populating alpha-decay states in $^{12}C$ indicate that there may be broad $2^+$ strength in $^{12}C$ but at somewhat higher energies than suggested in the proton and alpha inelastic scattering measurements [13]. This contradiction between the two types of measurements remains to be reconciled and hence a firm conclusion regarding the excitation structure of $^{12}C$ above the alpha-decay threshold remains to be established.

It is abundantly clear that the resolution of the structure of the Hoyle-state and the nature of its excitations is of prime importance in nuclear physics as it is simultaneously central to both the understanding of clustering in light nuclear systems and testing ab initio models of nuclei. The renewed interest in this state has been driven by the authors of Reference [5] and in particular Peter Schuck, and the field owes them a great debt.

**References**

[1] Kanada En'yo Y, Horiuchi H 2001 *Prog. Theo. Phys. Suppl.* **142**, 205
[2] Wiringa R B, Pieper S C, Carlson J and Pandharipande V R 2000 *Phys. Rev.* C **62** 014001
[3] Hoyle F 1954 *The Astrophysical Journal, Supplement Series* Vol. 1 p. 12
[4] Chernykh M, et al. 2007 *Phys. Rev. Lett.* **98** 032501
[5] Tohsaki A, et al. 2001 *Phys. Rev. Lett.* **87** 192501
[6] Epelbaum E, Krebs H, Lee D and Meißner U-G 2011 *Phys. Rev. Lett.* **106** 192501
[7] Itoh M, et al. 2004 *Nucl. Phys.* A **738** 268
[8] Freer M, et al. 2009 *Phys. Rev.* C **80** 041303
[9] Gai M, et al. *submitted to Phys. Rev. C*
[10] Gai M for the UConn-Yale-Duke-Weizmann-PTB-UCL Collaboration 2011 *In press, APH Pol* B, March.
[11] Freer M, Almaraz-Calderon S, Aprahamian A, Ashwood N I, Barr M, Bucher B, Copp P, Couder M, Curtis N, Fang X, Jung F, Lesher S, Lu W, Malcolm J D, Roberts A, Tan W P, Wheldon C and Ziman V A 2011 *Phys. Rev.* C **83** 034314
[12] Aslanoglou X and Kemper K W 1986 Phys. Rev. C **34** 1649
[13] Hyldegaard S, et al. 2010 *Phys. Rev.* C **81** 024303

Journal of Physics: Conference Series **321** (2011) 012037 doi:10.1088/1742-6596/321/1/012037

# Lattice EFT calculation of thermal properties of low-density neutron matter

**T. Abe**[1] **and R. Seki**[2]

[1] Department of Physics, the University of Tokyo, Hongo, Tokyo 113-0033, Japan

[2] Department of Physics and Astronomy, California State University, Northridge, California 91330, USA

E-mail: `tabe@nt.phys.s.u-tokyo.ac.jp`

**Abstract.** Thermal properties of low-density neutron matter are investigated by lattice calculation with nuclear effective field theory without pions up to the next-to-leading order. The $^1S_0$ pairing gap is extracted near zero temperature at low densities. We find that the pairing gap is smaller than the BCS approximation with the conventional $NN$ potentials, but not as small as those by various many-body calculations beyond BCS approximation. Our result is consistent with the recent Green's Function Monte Carlo calculation within the statistical errors. The critical temperature of the normal-to-superfluid phase transition and the pairing temperature scale are also extracted at low densities, and the phase diagram is given. We find that the physics of low-density neutron matter is clearly identified as being BCS-BEC crossover.

*Dedicated to Peter Schuck, in celebration of his retirement*

## 1. Introduction

Even though the investigations over many years provided much understanding of the pairing gap $\Delta$, reliable quantitative information has not been fully available, even in the $^1S_0$ channel of the partial wave decomposition, which is the dominant component in low-density neutron matter [1]. For example, $\Delta$ had been firmly established *in the BCS approximation*, as evident in the fact that various conventional $NN$ potentials have provided nearly the same $\Delta$ as a function of density. Many-body calculations beyond the BCS approximation, however, have yielded $\Delta$ of various magnitudes, generally smaller than BCS $\Delta$. Quantum Monte Carlo calculations have also been used for the determination of $\Delta$. The Green's function Monte Carlo (GFMC) method has yielded $\Delta$ smaller than BCS $\Delta$ [2], but not as small as those obtained by the many-body calculations beyond the BCS approximation. Another method closely related to GFMC, the auxiliary field diffusion Monte Carlo (AFDMC) method has given $\Delta$ quite close to BCS $\Delta$ [3]. Our motivation is to determine $\Delta$ by a quantum Monte Carlo lattice calculation with the nuclear Effective Field Theory (EFT), which is motivated by the work of Ref. [4] and has been now widely applied to a few- and many-body systems [5, 6, 7, 8, 9, 10, 11, 12, 13]. This contribution of the proceedings is the summary based on our work in Refs. [14, 15].

## 2. Low-density neutron matter on the lattice

### *2.1. Lattice Hamiltonian up to the next-to-leading order of pionless EFT*

Nuclear EFT is a low-energy effective theory of Quantum Chromodynamics (QCD) [16, 17, 18, 19]. It holds all the symmetries of low-energy QCD such as the approximate chiral symmetry

Journal of Physics: Conference Series **321** (2011) 012037 doi:10.1088/1742-6596/321/1/012037

and its breaking. Below the chiral symmetry breaking scale $\Lambda_\chi \sim 1$ GeV, pions and nucleons are the relevant degrees of freedom to describe the low-energy theory. At low energies sufficiently below the pion mass, even the pions can be integrated out from the theory, the relevant degrees of freedom are only nucleons. At that energy scale, we can describe low-energy nuclear physics by a low-energy effective theory without the pion degrees of freedom referred to as pionless EFT.

Starting with the general second-quantized Hamiltonian with the contact and its derivative interaction terms in the continuous coordinate space, the pionless EFT Hamiltonian up to the next-to-leading order (NLO) can be discretized on the lattice as

$$\begin{aligned}\hat{H} &= -t\sum_{\langle i,j\rangle\sigma}\hat{c}^\dagger_{i\sigma}\hat{c}_{j\sigma} + 6t\sum_{i\sigma}\hat{c}^\dagger_{i\sigma}\hat{c}_{i\sigma} \\ &+ \frac{1}{a^3}\left[c_0(\Lambda) - \frac{6}{a^2}c_2(\Lambda)\right]\sum_i \hat{c}^\dagger_{i\uparrow}\hat{c}^\dagger_{i\downarrow}\hat{c}_{i\downarrow}\hat{c}_{i\uparrow} + \frac{1}{2a^5}c_2(\Lambda)\sum_{\langle i,j\rangle\sigma\sigma'}\hat{c}^\dagger_{i\sigma}\hat{c}_{i\sigma}\hat{c}^\dagger_{j\sigma'}\hat{c}_{j\sigma'},\end{aligned} \tag{1}$$

with the hopping amplitude $t \equiv 1/(2Ma^2)$, the neutron mass $M$, the spatial lattice spacing $a$, and the momentum cutoff scale $\Lambda$ set to be $\pi/a$. $\langle i,j\rangle$ indicates that the summation is restricted to the neighboring pairs of $i$ and $j$. $\hat{c}^\dagger_{i\sigma}$ ($\hat{c}_{i\sigma}$) is the creation (annihilation) operator of the neutron with spin $\sigma = \uparrow, \downarrow$ at the $i$th site.

$c_0$ and $c_2$ are the interaction strengths of the contact and its derivative terms, respectively. As explained later in Sec. 2.2, these coupling constants are determined by solving the coupled equations, Eqs. (3) and (4), so as to reproduce the experimental values of the scattering length and effective range.

The lattice Hamiltonian, Eq. (1), corresponds to the extended Hubbard-model Hamiltonian in condensed matter physics [20]. We can apply the Determinantal Quantum Monte Carlo (DQMC) method of the attractive Hubbard model to the low-density neutron matter system on the lattice. After determining the coupling constants, we can perform the DQMC lattice calculation to include the non-perturbative effects or neutron loop contributions as introduced briefly in Sec. 2.3.

### *2.2. Determination of the coupling constants, $c_0$ and $c_2$*

The cutoff-dependent neutron-neutron interaction parameters, $c_0(\Lambda)$ and $c_2(\Lambda)$, in Eq. (1) are determined from the neutron-neutron scattering phase shift, using the effective range expansion up to the $p^2$ terms in $^1S_0$ channel,

$$p\cot\delta_0(p) = -\frac{1}{a_0} + \frac{1}{2}r_0p^2 + \mathcal{O}(p^4), \tag{2}$$

with the $^1S_0$ phase shift $\delta_0$, scattering length $a_0$, effective range $r_0$, and relative momentum of two neutrons $p$.

$a_0$ and $r_0$ can be related to $c_0$ and $c_2$ through the coupled equations [21],

$$\frac{M}{4\pi}\frac{1}{a_0} = \left[\frac{1}{c_0(\Lambda)} + \frac{M}{2\pi^2}L_1(\Lambda)\right] + \frac{M}{\pi^2}L_3(\Lambda)\frac{c_2(\Lambda)}{c_0(\Lambda)} \tag{3}$$

$$\frac{M}{16\pi}r_0 = \frac{c_2(\Lambda)}{c_0^2(\Lambda)} - \frac{M}{4\pi^2}\frac{1}{\Lambda}R\left(\frac{p^2}{\Lambda^2}\right), \tag{4}$$

where $L_1$, $L_3$, and $R$ are the quantites related to the choice of the regulator in the momentum integrations. By solving the coupled equations, Eqs. (3) and (4), we can obtain the interaction parameters on the lattice, $c_0$ and $c_2$, from the experimental values of $a_0$ and $r_0$. Note that $c_2$ should be treated perturbatively to keep the power counting rule. The details of the determination of $c_0$ and $c_2$ can be found in Ref. [21].

### *2.3. Determinantal Quantum Monte Carlo*

For the evaluation of the observables, we compute the correlation functions by DQMC method [22]. The essential idea of DQMC is to describe and to compute the observables by the Green's function. The Green's function is the fundamental tool of DQMC method, not only for the computation of the observables, but also for the update process of Monte Carlo sampling. For the sake of simplicity, we restrict our discussion to the leading order (LO) case. The concrete introduction including the NLO terms perturbatively can be found in Ref. [14].

By applying the Hubbard-Stratonovich transformation, the expectation value of the observable is written in the form of the functional integral over the auxiliary fields $\chi$,

$$\langle \hat{O} \rangle \equiv \frac{1}{Z} \mathrm{Tr}\left(\hat{O}\hat{U}\right) = \frac{1}{Z}\int d[\chi]\langle \hat{O}(\chi)\rangle G(\chi)\xi(\chi), \tag{5}$$

with the grand partition function,

$$Z \equiv \mathrm{Tr}(\hat{U}) = \int d[\chi] G(\chi)\xi(\chi), \tag{6}$$

and the evolution operator,

$$\hat{U} \equiv \exp\left[-\beta\left(\hat{H} - \mu\hat{N}\right)\right] \approx \left\{\exp\left[-\Delta\beta\left(\hat{H} - \mu\hat{N}\right)\right]\right\}^{N_t} = \int d[\chi] G(\chi)\hat{U}_{\chi}, \tag{7}$$

with the inverse temperature $\beta \equiv 1/T$, the chemical potential $\mu$, and the number operator $\hat{N}$. In the second equality, $\hat{U}$ (and also $\hat{U}_{\chi}$) is discretized into $N_t \equiv \beta/\Delta\beta$ time slices by using the Torroter-Suzuki approximation. $G(\chi)$ is the Gaussian factor. $\xi(\chi)$ is the trace of the one-body evolution operator $U_{\chi}$, and is related to the Green's function $G_{ij}$ through $U_{\chi}$ in matrix form as $(\mathbf{U}_{\chi})_{ij} = (\mathbf{G}^{-1})_{ij}$. $\langle \hat{O}(\chi)\rangle$ is defined as $\mathrm{Tr}(\hat{O}\hat{U}_{\chi})/\xi(\chi)$.

The functional integral, Eq. (5), is evaluated by Monte Carlo integration with the Metropolis algorithm, regarding $G(\chi)\xi(\chi)$ as the weight,

$$\langle \hat{O} \rangle = \frac{\sum_{i=1}^{N}\langle \hat{O}\rangle_i \Phi_i}{\sum_{i=1}^{N}\Phi_i}, \tag{8}$$

with the sign of the weight $\Phi_i \equiv G(\chi)\xi(\chi)/|G(\chi)\xi(\chi)|$ and the number of samples $N$.

## 3. Results

### *3.1. Setup*

The DQMC framework is based on the grand canonical ensemble. To measure the observables at some fixed neutron number, the neutron number for the observables that we compute is determined by interpolation of the observables for various values of chemical potential. To take the thermodynamic and continuum limits, we perform the calculations with the spatial lattice sizes $N_s$ of $4^3$, $6^3$, $8^3$, and $10^3$, and the filling fraction of neutrons in the cubic box $n$ ranging from 1/16 to 1/2. By using these results in $N_s$ and $n$, the final observables are extrapolated in the thermodynamic ($N_s \to \infty$) and continuum limits ($n \to 0$). To observe the phase transition, we set the temporal lattice sizes $N_t$ ranging from 2 to 128.

### *3.2. $^1S_0$ pairing gap near zero temperature*

$\Delta$ is extracted directly from the off-diagonal long-range order of the spin pair-pair correlation function $P_s$ [23],

$$P_s(R) = \frac{1}{N_s}\sum_i \langle \hat{\Delta}^{\dagger}_{i+R}\hat{\Delta}_i\rangle = \frac{1}{N_s}\sum_{i,\, j=i+R}(\delta_{ij} - G_{ji})^2, \tag{9}$$

where $\hat{\Delta}_i \equiv \hat{c}_{i\uparrow}\hat{c}_{i\downarrow}$ is the two-neutron spin-pairing operator at the $i$-th site, and $R$ is the separation of the neutron pairs in the lattice spacing unit. $G_{ij} \equiv G^{\sigma}_{ij} = \langle \hat{c}_{i\sigma}\hat{c}^{\dagger}_{j\sigma}\rangle$ is the equal-time Green's function for $\sigma = \uparrow, \downarrow$. From the diminishing asymptotic behavior of $P_s$ at sufficiently large $R$, $\Delta$ near zero temperature is evaluated by $\Delta = \frac{|c_0|}{a^3}\sqrt{P_s(T \approx 0, R \gg 1)}$.

Figure 1 shows comparison of our $\Delta$ at low densities with other selected calculations in the literature. In the figure, the density $\rho$ is described by the Fermi momentum $k_F$ through the relation in the system of non-interacting neutrons, $k_F = (3\pi^2\rho)^{1/3}$. The solid curve is the typical BCS result with realistic two-nucleon interactions. Note that all BCS results with various realistic $NN$ forces fall into this curve. The other curves are some selected many-body calculations beyond BCS approximation. Calculations in random phase approximation are labeled by R1 [24], R2 [25] and R3 [26], those with correlated basis functions by C1 [27] and C2 [28], and the renormalization group method by RG [29]. These results are scattered, but show the trend of the drastic suppression of $\Delta$ from the BCS results, except for R3 [26] and C2 [28]. Symbols are the quantum Monte Carlo results. The open squares are the recent GFMC [2], the open circles are the AFDMC [3], and the open diamonds are our DQMC lattice results [14]. It has been pointed out that the AFDMC results may not have been converged to the ground state and would become consistent with the GFMC and our results by using an appropriate initial wave function [30]. These quantum Monte Carlo results imply that $\Delta$ is suppressed with respect to the BCS results, but is not much smaller than BCS results as predicted by most of the other many-body calculations beyond BCS approximation. Note that the recent path integral Monte Carlo calculation seems to confirm our $\Delta$ at $k_F \sim 0.46$ fm$^{-1}$ [31].

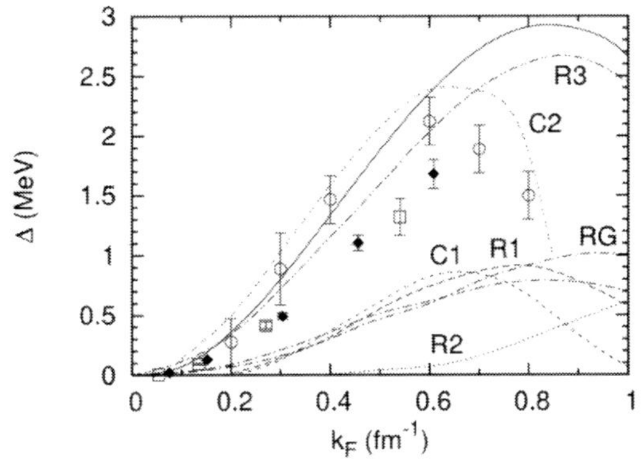


**Figure 1.** Comparison of our Monte Carlo $\Delta$ to other calculations as a function of the neutron matter density (represented by the Fermi momentum $k_F$). The solid diamonds show our results, with statistical uncertainties. The other calculations consist of three types: quantum Monte Carlo (shown by symbols with statistical uncertainties), BCS (shown by the solid curve), and the many-body calculations beyond BCS approximation (R's, C's, and RG; shown by dotted and dashed curves). See text for the description of each calculation shown.

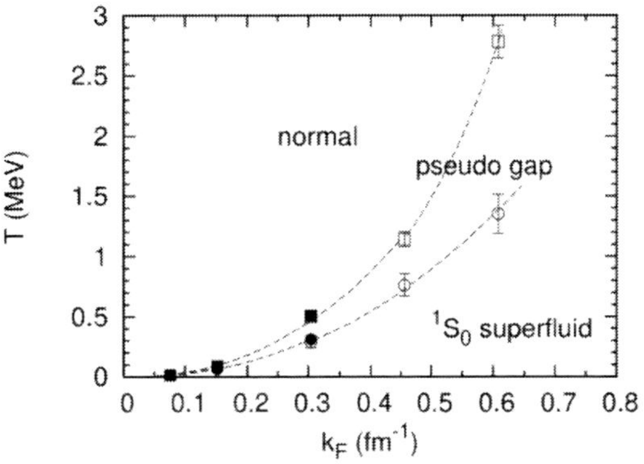


**Figure 2.** The $^1S_0$ phase diagram of low-density neutron matter. The solid and open symbols with statistical uncertainties show the LO and NLO results, respectively. The dashed curves for $T_c$ and $T^*$ are drawn by interpolation. Neutron matter is in the superfluid phase below the critical temperature $T_c$ of the second-order phase transition. Above $T_c$, neutron matter is in the pseudo-gap phase, in which pairing remains locally without forming long-range order, and undergoes a smooth transition from the pseudo-gap phase to the normal phase around $T^*$, as pairing gets much less.

### *3.3. Phase diagram at finite temperature*

To investigate the phase diagram of low-density neutron matter, we evaluate the critical temperature $T_c$ of the normal-to-superfluid phase transition and the pairing temperature scale $T^*$.

$T_c$ is measured from the spin pair-pair correlation sum [32],

$$C_\Delta(T) = \frac{1}{N_s}\sum_{i,j}\langle \hat{\Delta}_i\hat{\Delta}_j^\dagger + \hat{\Delta}_i^\dagger\hat{\Delta}_j\rangle = \frac{1}{N_s}\sum_{i,j}\left[(G_{ij})^2 + (\delta_{ij} - G_{ji})^2\right]. \tag{10}$$

$T_c$ is then extracted from the inflexion point of $C_\Delta(T)$ by fitting an assumed functional form near the inflexion point.

$T^*$ is extracted from the Pauli spin susceptibility $\chi_P$, which is given by

$$\chi_P(T, N_s) = \frac{1}{T}\frac{1}{N_s}\sum_{i,j}\langle \boldsymbol{S}_i \cdot \boldsymbol{S}_j\rangle = \frac{1}{T}\frac{1}{N_s}\sum_{i,j} 2G_{ij}\left(\delta_{ij} - G_{ji}\right), \tag{11}$$

where $\boldsymbol{S}_i = \sum_{\mu,\nu=\uparrow,\downarrow} c_{i\mu}^\dagger \boldsymbol{\sigma}_{\mu\nu} c_{j\nu}$ and $\boldsymbol{\sigma}$ is the vector of Pauli matrices. $T^*$ is evaluated by identifying the maximum point of $\chi_P$ as a function of $T$ [32].

Figure 2 shows the phase diagram. As in Figure 1, the density is described by the Fermi momentum. The solid and open symbols with statistical uncertainties show the LO and NLO results. The dashed curves are drawn for $T_c$ and $T^*$. Note that $T^*$ is just the pairing temperature scale, not the signature of the phase transition, and is related to the pseudo gap phase.

Table 1 summarizes our results of the $^1S_0$ pairing gap $\Delta$ near zero temperature, the critical temperature of the normal-to-superfluid phase transition $T_c$, and the pairing temperature scale $T^*$. In the table, the first and second columns are for the Fermi momenta $k_F$ and the correspoinding densities $\rho$, respectively. $\rho_0$ is the normal nuclear density, $\rho_0 = 0.16$ fm$^{-3}$. Our results of $\Delta$ at $T \approx 0$, $T_c$, and $T^*$ are listed in the third, fourth, and fifth columns, respectively. These results have been extrapolated to the thermodynamic ($N_s \to \infty$) and continuum limits ($n \to 0$). The values in the parenthese are only the statistical errors. (The systematic and theoretical errors are not included here.)

**Table 1.** Our results of the $^1S_0$ pairing gap $\Delta$ near zero temperature, the critical temperature of the normal-to-superfluid phase transition $T_c$, and the pairing temperature scale $T^*$ in the thermodynamic and continuum limits. Uncertainties shown are statistical only.

| $k_F$ (MeV) | $\rho$ ($\rho_0$) | $\Delta$ (MeV) | $T_c$ (MeV) | $T^*$ (MeV) |
|---|---|---|---|---|
| 15 | $9 \times 10^{-5}$ | 0.021(1) | 0.014(3) | 0.014(1) |
| 30 | $7 \times 10^{-4}$ | 0.13(1) | 0.067(5) | 0.091(9) |
| 60 | $6 \times 10^{-3}$ | 0.49(3) | 0.29(5) | 0.45(5) |
| 90 | $2 \times 10^{-2}$ | 1.10(7) | 0.76(9) | 1.1(1) |
| 120 | $5 \times 10^{-2}$ | 1.7(1) | 1.4(2) | 2.8(1) |

## 4. Summary

In summary, we have investigated thermal properties of low-density neutron matter by determinantal quantum Monte Carlo lattice calculations with the pionless EFT up to the NLO. The $^1S_0$ pairing gap at $T \approx 0$, the critical temperature of normal-to-superfluid phase transition, and the pairing temperature scale have been extracted directly from the correlation functions, and have provided the phase diagram for densities of $(10^{-4} - 10^{-1})\rho_0$. The thermodynamic limit was taken, and the continuum limit was examined in the determination. The pairing gap was found to be suppressed by approximately one third from the BCS value. The physics of neutron matter in this density region has clearly been identified as a BCS-BEC crossover.

## Acknowledgments

T. A. acknowledges all the organizers of the International EFES-IN2P3 Conference "Many body correlations from dilute to dense nuclear system". T. A. would like to thank also a financial support by the JSPS Core-to-Core Program EFES for attending at the conference. This work was supported by the U.S. Department of Energy under Grant No. DE-FG02-87ER40347 at CSUN. The calculations were carried out on Seaborg, Bassi, and Franklin at the National Energy Research Scientific Computing Center, which was supported by the Office of Science of the U.S. Department of Energy under Contract No. DE-AC03-76SF00098, and on Titech Grid and TSUBAME, Tokyo Institute of Technology, Japan.

## References

[1] Dean D J and Hjorth-Jensen M 2003 *Rev. Mod. Phys.* **75** 607
[2] Gezerlis A and Carlson J 2008 *Phys. Rev.* C **77** 032801(R)
[3] Gandolfi S, Illarionov A Y, Fantoni S, Pederiva F and Schmidt K E 2008 *Phys. Rev. Lett.* **101** 132501
[4] Müller H M, Koonin S E, Seki R and van Kolck U 2000 *Phys. Rev.* C **61** 044320
[5] Lee D 2009 *Prog. Part. Nucl. Phys.* **63** 117
[6] Lee D, Borasoy B and Schafer T 2004 *Phys. Rev.* C **70** 014007
[7] Lee D and Schafer T 2005 *Phys. Rev.* C **72** 024006
[8] Lee D and Schafer T 2006 *Phys. Rev.* C **73** 015201
[9] Lee D and Schafer T 2006 *Phys. Rev.* C **73** 015202
[10] Borasoy B, Epelbaum E, Krebs H, Lee D and Meissner U.-G. 2008 *Eur. Phys. J.* A **35** 357
[11] Epelbaum E, Krebs H, Lee D and Meissner U.-G. 2009 *Eur. Phys. J.* A **40** 199
[12] Epelbaum E, Krebs H, Lee D and Meissner U.-G. 2010 *Phys. Rev. Lett.* **104** 142501
[13] Epelbaum E, Krebs H, Lee D and Meissner U.-G. 2011 *Phys. Rev. Lett.* **106** 192501
[14] Abe T and Seki R 2009 *Phys. Rev.* C **79** 054002
[15] Abe T and Seki R 2009 *Phys. Rev.* C **79** 054003
[16] Bedaque P F and van Kolck U 2002 *Ann. Rev. Nucl. Part. Sci.* **52** 339
[17] Epelbaum E 2006 *Prog. Part. Nucl. Phys.* **57** 654
[18] Epelbaum E, Hammer H.-W. and Meissner U.-G. 2009 *Rev. Mod. Phys.* **81** 1773
[19] Machleidt R and Entem D R 2011 *Phys. Rept.* **503** 1
[20] Abe T, Seki R and Kocharian A N 2004 *Phys. Rev.* C **70** 014315, 2005 *Phys. Rev.* C **71** 059902(E)
[21] Seki R and van Kolck U 2006 *Phys. Rev.* C **73** 04400
[22] dos Santos R R 2003 *Braz. J. Phys.* **33** 36
[23] Guerrero M, Ortiz G and Gubernatis J E 2000 *Phys. Rev.* B **62** 600
[24] Wambach J, Ainsworth T L and Pines D 1993 *Nucl. Phys.* A **555** 128
[25] Schulze H J, Cugnon J, Lejeune A, Baldo M and Lombardo U 1996 *Phys. Lett.* B **375** 1
[26] Cao L G, Lombardo U and Schuck P 2006 *Phys. Rev.* C **74** 064301
[27] Chen J M C, Clark J W, Davé R D and Khodel V V 1993 *Nucl. Phys.* A **555** 59
[28] Fabrocini A, Fantoni S, Illarionov A Y and Schmidt K E 2005 *Phys. Rev. Lett.* **95** 192501
[29] Schwenk A, Friman B and Brown G E 2003 *Nucl. Phys.* A **713** 191
[30] Gezerlis A and Carlson J 2010 *Phys. Rev.* C **81** 025803
[31] Wlazlowski G and Magierski P 2011 *Phys. Rev.* C **83** 012801
[32] Sewer A, Zotos X and Beck H 2002 *Phys. Rev.* B **66** 140504(R)

Journal of Physics: Conference Series **321** (2011) 012038 doi:10.1088/1742-6596/321/1/012038

# Role of short-range and tensor correlations in nuclei

**A Rios[1], W H Dickhoff[2] and A Polls[3]**

[1]Department of Physics, Faculty of Engineering and Physical Sciences, University of Surrey, Guildford, Surrey GU2 7XH, United Kingdom

[2]Department of Physics, Washington University, St. Louis, Missouri 63130, USA

[3]Departament d'Estructura i Constituents de la Matèria and Institut de Ciències del Cosmos, Universitat de Barcelona, Avda. Diagonal 647, E-08028 Barcelona, Spain

E-mail: a.rios@surrey.ac.uk

**Abstract.** The role of short-range and tensor correlations in nuclei can be investigated through a careful comparison with microscopic nuclear matter calculations. We focus on the momentum distribution, a one-body operator which is particularly sensitive to correlations. We identify a depletion of the population of hole states of around 15% in symmetric matter and a significant isospin dependence due to the tensor force. The increased role of short-range and tensor correlations for the minority species makes the case for further experimental scrutiny of nuclei with large neutron excess.

The nucleon-nucleon (NN) interaction is extremely repulsive at short distances [1]. As a consequence, the probability of finding two nucleons close to each other is substantially suppressed. In turn, such a decrease in the configuration space components implies the population of high-momentum components in the many-body wave function. These significant modifications cannot be described within an independent particle model and are generically referred to as *correlations* [2]. An additional mechanism that provides high-momentum components is associated with the NN tensor force, which is required, among other things, to obtain the correct quadrupole moment of the deuteron [1]. Particle number conservation requires the high-momentum components to be accompanied by a corresponding depletion of the population of states within the nuclear Fermi sea [3]. A characteristic feature of this depletion in nuclear matter is its essentially momentum-independent character [4]. For realistic NN interactions, different many-body techniques consistently predict a depletion of the nuclear Fermi sea of a little over 15%, of which around a third is caused by tensor correlations. An $(e, e'p)$ experiment on $^{208}$Pb at NIKHEF in a large domain of missing energy and a momentum range corresponding to the mean-field Fermi sea confirmed that a global depletion between 15 and 20% of proton orbits below the Fermi energy explains all the measured coincidence cross sections [5]. The differences with respect to nuclear matter calculations can be largely attributed to long-range correlations [6].

A particularly well-suited technique to study the depletion of nuclear systems is the SCGF method [2]. Within this approach, short-range (SRC) and tensor correlations (TC) are accounted for by a fully self-consistent treatment of ladder diagrams for the interaction between particles that propagate with respect to a correlated (by short-range and tensor effects) ground state [7]. The intermediate particle-particle and hole-hole states are fully dressed, with off-shell effects consistently taken into account. A new generation of SCGF numerical calculations have recently become available at finite temperature [7, 8]. Thermal effects are expected to smooth out the

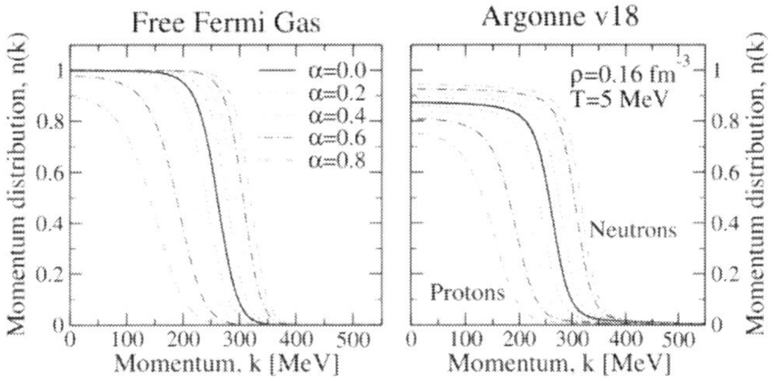


**Figure 1.** Isospin asymmetry dependence of the Free Fermi gas (left panel) and the correlated momentum distribution from Av18 (right panel). The density ($\rho = 0.16$ fm$^{-3}$) and temperature ($T = 5$ MeV) are fixed and the results for different asymmetries correspond to different line-styles.

momentum distribution near the Fermi momentum, but they hardly affect the small and very high-momentum content of the ground state [3].

Interaction-induced correlations have a distinctive signature in the momentum distribution, removing strength at momenta below the Fermi surface and shifting it to high momenta. The momentum distribution for symmetric and isospin asymmetric matter, obtained with the SCGF method, are shown in the right panel of Fig. 1 for a fixed density, $\rho = 0.16$ fm$^{-3}$, and temperature, $T = 5$ MeV. For comparison, we show equivalent free Fermi gas (FFG) momentum distributions in the left panel. At such low temperature, the Fermi-Dirac distribution of the symmetric matter FFG (solid line) deviates very little from a step function, $\Theta(k_F - k)$. This indicates that thermal effects are small in the deep interior of the Fermi sea under these conditions. In contrast, for the correlated case (right panel), $n(k)$ for symmetric matter presents a sizable depletion and it is rather flat below $k_F$. This depletion is mainly caused by dynamical SRC and TC. As a consequence, the contribution to the density sum-rule of the states below $k_F$ is reduced to 75%, *i.e.* 10% of the strength is shifted to higher momenta due to NN correlations. The high-momentum tail in $n(k)$ for symmetric matter provides 52% of the total kinetic energy per particle, to be compared to the much smaller 25% in the FFG.

For a fixed total density, one can switch from symmetric to neutron matter by modifying the relative concentration of neutrons and protons, defined by the asymmetry parameter, $\alpha = \frac{\rho_n - \rho_p}{\rho_n + \rho_p}$. The SCGF method within the ladder approximation can be generalized to the case of partially isospin-polarized matter [9, 10]. Since the approach accounts for both the SRC and TC associated with the underlying NN interaction, it can be used to generate quantitative predictions for the importance of different types of correlations in isospin asymmetric systems. The study of the asymmetry dependence of SRC and TC is motivated to a large extent by the future study of rare isotopes with large neutron excess [11].

In Fig. 1, the momentum distributions for different asymmetries, from symmetric matter ($\alpha = 0$) to extremely neutron-rich matter ($\alpha = 0.8$), correspond to different line styles. As asymmetry increases, the system becomes more neutron-rich and, consequently, the Fermi momentum of neutrons (protons) increases (decreases). This trivial effect can be observed in both the left and right panel. For the FFG, the neutron momentum distribution does not change substantially with asymmetry. For protons, however, $n(k)$ is substantially affected by asymmetry and it is modified at all momenta as the system becomes more neutron-rich. This is a direct consequence of the fact that protons become classical as their density decreases. Consequently, this effect should be ascribed to thermal correlations rather than dynamical correlations [3].

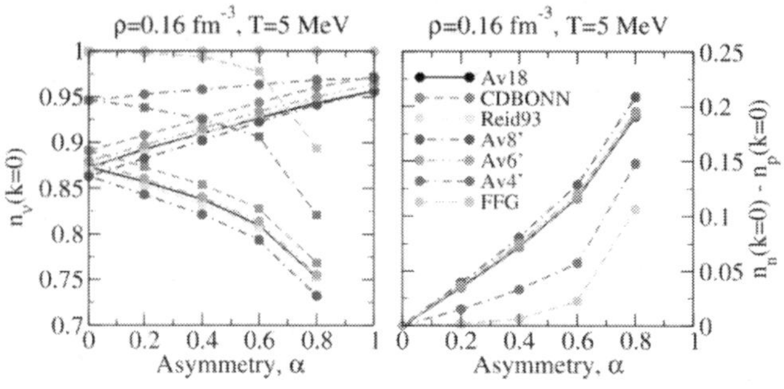


**Figure 2.** Left panel: lowest momentum occupation for neutrons (circles) and protons (squares) as a function of isospin asymmetry. Right panel: difference between neutron and proton occupations at $k = 0$ as a function of isospin asymmetry. The two panels show results at $T = 5$ MeV and $\rho = 0.16$ fm$^{-3}$ for different NN interactions.

For the correlated case, the most abundant component (neutrons for positive $\alpha$) gets less depleted when the asymmetry increases, *i.e.* neutrons become "less correlated". This is in contrast to the FFG results, which, for neutrons, exhibit no change inside the Fermi sea. This behavior can be explained in simple terms as follows. Although the total number of pairs is the same when increasing the asymmetry at constant density, some of the proton-neutron pairs are replaced by neutron-neutron pairs. The latter correlations are weaker than the proton-neutron ones, due to the absence of tensor effects, and therefore neutrons become less correlated at large $\alpha$'s. Conversely, the momentum distributions of the less abundant species (protons) become more depleted with asymmetry.

The information about the isospin dependence of the depletion is summarized in the left panel of Fig. 2, where $n_\nu(0)$ is plotted as a function of the asymmetry. The FFG results (orange lines) are compared to those obtained with a wide range of realistic NN interactions. The FFG provides a measure of thermal effects in the depletion. For the non-interacting case, $n(0)$ for neutrons does not change when the asymmetry increases, indicating that they are totally degenerate. The corresponding occupation for protons (the less abundant component) is close to 1 only at small asymmetries. As a matter of fact, in the limit $\alpha \to 1$, protons become an impurity gas in a Fermi sea of neutrons, thus behaving as a classical gas with $n_p(0) \to 0$. The steepness of the change in $n_p(0)$ would be smaller for higher densities, *i.e.* more degenerate systems.

For the correlated depletion, the occupation of the zero momentum state is an increasing (decreasing) function of the asymmetry for the more (less) abundant component. The behavior is similar for all NN interactions, with a changing offset due to the different predictions for the depletion of symmetric matter. Asymmetries of stable nuclei belong to the lower range of asymmetries, *i.e.* $\alpha \sim 0.2$ for $^{208}$Pb. The SCGF predictions should be valid in this range, where thermal effects are unimportant. Unfortunately, the difference between the occupation of protons at this asymmetry and the symmetric case is only $\sim 2\%$, too small to allow experimental verification. Nuclei at larger asymmetries produced at future rare isotope facilities may provide a better testing ground. Moreover, integrated effects over the whole neutron and proton density profiles might also enhance the effect of asymmetry. Let us note that, for finite nuclei, one should also take into account surface properties and their isospin dependence [12].

A surprising feature arises when comparing the different NN interaction results in the left panel of Fig. 2. In spite of the differences observed in both symmetric and pure neutron matter for the different potentials, the asymmetry dependence of $n_\nu(0)$ is relatively similar.

This is intriguing because these forces have a rather different short-range behavior and tensor structure. Given the almost linear dependence of $n_\nu(0)$ with asymmetry, a better insight into these differences can be gained by plotting the difference $n_n(0) - n_p(0)$. In the right panel of Fig. 2, we present this "iso-depletion" for different interactions as a function of the asymmetry at $T = 5$ MeV and $\rho = 0.16$ fm$^{-3}$. The iso-depletion is the same for a wide variety of modern NN potentials, independently of their short-range or operatorial structure. This suggests that the isospin dependence of $n(0)$ is fixed by the phase-shifts, most probably via their isospin dependence.

The right panel of Fig. 2 shows two exceptions to this common behaviour. On the one hand, the iso-depletion of the FFG falls below the other results. This is expected, because thermal effects are comparatively smaller than dynamical correlations for this particular density and temperature. On the other hand, immediately below the majority of the interactions, one finds the results corresponding to the Av4' potential. This interaction is an oversimplified version of Av18, with an unrealistically simple operatorial structure that only has a spin-isospin part and no tensor components [13]. The fact that the Av4' results lie significantly below the others demonstrates that correlations in asymmetric matter are largely driven by tensor effects, *i.e.* the tensor force increases substantially the difference between neutron and proton momentum distributions with isospin asymmetry. Moreover, the comparison with Av6' and Av8' suggests that, once the tensor components are included in a force, the iso-depletion becomes universal. Let us stress the fact that such an agreement for different potentials is surprising in view of their different momentum-space and operatorial structures.

We close this discussion by noting that the employed interactions appear to slightly underestimate the depletion of the experimentally determined depletion of the deep proton mean-field orbits in $^{208}$Pb [5]. While long-range correlations have an additional influence on the depletion of the Fermi sea, it is expected that they are more important near the Fermi energy. For a proper comparison with finite nuclei, however, these long-range volume effects are presumably irrelevant, since they must anyway be replaced by the surface dominated physics of low-lying nuclear states. Nevertheless, while there remains a few % uncertainty as to the exact amount that is experimentally required, it is also clear that different interactions, including modern ones, lead to very similar predictions for the depletion of the nuclear Fermi sea within the SCGF method.

## Acknowledgments

This work has been supported by a Marie Curie Intra European Fellowship within the $7^{th}$ Framework programme, STFC grant ST/F012012, by the U.S. National Science Foundation under grants PHY-0652900 and PHY-0968941, by grant No. FIS2008-01661 (MEC, Spain) and grant No. 2009SGR1289 from Generalitat de Catalunya.

## References

[1] Bethe H A 1971 *Annu. Rev. Nucl. Sci.* **21** 93
[2] Dickhoff W H and Van Neck D 2008 *Many-Body Theory Exposed!* 2nd ed (Singapore: World Scientific)
[3] Rios A, Polls A and Dickhoff W H 2009 *Phys. Rev. C* **79** 064308
[4] Vonderfecht B E, Dickhoff W H, Polls A and Ramos A 1991 *Phys. Rev. C* **44** R1265
[5] van Batenburg M F 2001 Ph.D. thesis University of Utrecht
[6] Barbieri C 2009 *Phys. Rev. Lett.* **103** 202502
[7] Frick T and Müther H 2003 *Phys. Rev. C* **68** 034310
[8] Rios A, Polls A and Vidaña I 2009 *Phys. Rev. C* **79** 025802
[9] Bożek P 2004 *Phys. Lett. B* **586** 239
[10] Frick T, Müther H, Rios A, Polls A and Ramos A 2005 *Phys. Rev. C* **71** 014313
[11] Gade A *et al.* 2004 *Phys. Rev. Lett.* **93** 042501
[12] Charity R J, Sobotka L G and Dickhoff W H 2006 *Phys. Rev. Lett.* **97** 162503
[13] Wiringa R B and Pieper S C 2002 *Phys. Rev. Lett.* **89** 182501

Journal of Physics: Conference Series **321** (2011) 012039 doi:10.1088/1742-6596/321/1/012039

# Gorkov self-consistent Green's function calculations of semi-magic nuclei

**V. Somà**

CEA-Saclay, IRFU/Service de Physique Nucléaire, F-91191 Gif-sur-Yvette, France

E-mail: vittorio.soma@cea.fr

**T. Duguet**

CEA-Saclay, IRFU/Service de Physique Nucléaire, F-91191 Gif-sur-Yvette, France
National Superconducting Cyclotron Laboratory and Department of Physics and Astronomy, Michigan State University, East Lansing, MI 48824, USA

**C. Barbieri**

Department of Physics, University of Surrey, Guildford GU2 7XH, UK

**Abstract.** The first nuclear structure application of the newly developed Gorkov self-consistent Green's function method is presented. The approach aims to describe many-nucleon systems from an ab-initio standpoint featuring an explicit treatment of pairing correlations. In the present work calculations of binding energies of calcium isotopes are reported and compared with experimental data and other theoretical references.

## 1. Introduction

Although considerable progress has been made in recent years, e.g. using coupled-cluster (CC) [1] or Dyson self-consistent Green's function [2] (Dyson-SCGF) methods, ab-initio calculations of medium-mass nuclei are still limited to a few tens of (doubly-magic $\pm 1$) nuclei. Besides the challenging numerical scaling of such ab-initio methods with nucleon number, there exists an intrinsic difficulty in going to truly open-shell nuclei. Indeed, doing so requires either to use a multi-reference based method, e.g. multi-reference CC [3], or to build the correlated state onto a zeroth-order reference state that already accounts for pairing correlations.

Following the second option, we are currently developing [4] a many-body method based on the self-consistent Green's function theory in the Gorkov formalism that allows for an explicit treatment of pairing correlations and that is therefore applicable to several hundreds of (semi-magic $\pm 1$) nuclei, including systems up to, e.g., the tin isotopic chain. The present implementation of the method starts from realistic low-momentum interactions and consists in the self-consistent solution of Gorkov's equations on the basis of first- and second-order self-energies.

Journal of Physics: Conference Series **321** (2011) 012039 doi:10.1088/1742-6596/321/1/012039

## 2. Gorkov Green's function theory

*2.1. Elements of formalism*

Let $\{c_a^\dagger\}$ be a basis of the one-body Hilbert space $\mathcal{H}_1$ that can be divided into two blocks mapped into each other by the time-reversal transformation. In our case, such a basis will correspond to the harmonic oscillator basis. To any state $a$ belonging to the first block is associated a single-particle state $\bar{a}$ belonging to the second block and having the same quantum numbers as $a$, except for the sign of the third component of the j-coupled angular momentum. With that in mind one can define a basis $\{\bar{c}_a^\dagger\}$ dual to the basis $\{c_a^\dagger\}$ through

$$\bar{c}_a^\dagger(t) \equiv \eta_a c_{\bar{a}}^\dagger(t)\,, \qquad \bar{c}_a(t) \equiv \eta_a c_{\bar{a}}(t)\,, \tag{1}$$

which correspond to exchanging the state $a$ by its partner $\bar{a}$ up to the phase $\eta_a$. By convention $\bar{\bar{a}} = a$ with $\eta_a\,\eta_{\bar{a}} \;=\; -1$.

In Gorkov formalism, one targets the ground state $|\Psi_0\rangle$ of the grand-canonical-like potential $\Omega = H - \mu N$, where $\mu$ is the chemical potential and $N$ the particle-number operator, having the number $N = \langle\Psi_0|N|\Psi_0\rangle$ of particles in average. The state $|\Psi_0\rangle$ is a priori not an eigenstate of $N$, i.e. it is likely to break particle number symmetry. In order to access complete one-body information contained in $|\Psi_0\rangle$, one must introduce a set of four Green's functions, known as Gorkov propagators [5]. Gorkov's propagators can be conveniently grouped into a matrix representation, first introduced by Nambu [6]. After defining an "annihilation" column vector made of annihilation and creation operators

$$\mathbf{C}_a(t) \equiv \begin{pmatrix} c_a(t) \\ \bar{c}_a^\dagger(t) \end{pmatrix}, \tag{2a}$$

and a "creation" row vector

$$\mathbf{C}_a^\dagger(t) \equiv \begin{pmatrix} c_a^\dagger(t) & \bar{c}_a(t) \end{pmatrix}, \tag{2b}$$

one can write the four propagators through

$$i\,\mathbf{G}_{ab}(t,t') \;\equiv\; \langle\Psi_0|T\left\{\mathbf{C}_a(t)\mathbf{C}_b^\dagger(t')\right\}|\Psi_0\rangle = i\begin{pmatrix} G_{ab}^{11}(t,t') & G_{ab}^{12}(t,t') \\ G_{ab}^{21}(t,t') & G_{ab}^{22}(t,t') \end{pmatrix}. \tag{3}$$

Similarly to the Dyson case, self-consistent, i.e. *dressed*, Gorkov propagators are solution of an equation of motion taking, in Nambu's notation and in the energy representation, the typical form

$$\mathbf{G}_{ab}(\omega) = \mathbf{G}_{ab}^{(0)}(\omega) + \sum_{cd} \mathbf{G}_{ac}^{(0)}(\omega)\,\tilde{\mathbf{\Sigma}}_{cd}(\omega)\,\mathbf{G}_{db}(\omega)\,. \tag{4}$$

In Eq. (4) $\mathbf{G}_{ab}^{(0)}(\omega)$ denotes the unperturbed Gorkov Green's function associated with a reference state of the Bogoliubov type, whereas $\tilde{\mathbf{\Sigma}}_{cd}(\omega)$ defines normal and anomalous irreducible self-energies through

$$\tilde{\mathbf{\Sigma}}_{ab}(\omega) \equiv \begin{pmatrix} \tilde{\Sigma}_{ab}^{11}(\omega) & \tilde{\Sigma}_{ab}^{12}(\omega) \\ \tilde{\Sigma}_{ab}^{21}(\omega) & \tilde{\Sigma}_{ab}^{22}(\omega) \end{pmatrix}. \tag{5}$$

*2.2. Calculation scheme*

Proceeding to an actual calculation and solving Eq. (4) relates to truncating the diagrammatic expansion of the irreducible self-energies. The latter expansion, originally based on perturbation theory, is re-expressed in terms of skeleton diagrams, i.e. in terms of dressed propagators solution of Eq. (4). This key feature of *self-consistent* Green's function methods allows the resummation of self-energy insertions to all orders and makes the method intrinsically iterative.

In a first step, our objective is to perform self-consistent second-order calculations, i.e. include first- and second-order diagrams in the computation of the self-energies and solve Eq. (4) iteratively to convergence. Figure 1 (2) displays the corresponding diagrammatic contributions to the normal (anomalous) self-energy. In such diagrams, double lines denote self-consistent normal (two arrows in the same direction) and anomalous (two arrows in opposite directions) propagators, solutions of Eq. (4). Notice that this choice of diagrams constitutes a Kadanoff-Baym $\Phi$-derivable approximation, which automatically ensures the exact fulfillment of conservation laws [7]. Extensive details regarding both the formalism and the computational scheme will be reported in a forthcoming publication [4].

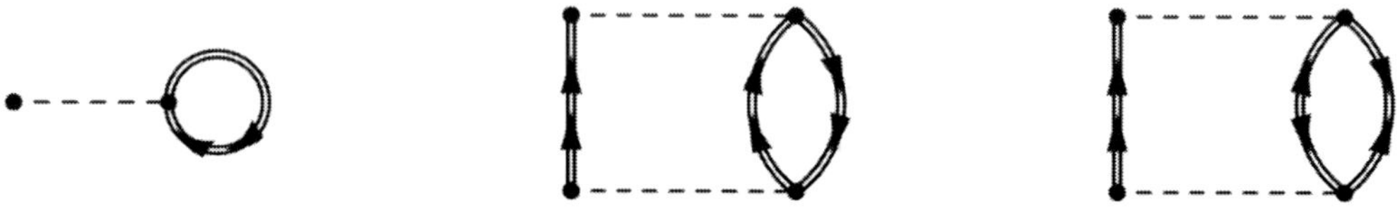

**Figure 1.** First- and second-order contributions to the normal self-energy $\Sigma^{11}$.

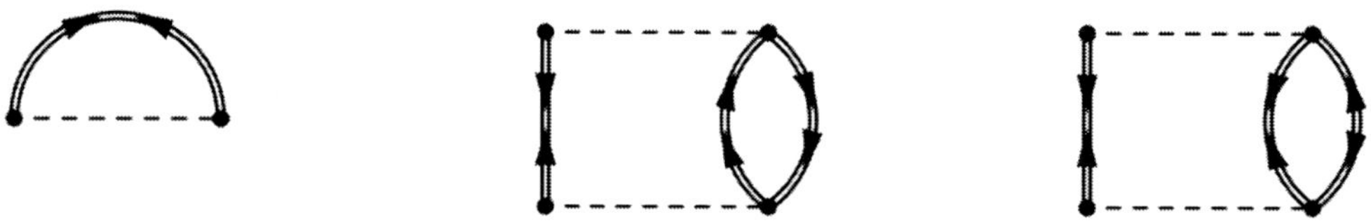

**Figure 2.** First- and second-order contributions to the anomalous self-energy $\Sigma^{21}$.

## 3. Results

A large set of observables are accessible as an outcome of Gorkov calculations, ranging from the binding energy and all one-body observables of the targeted $A$-body ground-state to one-nucleon separation energies to eigenstates of $A \pm 1$ systems. In the present paper, we simply wish to highlight the potential of this newly developed ab-initio method by reporting on one illustrative example, i.e. the binding energy evolution in the calcium isotopic chain. According to the Koltun-Galitskii sum rule [8], the total energy associated with a two-body operator can be accessed from the one-body (normal) dressed propagator, i.e.

$$E_0 = \sum_{ab} \frac{i}{4\pi} \int_{C\uparrow} d\omega \, G^{11}_{ba}(\omega) \left[t_{ab} + \omega \, \delta_{ab}\right] , \tag{6}$$

where the integration is performed over a closed contour in the upper half of the complex plane and where $t_{ab}$ denotes the matrix elements of the kinetic energy operator.

Figure 3 displays the binding energy from $^{40}$Ca to $^{48}$Ca. Results obtained using various ab-initio methods and approximation schemes are compared, i.e. (i) first-order Gorkov self-consistent Green's function (GGF(1)), equivalent to self-consistent Hartree-Fock-Bogoliubov (HFB), (ii) second-order perturbation theory (MBPT(2)), (iii) second-order Gorkov self-consistent Green's function (GGF(2)) and (iv) coupled cluster in singles and doubles approximation (CCSD) [9] methods. All calculations are performed using the same $\Lambda$ = 500 MeV chiral two-nucleon interaction [10], further evolved down to $\Lambda = 2.1$ fm$^{-1}$ through

Journal of Physics: Conference Series **321** (2011) 012039 doi:10.1088/1742-6596/321/1/012039

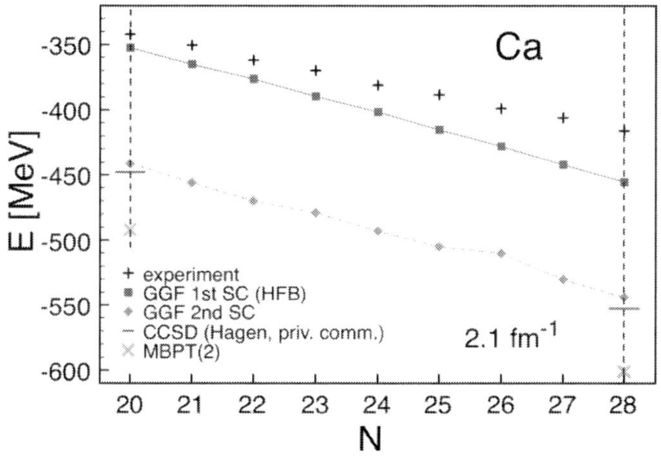


**Figure 3.** Binding energy in Ca isotopes from $N = 20$ to $N = 28$. Results of calculations from (i) HFB, (ii) MBPT(2), (iii) GGF(2) and (iv) CCSD [9] methods are reported and compared with experimental values. Calculations are performed using the $\Lambda = 500$ MeV chiral two-nucleon interaction [10] evolved down to $\Lambda = 2.1$ fm$^{-1}$ through renormalization group technique [11]. Three-nucleon forces are omitted.

renormalization group technique [11]. Three-nucleon forces (TNF) are omitted in the present calculations.

As a consequence of the softness of the interaction and the omission of TNF the Hamiltonian employed is not expected to reproduce the experimental energies. All sets of results overbind the targeted nuclei. In this respect experimental values have to be considered a qualitative rather than quantitative reference. An additional signal for the need of TNF comes from the fact that, in all calculations, this overbinding increases with the mass number $A$.

Self-consistent second-order calculations correct MBPT(2) estimations reducing the overbinding of about 10%. This significant difference evidences the importance of a self-consistent treatment of the dynamical correlations introduced by second-order self-energy diagrams.

Binding energies from GGF(2) calculations are in the range of CCSD results in the doubly-closed shell $^{40}$Ca and $^{48}$Ca (note the large vertical scale). This rudimentary agreement, although in a small model space and with a relatively soft interaction, is a promising indication of the possible capabilities of the present method. It is in particular essential to note the qualitative improvement of GGF(2) over second order many-body perturbation theory that provides a clear overbinding compared to a CC calculation. This underlines the key role played by the self-consistent nature of GGF method and of the associated resummation of self-energy insertions to all orders. Starting from there, GGF approach allows us to address the trend of binding energy along the full isotopic chain. A thorough comparison between Dyson- and Gorkov-GF and CC approaches is in progress and will be object of a future publication [12].

## 4. Conclusions

Binding energies from ab-initio calculations of several calcium isotopes have been presented. In particular, first illustrative results from the newly developed Gorkov self-consistent Green's function method have been shown and compared with other many-body approaches. In this precursory application the GGF(2) approximation scheme shows to be in the range of CCSD results that employ the same interaction and model-space parameters. Thanks to the ability of tackling truly open-shell systems Gorkov self-consistent Green's function techniques represent a promising candidate for the first-time ab-initio description of long isotopic chains in the mid-mass region of the nuclear chart.

## Acknowledgments
The authors wish to thank G. Hagen for providing them with unpublished results of CCSD calculations. Gorkov-SCGF calculations have been performed at the Centre de Calcul Recherche et Technologie (CCRT) of the Commissariat à l'Energie Atomique et aux Energies Alternatives (CEA), France. C.B. acknowledges support from the United Kingdom Science and Technology Facilities Council (STFC) through grant No. ST/I003363.

## References
[1] G. Hagen, T. Papenbrock, D. J. Dean, M. Hjorth-Jensen, Phys. Rev. C82, 034330 (2010).
[2] C. Barbieri, M. Hjorth-Jensen, Phys. Rev. C79, 064313 (2009).
[3] B. Jeziorski, H. J. Monkhorst, Phys. Rev. A24, 1668 (1981).
[4] V. Somà, T. Duguet, C. Barbieri, *unpublished* (2011).
[5] L. P. Gorkov, Sov. Phys. JETP 7, 505 (1958).
[6] Y. Nambu, Phys. Rev. 117, 648 (1960).
[7] G. Baym, Phys. Rev. 127, 1391 (1962).
[8] D. S. Koltun, Phys. Rev. Lett. 28, 182 (1972).
[9] G. Hagen, *private communication* (2011).
[10] D. R. Entem, R. Machleidt, Phys. Rev. C68, 041001 (2003).
[11] S. K. Bogner, T. T. S. Kuo, A. Schwenk, Phys. Rept. 386, 1 (2003).
[12] C. Barbieri, T. Duguet, G. Hagen, V. Somà, *unpublished* (2011).

# Symmetry-adapted Ab Initio Theory for Many-body Correlations in Nuclei

**J P Draayer[1], T Dytrych[1], K D Launey[1], and D Langr[2]**
[1]Department of Physics and Astronomy, Louisiana State University, Baton Rouge, LA 70803, USA
[2]Department of Computer Systems, Faculty of Information Technology, Czech Technical University in Prague, Thákurova 9, 160 00 Prague, Czech Republic

E-mail: draayer@lsu.edu

**Abstract.** We demonstrate that no-core shell-model results for low-lying states of light and medium mass nuclei, whether they are dilute or dense systems, reveal a strong dominance of low-spin and high-deformation configurations. This result is independent of whether the system Hamiltonian is phenomenological in nature or derived from a realistic interaction. It implies that only a small fraction of the complete model space is required for a description of such states, and this in turn points to the importance of using a symmetry-adapted, no-core shell-model framework for describing such nuclei, one based on an LS coupling scheme with the associated spatial configurations organized according to deformation.

These results confirm that the pioneering work of early developers of the field, J. P. Elliott with his SU(3) model and M. Moshinsky with his U(3) many-body oscillator work, extends to more open, multi-shell environments. Specifically, algebraic methods are both relevant in such an environment and they can be used to quell the combinatorial growth in dimensionality that comes with the addition of oscillator shells to a model space. Indeed, our findings demonstrate the utility of a symmetry-adapted, no-core shell-model approach, one that takes advantage of group theoretical as well as advanced computational methods. And importantly, what at first glance appear to be a daunting task – casting complex algebraic expressions of a symmetry-adapted scheme into a user-friendly and efficient shell-mode code, turns out to be not only doable, but a logical framework that embraces constructs that can be made to execute efficiently on massively parallel, multi-processor (and core) systems. Early results for some light p-shell nuclei are presented. In addition, we will show that the method can be extended to heavier nuclei of the sd-shell and beyond, including some cases of special astrophysical interest in the upper fp- and lower gds-shells, like isotopes of Ge, Se, and even Kr.

## 1. Introduction

Many-body correlations play an important role in nuclear dynamics and are key to understanding a variety of interesting phenomena observed in nuclear structure. Among these are alpha-cluster substructures and collective deformations that can strongly influence energy spectra and other properties of nuclei. Ironically, even these relatively easy-to-recognize features continue to resist microscopic solutions, challenging even the most sophisticated of *ab initio* shell model theories (e.g., see Refs. [1, 2, 3]). Such models, with their current relatively limited applicability to light nuclei and low-excitation structures, build on fundamental principles and should therefore offer predictive capabilities essential for a description of unstable and exotic nuclei, many of which are of high interest, e.g. in nucleosynthesis, but remain inaccessible even to experiment.

Journal of Physics: Conference Series **321** (2011) 012040 doi:10.1088/1742-6596/321/1/012040

We offer a novel model, the *ab initio* symmetry-adapted no-core shell model (SA-NCSM), that adapts the *ab initio* (first principles) concept and in addition, utilizes the symmetries known to dominate the nuclear dynamics. Hence, the model holds promise to expand dramatically the reach as well as the impact of current *ab initio* shell models toward describing – with unprecedented accuracy – heavier mass nuclei together with highly deformed and cluster substructures. This is achieved by recognizing that the choice of coordinates, especially when deformed nuclear shapes dominate, is crucial, and that the SA-NCSM affords solution in terms of coordinates that reflect symmetries inherent to nuclear systems. We present first results of the SA-NCSM that employs the SU(3) symmetry, which underpins the successful Elliott model [4, 5, 6] and which is a subgroup of the Sp(3, $\mathbb{R}$) of the symplectic model [7]. Indeed, these results confirm that the pioneering work of early developers of the field, J. P. Elliott with his SU(3) model [8] and M. Moshinsky with his U(3) many-body oscillator work [9], extends to more open, multi-shell environments. While the SA-NCSM states can be obtained through a unitary transformation from the $m$-scheme basis used in the NCSM – and hence, span the entire space, the growth of the model space within the SA-NCSM framework can be managed by winnowing to only physically relevant states as determined through symmetry considerations.

## 2. *Ab initio* symmetry-adapted no-core shell model

Using conventional labels, the SU(3)×SU$_S$(2) basis of the SA-NCSM is given as, $|\alpha(\lambda\,\mu)\kappa(LS)JM_J\rangle$ with $\alpha$ labeling the many-body configuration and $\kappa$ distinguishing multiple occurrences of the same $L$ value in the parent configuration ($S$ and $J$ denote spin and total angular momentum, respectively). The deformation-related $(\lambda\,\mu)$ set of quantum numbers labels SU(3) irreducible representations, irreps, and bring forward important information about nuclear shapes and deformation, for example, $(0\,0)$, $(\lambda\,0)$ and $(0\,\mu)$ describe spherical, prolate and oblate shapes, respectively.

The SA-NCSM implements fast methods for calculating matrix elements of arbitrary (currently up to two-body, but expandable to higher ranks) operators in the SU(3)×SU$_S$(2) symmetry-adapted basis. This facilitates both the evaluation of the Hamiltonian matrix elements and the use of the resulting eigenvectors to evaluate other experimental observables. Any two-body operator can be expanded in terms of SU(3)-coupled fermion creation and annihilation operators,

$$\left[\{a^\dagger_{\eta_i}\otimes a^\dagger_{\eta_j}\}^{(\lambda_{of}\,\mu_{of})}_{S_{of}}\otimes\{a_{\eta_l}\otimes a_{\eta_k}\}^{(\mu_{oi}\,\lambda_{oi})}_{S_{oi}}\right]^{\alpha_o(\lambda_o\,\mu_o)}_{\kappa_o L_o S_o J_o M_{oJ}}. \tag{1}$$

Here $a^\dagger_\eta$ and $a_\eta$ denote fermion creation and annihilation operators in the $\eta$ major oscillator shell with SU(3) symmetry $(\eta\,0)$ and $(0\,\eta)$, respectively. It is a straightforward matter to generalize this expansion to three-body (and higher ranks, if necessary) operators.

Exploiting the SU(3)×SU$_S$(2) symmetry of the SA-NCSM basis states, it suffices to evaluate the following SU(3)-reduced matrix elements, r.m.e. ('$i$' for 'initial', '$f$' for 'final', 'o' for 'operator'):

$$\langle\alpha_f(\lambda_f\,\mu_f)S_f|||\left[\{a^\dagger_{\eta_i}\otimes a^\dagger_{\eta_j}\}^{(\lambda_{of}\,\mu_{of})}_{S_{of}}\otimes\{a_{\eta_l}\otimes a_{\eta_k}\}^{(\mu_{oi}\,\lambda_{oi})}_{S_{oi}}\right]^{\alpha_o(\lambda_o\,\mu_o)}_{S_o}|||\alpha_i(\lambda_i\,\mu_i)S_i\rangle. \tag{2}$$

As the SA-NCSM SU(3) configurations are constructed by coupling a sequence of single-shell SU(3) irreps that are physically allowed, the reduced matrix elements (2) can be extracted from matrix elements between the highest-weight states of the single-shell SU(3) irreps (a SU(3) highest-weight state, which is annihilated by the raising SU(3) generators, is analogous to a $|J, M_J = J\rangle$ state for the case of the SU(2) group).

This factorization of matrix elements into the product of single-shell SU(3) r.m.e. and the associated SU(3) coupling coefficient, which can be calculated 'on the fly', reduces the number of key pieces of information required to the single-shell SU(3) r.m.e., and these track with the number of single-shell SU(3) highest-weight states. Clearly, compared to the explosive growth of the number of NCSM many-body configurations with increasing $N_{\max}$ cutoff (Fig. 1a), the number of single-shell highest-weight states of the SA-NCSM (Fig. 1b) is manageable even, e.g., for '$pf$-shell' nuclei and beyond.

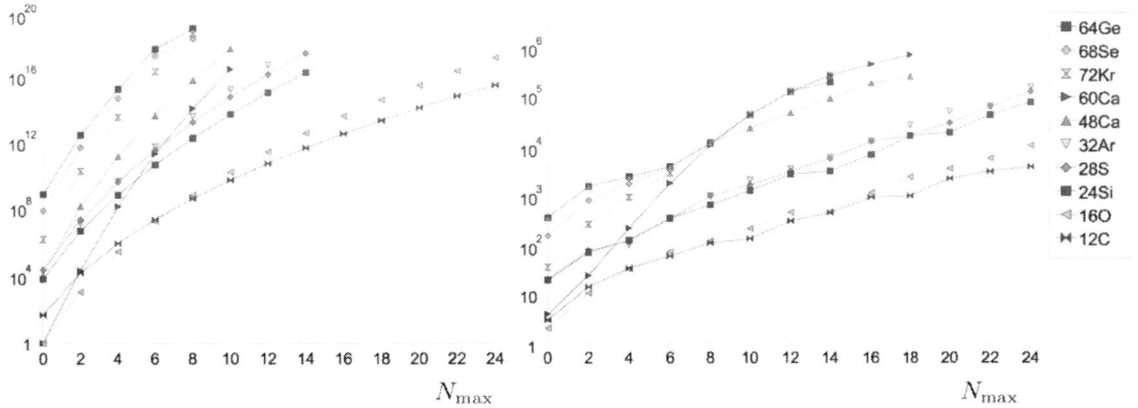


**Figure 1.** Number of (a) NCSM many-body states and (b) all single-shell SU(3) $\times$ SU$_S$(2) highest-weight states of the SA-NCSM many-body model space as a function of the $N_{\max}$ many-body cutoff for several representative nuclei in the $p$, $sd$, and $pf$ shells.

Furthermore, within the SA-NCSM framework, the growth of the model space can be managed by winnowing the model space to only physically relevant states as determined through symmetry considerations. The underlying concept of this framework is illustrated in our proof-of-principle study [10, 11] that exploits symplectic Sp(3, $\mathbb{R}$) symmetry and its SU(3) subgroup symmetry in an analysis of large-scale nuclear physics applications for $^{12}$C and $^{16}$O. What one learns from the outcome of these studies is that typically a small fraction of the full model space, several orders of magnitude less than that of the corresponding NSCM approach, suffices to represent most of the physics – typically 90% or more as measured by projecting NCSM results onto a symmetry-adapted equivalent basis and noting that only a small subset of the full space contributes to the low-energy dynamics.

## 3. Symmetry-guided framework

The SA-NCSM model offers a natural organization of the model space (Fig. 2). In particular, in the SA-NCSM structured approach, the space is first separated into proton and neutron subspaces and then, in keeping with the $LS$ coupling foundation of the SA-NCSM, this is further organized according to the total proton spin ($S_\nu$) and neutron spin ($S_\pi$) coupled to total spin ($S$). These spin spaces can be further organized into SU(3) structures, each of which realizes a nuclear shape deformation given by $(\lambda\,\mu)$.

This SA-NCSM framework allows users to down-select from all possible configurations to a subset that tracks with an inherent preference of a system towards low-spin and high-deformation dominance as revealed in realistic NCSM wavefunctions. For example, for the $0^+$ ground state in $^{12}$C, calculated by NCSM with JISP16 interaction for $\hbar\Omega = 15$ MeV [11], the SU(3) states that correspond to the most deformed oblate shape realize the major component of this state. As for the low-spin dominance, our recent study [12] showed the preponderance of only proton (neutron) spin values $S_\pi$ $(S_\nu) = 0$ and 1 for the $6\hbar\Omega$ NCSM results for $^{12}$C using the effective

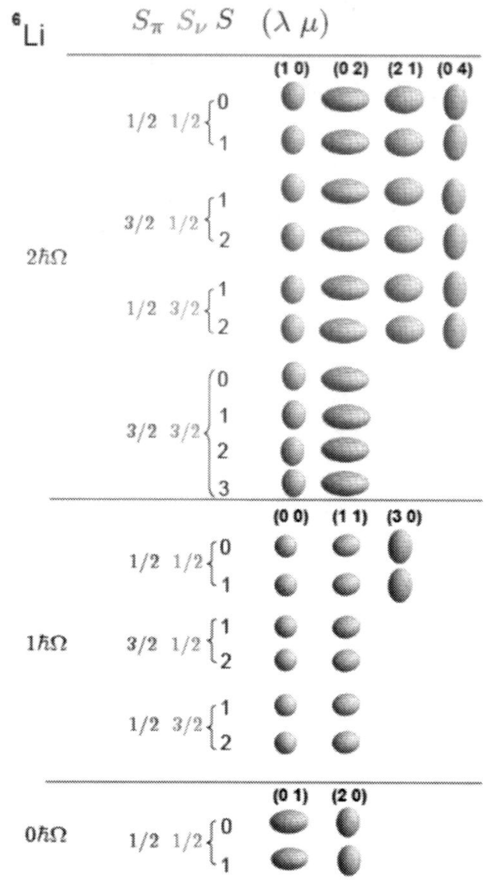

**Figure 2.** Many-body SA-NCSM configurations for $^6$Li organized into proton-, neutron- and total-spin [$SU_S(2)$] and shape [SU(3)] sectors up through $N_{\max} = 2$.

N$^3$LO interaction (and similarly for other interaction choices, such as JISP16) with $\hbar\Omega$=15 MeV. Specifically, $S_\pi \times S_\nu \to S$ with $0 \times 0 \to 0$, $0 \times 1 \to 1$, $1 \times 0 \to 1$, and $1 \times 1 \to 2$ are sufficient to describe 98.75%, 98.72%, and 98.82% of the converged $J = 0^+_{\rm gs}$, $2^+_1$, and $4^+_1$ $^{12}$C NCSM eigenvectors, respectively, with a dominance of about 80% of the $0 \times 0 \to 0$ configurations. The residual roughly 1% of the wavefunctions involves time/memory consuming higher-spin and consequently less relevant configurations. Retaining only the most relevant proton (neutron) spin values allows one, with the same computer resources, to accommodate the full basis within the selected spin spaces up through higher $\hbar\Omega$ values. And since typically configurations of maximum spatial deformation also dominate, only a percentage of the proton(neutron)/spin subspace need be further considered to accommodate the physically most significant configurations, including particle-hole excitations that are important for a description of clustering modes, in higher $\hbar\Omega$ model spaces. Beyond these, the Sp(3, $\mathbb{R}$) symplectic symmetry can guide an extension of model spaces to incorporate even higher $\hbar\Omega$ values [10] required to gain convergence of the lowest bound $0^+$ states in light nuclei.

As an illustration, we note that the *ab initio* SA-NCSM and the symmetry-guided framework was employed for the $1^+$ ground state of $^6$Li with JISP16 ($\hbar\Omega$=11 MeV). The model space included all the configurations up through $N_{\max} = 4$ (full space) with the $N_{\max} = 6$ subspace restricted to only 19 SU(3) $\times$ $SU_S(2)$ configurations. These first examples illustrate the remarkable result that only a fraction of the model space yields high (99.6% for $^6$Li) overlap with the corresponding NCSM wavefunctions and most (98.7%) of the corresponding binding energy.

The symmetry-guided framework of the SA-NCSM is especially crucial for reaching heavier nuclei. An illustrative examples is shown in Fig. 3 for the case of the $N \sim Z$ $pf$-nucleus $^{64}$Ge of astrophysical significance. The figure shows the combinatorial (near-exponential) growth of the full NCSM space (blue squares). Employing spin considerations, the full space can be reduced roughly by an order of magnitude while achieving considerable decrease in computing intensity because low-spin configurations typically have a simpler structure than for higher ones. In the example of Fig. 3, we chose to retain only proton (neutron) spin values $S_\pi = 0$ and 1 ($S_\nu = 0$ and 1) coupled to total spin $S = 0$, 1 and 2 (no restrictions, red diamonds). Further reductions of several order of magnitude can be achieved by selecting SU(3) proton (neutron) modes according to their shape deformation with the most deformed structures playing the foremost role in nuclear collectivity. As for the spin, the selected proton SU(3) configurations are coupled to the selected neutron configurations to yield the total number of many-body states (yellow, green and purple triangles). Beyond these winnowing considerations, only reduced matrix elements of the associated highest-weight states need to be calculated and stored because the associated coupling coefficients can be computed 'on the fly'.

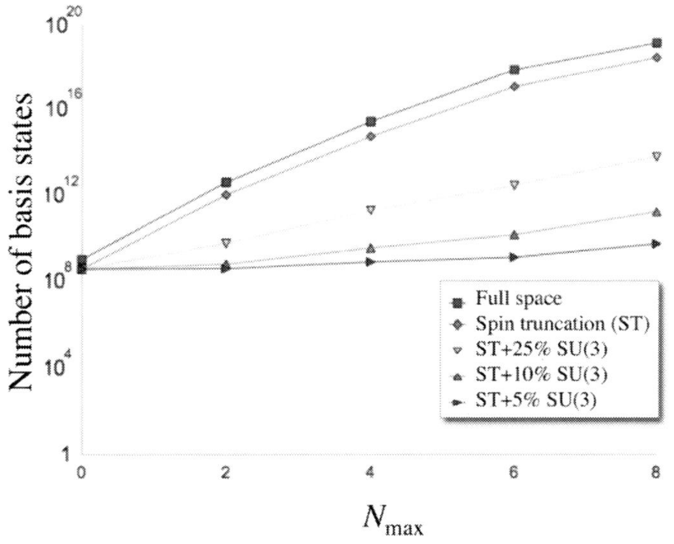


**Figure 3.** Dimensions of various model spaces for $^{64}_{32}Ge_{32}$: conventional NCSM model spaces (blue squares) and reduced model space dimensions when only proton (neutron) $S_\pi$ ($S_\nu$) $= 0$ and 1 values coupled to total spin 0, 1 and 2 are considered (red diamonds) and when beyond this only 25%, 10% and 5% of the most deformed proton (neutron) SU(3) configurations are retained (yellow, green and purple triangles, respectively) as a function of $N_{\max}$.

In short, the SA-NCSM builds upon a proton-neutron formalism in a $LS$ coupling scheme with the proton/neutron spaces organized into subspaces of definite spin and further into spatial SU(3) representations that key to deformation, and for even higher $\hbar\Omega$ model spaces, into symplectic $Sp(3,\mathbb{R}) \supset SU(3)$ structures. The symmetry-guided framework utilizes the physical relevance at each level of the structured space, namely, low-spin and high-deformation dominance together with most important patterns of symplectic excitations. This approach will open up an entire

Journal of Physics: Conference Series **321** (2011) 012040 doi:10.1088/1742-6596/321/1/012040

region of the periodic table to investigation with *ab initio* methods with forefront predictive capabilities.

## 4. Conclusion

The SA-NCSM advances an extensible microscopic framework for studying nuclear structure and reaction processes – strong as well as weak interaction dominated, that capitalizes on advances being made in *ab initio* methods while exploiting symmetries – exact and partial, known to dominate the dynamics. We have developed a symmetry-adapted shell model with the view toward exploring the properties of nuclei far from stability using externally provided realistic interactions derived from Quantum Chromodynamics (QCD) considerations.

Winnowing considerations of the type shown above illustrate that the SA-NCSM offers a systematic framework for down-selecting to physically relevant and manageable subspaces associated with full NCSM based on spin and deformation selection, which are complementary and mutually reinforcing. The method is applied first for light systems so the currently available *ab initio* methods can be used to guide the development, but ultimately pushing toward heavier ones that require symmetry guided winnowing decisions.

## Acknowledgement

Supported by the U.S. National Science Foundation (PHY-0500291 & OCI-0904874), the U.S. Department of Energy (DE-SC0005248), and the Southeastern Universities Research Association (http:/www.sura.org).

## References

[1] Navrátil P, Vary J and Barrett B 2000 *Phys. Rev. Lett.* **62** 054311
[2] Wloch M, Dean D J, Gour J R, Hjorth-Jensen M, Kowalski K, Papenbrock T and Piecuch P 2005 *Phys. Rev. Lett.* **94** 212501
[3] Navrátil P, Quaglioni S, Stetcu I and Barrett B 2009 *J. Phys. G: Nucl. Part.* **36** 083101
[4] Elliott J P 1958 *Proc. Roy. Sot. A* **245** 128
[5] Elliott J P 1958 *Proc. Roy. Sot. A* **245** 562
[6] Elliott J P and Harvey M 1962 *Proc. Roy. Sot. A* **272** 557
[7] Rosensteel G and Rowe D 1977 *Phys. Rev. Lett.* **38** 10
[8] 1992 *Proceedings of the international symposium in honor of K.T. Hecht, Ann Arbor, MI, 19-21 September 1991* ed Draayer J and Janecke J (World Scientific)
[9] 2004 *Proceedings of the Symposium in Honor of Jerry P. Draayer's 60th Birthday, Playa del Carmen, Mexico, 18-21 February 2003* ed Escher J, Castaños O, Hirsch J G, Pittel S and Stoicheva G (World Scientific)
[10] Dytrych T, Sviratcheva K, Bahri C, Draayer J and Vary J 2007 *Phys. Rev. Lett.* **98** 162503
[11] Dytrych T, Sviratcheva K D, Draayer J P, Bahri C and Vary J P 2008 *J. Phys. G: Nucl. Part. Phys.* **35** 123101
[12] Dytrych T, Draayer J P, Sviratcheva K D, Bahri C and Vary J P 2009 *"Nuclear Structure And Dynamics '09": Proceedings of the International Conference, Dubrovnik, Croatia, 4-8 May 2009* vol 1165 ed Milin M, Niksic T, Szilner S and Vretenar D (AIP Conference Proceedings)

Journal of Physics: Conference Series **321** (2011) 012041 doi:10.1088/1742-6596/321/1/012041

# Spin Modes in Nuclei and Astrophysical Processes based on New Shell-Model Hamiltonians

**Toshio Suzuki**

Department of Physics, College of Humanities and Sciences, Nihon University
Sakurajosui 3-25-40, Setagaya-ku, Tokyo 156-8550, Japan
Center for Nuclear Study, University of Tokyo, Hirosawa, Wako-shi, Saitama 351-0198, Japan
National Astronomical Observatory of Japan, Mitaka, Tokyo 181-8588, Japan

E-mail: suzuki@chs.nihon-u.ac.jp

**Abstract.** Neutrino-nucleus reactions on $^{12}$C and $^{16}$O are investigated by shell model calculations with the use of new shell model Hamiltonians for $p$-shell and $p-sd$-shell. The new Hamiltonians can describe spin modes in nuclei quite successfully, for example, the Gamow-Teller (GT) transitions in $^{12}$C and $^{14}$C. New reaction cross sections are compared with previous calculations. Possible implications on nucleosynthesis in supernova explosions are discussed. Electron capture reaction rates are evaluated by using new shell model Hamiltonians in $fp$-shell, which can describe magnetic dipole transitions and GT transitions quite nicely. Capture rates for Ni and Co isotopes in stellar environments are evaluated. The calculated rates in $^{58}$Ni and $^{60}$Ni are found to reproduce well the rates obtained by using experimental GT strengths.

## 1. Introduction

We study neutrino-nucleus reactions and electron capture reactions, which are dominantly induced by excitations of spin modes in nuclei. Recently, successful descriptions of spin responses in nuclei have been obtained by shell model investigations with the use of new shell model Hamiltonians in $p$-shell, SFO [1] and SFO-tls [2], and $fp$-shell, GXPF1 [3]. Neutrino induced reaction cross sections and electron capture rates are evaluated by shell model calculations with the new Hamiltonians, and applied to astrophysical processes in stars.

Neutrino nucleus reactions on $^{12}$C and $^{16}$O are studied in Sect. 2. Electron capture reactions on Ni and Co isotopes in stellar environments are discussed in Sect. 3.

## 2. Neutrino-Nucleus Reactions

### *2.1. Neutrino-Induced Reaction on $^{12}C$*

We first explain some important points of the recent advances in the shell model calculations. Important roles of spin dependent interactions, especially, the tensor force are properly taken into account in the new Hamiltonians. The robust sign rule for the monopole terms of the tensor interaction is essential to explain shell evolutions and change of magic numbers toward drip-lines [4]. Magnetic moments of $p$-shell nuclei and Gamow-Teller (GT) transitions in $^{12}$C and $^{14}$C are well reproduced by SFO [1]. The model space including up to 2-3 $\hbar\omega$ excitations is used with a small quenching for the axial-vector coupling constant, $g_A^{eff}/g_A$ =0.95.

Cross sections for the exclusive charge-exchange $\nu$-induced reaction, $^{12}$C ($\nu$, $e^-$) $^{12}$N$_{g.s.}$, are shown in Fig. 1(a). The reaction is induced by the GT transition. The experimental values of the cross section [5] are well reproduced by shell model calculations with the use of SFO. The calculated cross sections are enhanced compared to those of conventional ones obtained by PSDMK2 [6, 7]. Neutral current reaction cross sections induced by supernova neutrinos with temperature $T$ (MeV) are also shown in Fig. 1(b). The enhancement of the neutrino-nucleus reaction cross sections in $^{12}$C as well as in $^{4}$He is found to lead to the enhancement of the production yields of light elements, $^{11}$B and $^{7}$Li, in supernova explosions [8, 9].

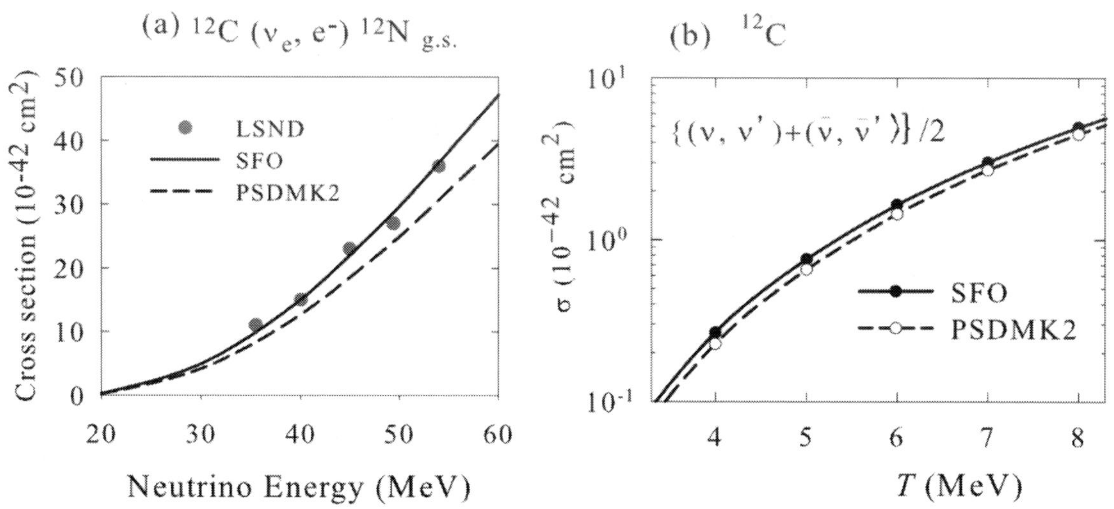


**Figure 1.** (a) Neutrino-induced reaction cross sections for $^{12}$C ($\nu$, $e^-$) $^{12}$N$_{g.s.}$ obtained by shell model calculations with the use of SFO and PSDMK2 as well as the experimental data [5]. (b) Cross sections for the neutral current reaction, $^{12}$C ($\nu$, $\nu'$) $^{12}$C, induced by supernova neutrinos with temperature $T$.

### *2.2. Spin-Dipole Strength in $^{16}$O and $\nu$-induced Reactions*

Next, we discuss spin-dipole strengths in $^{16}$O and $\nu$-induced reactions on $^{16}$O. The $^{16}$O nucleus is contained in water, which is an important target for the detection of neutrinos.

The SFO is modified to take full account of the important roles of the tensor force in the $p-sd$ cross shell part. The tensor and the two-body spin-orbit components of the Millener-Kurath interaction [6] are replaced by those of $\pi + \rho$ meson exchanges and $\sigma + \omega + \rho$ meson exchanges, respectively [2].

Spin dipole strengths in $^{16}$O are studied by using the modified Hamiltonian, SFO-tls [2]. Calculated spin-dipole strengths with the rank $\lambda$

$$SD(\lambda) = \frac{1}{2J_i+1} |\langle J_f \parallel r[Y^1 \times \vec{\sigma}]^{\lambda} t_- \parallel J_i \rangle|^2 \tag{1}$$

are shown in Fig. 2 for the transition; $^{16}$O $\rightarrow$ $^{16}$F. The observed energy levels of the spin-dipole states, $0^-, 1^-$ and $2^-$, are rather well reproduced by SFO-tls. The strength is shifted toward lower energy region for SFO-tls compared to SFO as shown in Fig. 2(b).

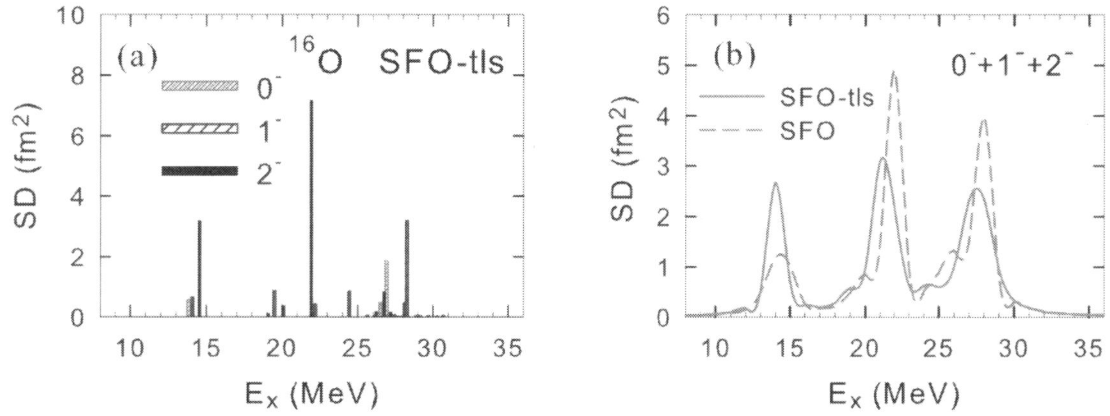


**Figure 2.** (a) Spin-dipole strengths for $^{16}O \rightarrow {}^{16}F$ transition obtained by using SFO-tls. (b) Comparison of the spin-dipole strengths in $^{16}O$ obtained for SFO-tls and SFO.

The $\nu$-induced reaction cross sections, whose main contributions come from the spin-dipole transitions, are therefore enhanced for SFO-tls as shown in Fig. 3(a) for $^{16}O$ ($\nu$, $e^-$) $^{16}F$. The enhancement factors are 1.60, 1.30 and 1.15 -1.06 at $E_\nu$ = 20, 30 and 40 -60 MeV, respectively. They are also enhanced compared to CRPA results [10] at low neutrino energies, 30 MeV $\leq E_\nu \leq$ 80 MeV, by about 10%. Cross sections for supernova neutrinos with temperature $T$ (MeV) are shown in Fig. 3(b). The calculated cross sections are enhanced for SFO-tls compared to SFO by about 17 -5 % for $T$ = 4 -10 MeV, and also compared to CRPA by about 41% and 17% for $T$ =4 and 8 MeV, respectively. Situations are similar for ($\bar{\nu}$, $e^+$) reaction as well as the neutral current reaction. It would be interesting to study effects of the enhancement of the cross sections for SFO-tls on nucleosynthesis and neutrino detections.

## 3. Electron Capture Reactions in *fp*-Shell Nuclei

Spin properties of *fp*-shell nuclei are well described by the new shell model Hamiltonian, GXPF1J [11]. Magnetic dipole strengths in *fp*-shell nuclei and $GT_-$ strength in $^{58}Ni$ are well reproduced by GXPF1J with the universal quenching factors of $g_A^{eff}/g_A$ =0.74 and $g_s^{eff}/g_s$ =0.75. Experimental $GT_+$ strengths in the $\beta_+$ channel in $^{58}Ni$ [12] and $^{60}Ni$ [13] are also well reproduced by GXPF1J [14].

Electron capture reactions in Ni isotopes, which are dominantly induced by the $GT_+$ transitions, are studied in stellar environments, that is, at high temperatures, $T = T_9 \times 10^9$ K with $T_9$ = 1~10, and at high densities, $\rho Y_e = 10^7 \sim 10^{10}$ g/cm$^3$ with $Y_e$ the lepton-to-baryon ration in stars. The capture rates obtained by shell model calculations with GXPF1J are found to reproduce well the rates in $^{58}Ni$ and $^{60}Ni$ evaluated by using the experimental $GT_+$ strengths [14]. The capture rates in $^{56}Ni$ for GXPF1J are found to be smaller compared to those for a conventional Hamiltonian KB3G [15] except for the case of $\rho Y_e$ =$10^7$ at low $T_9$. The difference can be attributed to larger fragmentation of the GT strength for GXPF1J compared to KB3G [8]. See Ref. [14] for more details as for Ni isotopes.

The capture rates in Co isotopes are also studied. Capture rates in $^{56}Co$, $^{58}Co$ and $^{60}Co$ are obtained for GXPF1J and compared with previous calculations of FFN [16] and LSSM calculations of Ref. [17]. The present results are consistent with those of Ref. [17] obtained

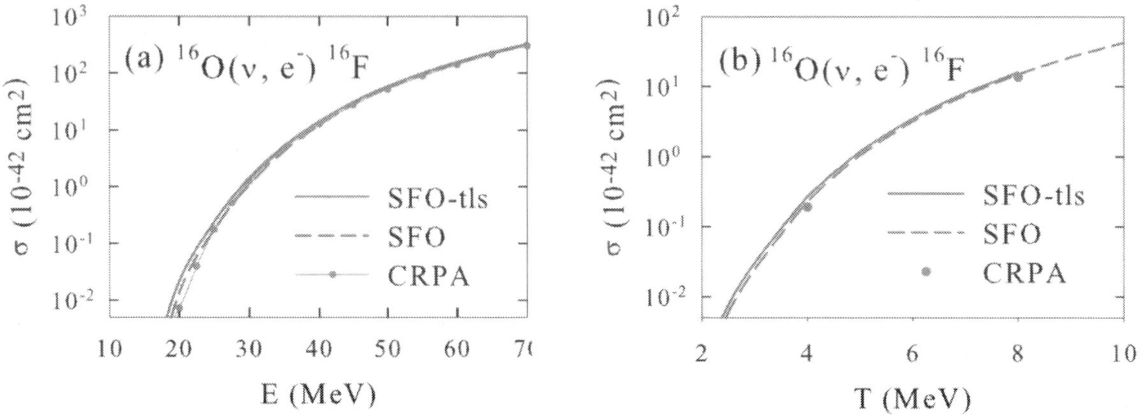


**Figure 3.** (a) Calculated reaction cross sections for $^{16}$O ($\nu$, $e^-$) $^{16}$F obtained by shell model calculations with the use of SFO-tls and SFO, as well as CRPA calculations. (b) The same as in (a) for supernova neutrinos with temperature $T$.

with KBF but much smaller than those of FFN. In $^{60}$Co, the present capture rates are enhanced compared to KBF at low $T_9$ for $\rho Y_e = 10^7$ g/cm$^3$ [18]. Use of the GXPF1J is quite promissing in $fp$-shell nuclei including neutron-rich isotopes [14]. The extension of the present work to various isotopes will be reported in future publications.

This work has been supported in part by Grants-in-Aid for Scientific Research (C)22540290 of the Ministry of Education, Culture, Sports, Science and Technology of Japan.

## References

[1] Suzuki T, Fujimoto R and Otsuka T 2003 *Phys. Rev.* C **67** 044302
[2] Suzuki T and Otsuka T 2008 *Phys. Rev.* **78** 061301(R)
[3] Honma M, Otsuka T, Brown B A and Mizusaki T 2002 *Phys. Rev.* C **65** 061301(R); 2004 ibid. **69** 034335
[4] Otsuka T, Suzuki T, et al. 2010 *Phys. Rev. Lett.* **104** 012501
[5] Athanassopoulos et al. 1997 *Phys. Rev.* C **55** 2078
[6] Millener D J and Kurath D 1975 *Nucl. Phys.* **A255** 315
[7] OXBASH; Brown B A, Etchegoyan A and Rae W D M 1986 *MSU Cyclotron Laboratory Report* No. 524
[8] Suzuki T, Chiba S, Yoshida T, Kajino T and Otsuka T 2006 *Phys. Rev.* C **74** 034307
[9] Yoshida T, Suzuki T et al., 2008 *Astrophys. J.* **686** 448
[10] Kolbe E, Langanke K and Vogel P 2002 *Phys. Rev.* D **66** 013007
[11] Honma M et al. 2005 *J. Phys. Conf. Ser.* **20** 7
[12] Hagemann M et al. 2004 *Phys. Lett.* B **579** 251
[13] Anantaraman et al. 2008 *Phys. Rev.* C **78** 065803
[14] Suzuki T, Honma M, Mao H, Otsuka T and Kajino T 2011 *Phys. Rev.* C **83** 044619
[15] Caurier E et al. 2005 Rev. Mod. Phys. **77** 427
[16] Fuller G M, Fowler W A and Newton M J 1982 *Astrophys. J.* **252** 715; 1985 ibid. **293** 1
[17] Langanke K and Martínez-Pinedo G 1999 **Phys. Lett.** B **453** 187
[18] Suzuki T, Honma M, Yoshida T, Mao H, Kajino T and Otsuka T 2011 *Prog. Part. Nucl. Phys.* **66** 385

Journal of Physics: Conference Series **321** (2011) 012042 doi:10.1088/1742-6596/321/1/012042

# Concluding Remarks

**Hisashi Horiuchi**

Research Center for Nuclear Physics, Osaka University, Ibaraki, Osaka 567-0047, Japan
International Institute for Advanced Studies, Kizugawa, Kyoto 619-0225, Japan

E-mail: horiuchi@rcnp.osaka-u.ac.jp

**Abstract.** Some remarks are given on many body correlations from dilute to dense nuclear systems, putting stress on three subjects; cluster-gas-like states, dineutron correlation, and relation of mean-field dynamics and clustering dynamics.

## 1. Introduction

I here give some remarks on many body correlations from dilute to dense nuclear systems, from the viewpoint of a theorist who is long involved in the study of nuclear clustering. Nuclear clustering is important especially for dilute nuclear systems. We can easily enumerate several subjects of clustering in relation with dilute nuclear density. (i) Light nuclei; Lots of cluster states exist as excited states in light nuclei which have large surface region. (ii) Very dilute excited states in light nuclei; Recently cluster-gas-like states typified by alpha-condensation-like states have been discussed to be existent widely. (iii) Neutron-rich nuclei: Cluster structure with the neutron molecular orbit forms dilute nuclei and dineutron correlation is discussed in neutron skin, (iv) Heavy ion collisions: Multifragmentation in the expansion phase of the collision. (v) Crust of neutron star: Formation of clusters and possible alpha particle condensation.

Our study of alpha-condensation-like states in finite nuclei statrted on the excursion boat of the cluster conference at Rab, Croatia, in 1999. Peter Schuck asked me whether there is any possiblity of alpha-condensation-like structure in finite nuclei. I recalled and mentioned 3-alpha cluster calculations [1, 2, 3] performed more than 30 years ago which concluded that the Hoyle state is a $3\alpha$ gas-like state with very dilute density, about 1/3 of the ground state. Later we found [4] actually that the fully microscopic cluster wave functions of $3\alpha$ RGM [2] and $3\alpha$ GCM [3] are almost 100 % equivalent to the $3\alpha$ THSR [5] wave function:

$$|\langle\Phi(3\alpha\mathrm{THSRw.f.})|\Phi(3\alpha\mathrm{RGM/GCMw.f.})\rangle|^2 \approx 100\%. \tag{1}$$

The structure study of the Hoyle state has now developed into the proposal and study of the novel type of nuclear structure, the cluster-gas-like structure, in wide region of nuclear chart, which is typified by alpha-condensation-like states.

The subjects discussed in this conference are wide and therefore, as I mentioned before, I give below some remarks on the following three problems. They are (1) cluster-gas-like states, (2) dineutron correlation, and (3) relation of mean-field dynamics and clustering dynamics.

Published under licence by IOP Publishing Ltd

Journal of Physics: Conference Series **321** (2011) 012042 doi:10.1088/1742-6596/321/1/012042

## 2. Cluster-gas-like states

We now know that the Hoyle state has a $3\alpha$-condensate-like structure, the lowest-energy state of cluster-gas-like states in $^{12}$C. It has long been discussed that the nucleus undergoes the liquid-gas phase transition, where the gas phase means the gas of nucleons. The excitation energy of the nucleon-gas state of mass-number $A$ nucleus is about $8A$ MeV and this high excitation energy makes the nucleon-gas state as the subject of nuclear matter and nuclear reaction rather than nuclear structure. On the other hand, the gas state of clusters is not so highly excited, and can be a discrete state accessible spectroscopically. This situation is shown in Fig. 1 for $^{12}$C.

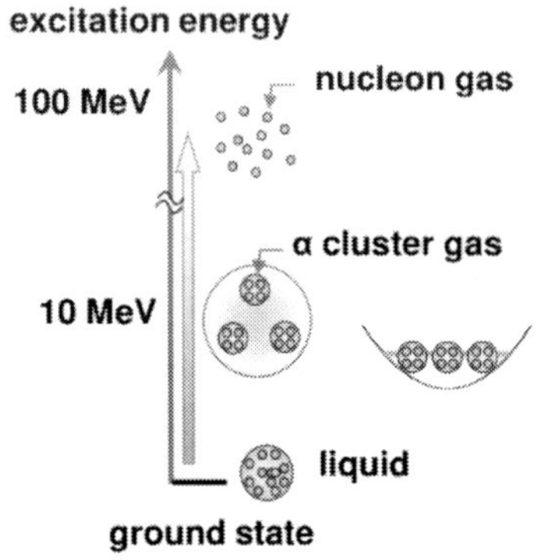


**Figure 1.** Excitation energies of $\alpha$-cluster gas state and nucleon gas state in the case of $^{12}$C.

If we can assume weak inter-cluster interaction, the lowest-energy state formed by $n$ clusters $C_1$, $C_2$, $\cdots$ $C_n$ is located near the breakup threshold of $C_1$, $C_2$, $\cdots$ $C_n$. The highest-energy breakup threshold is of course the full dissociation threshold into $A$ nucleons. The lowest-energy cluster structure of $C_1 + C_2 + \cdots C_n$ can not be liquid-like, because a liquid-like compact cluster structure is equivalent to a mean-field-type structure. The lowest-energy cluster states are zero-temperature states. When the temperature is raised from zero, two types of excitation occur: one is the mean-field-type excitation of individual clusters with liquid nature and the other is the excitation of inter-cluster relative motions. For the understanding of the configuration of nuclear system at finite temperature, the investigation of the caloric relation is useful. Fig. 2 gives the constant-pressure caloric curve of $^{36}$Ar ($N = Z = 18$) calculated with AMD [6]. We see the liquid-gas phase transition occurs through the region of negative heat capacity. Fig. 3 shows fragment mass distribution for the ensembles along the $P = 0.05$ MeV/fm$^3$ line. The caloric curve comes down towards the zero temperature line (abscissa) when the pressure $P$ becomes smaller. The multiplicity distributions of clusters along a caloric curve looks consistent with the distribution of breakup thresholds into clusters.

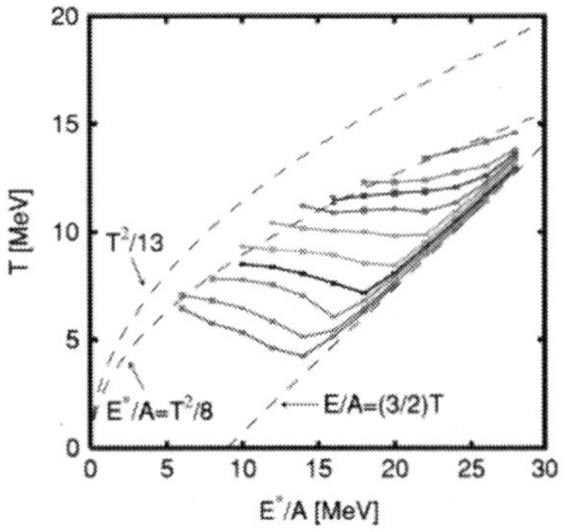


**Figure 2.** The constant pressure caloric curve for $^{36}$Ar obtained by AMD. The lines correspond to the pressure $P = 0.02$, 0.03, 0.05, 0.07, 0.10, 0.15, 0.20, 0.25, 0.30, and 0.40 MeV/fm$^3$ from the bottom. $E^*$ stands for the excitation energy and $A = 36$. The curves $E/A = T^2/(8\text{MeV})$ and $E/A = T^2/(13\text{MeV})$, and the line $E/A = (E^* + E_{\text{gas}})/A = (3/2)T$, are drawn for comparison.

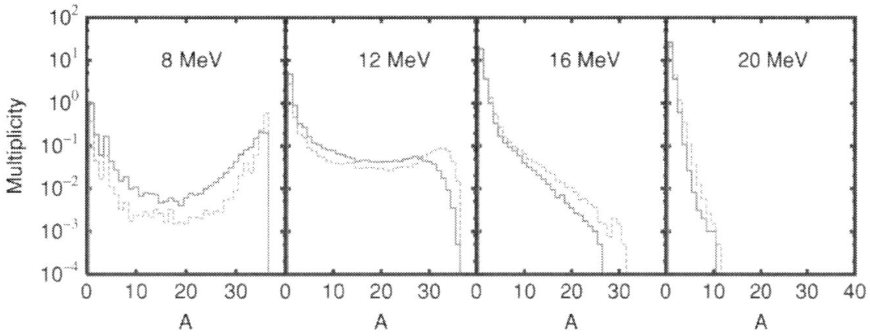


**Figure 3.** The fragment mass distributions along the $P = 0.05$ MeV/fm$^3$ line. Full lines are obtained with $r_{cl} = 2.5$ fm, while dashed lines are those with $r_{cl} = 3.0$ fm. If two nucleon seperation is smaller than $r_{cl}$, they are regarded as belonging to the same cluster.

## 3. Dineutron correlation

In order to study the halo phenomena of $^{11}$Li, Hansen and Jonson [7] introduced the dineutron picture for the two valence neutrons around $^9$Li core. Bertsch and Esbensen [8] solved the three-body problem of $^9$Li + $n$ + $n$ and found dineutron spatial correlation in the surface region in the two-neutron correlation density. They reported the necessity of the inclusion of high angular momentum of single particle states up to about 14, which is the reflection of spatial dineutron correlation. This point was shown also in the shell model study of Ref. [9] which reported rapid convergence of the calculation by the inclusiion of the basis state with Jacobi coordinate. If $^{11}$Li has '$^9$Li + dineutron' structure, Coulomb breakup strength is approximately estimated by $dB(E1)/dE_x \propto \sqrt{S_c}(E - S_c)^{3/2}/E^4$ which has the peak at $E(\text{peak}) = 1.6S_c$. If we use $S_c = S_{2n}$ = 0.3 MeV, we have a peak at low energy $E(\text{peak}) \approx 0.5$ MeV, which agrees with the result of Ref. [8]. Unfortunately, the discrepansies among the experimental results of Coulomb breakup obtained at GSI, RIKEN, and MSU, made it difficult to persue in more detail the validity of the di-neutron model. However, Nakamura and his group [10] remeasured the breakop reaction of $^{11}$Li with improved experimental procedure at RIKEN, and obtained new results which differ from previous data from the three institutions. The peak position of the new data is now much closer to $E(peak) \approx 0.5$ MeV and the whole $B(E1)$ distribution due to new data is now in good agreement with the Bertsch-Esbensen result.

Peter Schuck and his coworkers discussed coexistence of BCS and BEC-like pair structures in nuclei for clarifying the two-neutron structure of $^{11}$Li [11]. They used the solutions of the three-body model of *core* + $n$ + $n$ which, in the case of $^{11}$Li, is almost the same as the Bertsch-Esbensen model. They analysed the $L = 0$ radial wave function $f_{L=0}(r, R)$ of the $S = 0$ component of $^{11}$Li wave function, where $L$ and $S$ are orbital angular momentum and intrinsic spin of the system,respectively, and $\mathbf{r}$ and $\mathbf{R}$ are two-neutron relative coordinate and two–neutron center-of-mass coordinate measured from $^9$Li core, respectively. As shown in Fig. 4, $f_{L=0}(r, R)$ as a function of $r$ has an oscillatory behavior for smaller $R$ where the nuclear density is normal, while it becomes a well localized single peak for larger $R$ the nuclear density is dilute. They argue that these features in $^{11}$Li qualitatively correspond to the BCS and BEC-like structures of the pair wave function found in infinite nuclear matter. They also argue that, although the size of Cooper pair usually is very little influenced by pairing strength, in $^{11}$Li it is strongly influenced by pairing strength, which stems from that fact that two neutrons occupy very weakly bound $0p$ ans $1s$ orbits which spread out very far and make a halo structure.

In Ref. [11] the $^9$Li - $n$ potential $V_{^9\text{Li}-n}$ is deeper for positive parity than for negative parity.

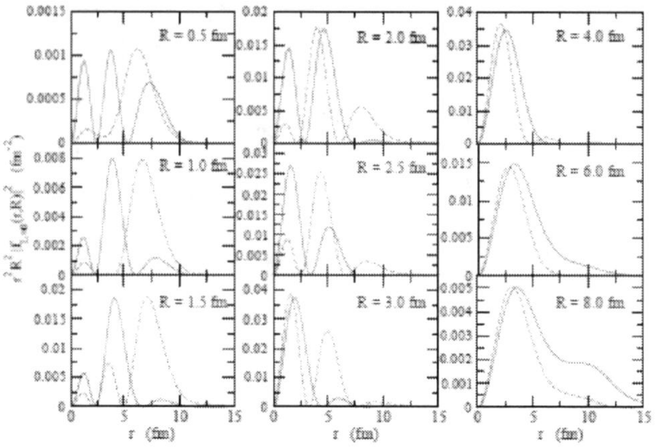


**Figure 4.** The ground state two-particle wave functions, $r^2R^2|f_{L=0}(r,R)|^2$. The solid lines correspond to the two-particle wave functions of $^{11}$Li, while the dashed lines denote those of $^{16}$C.

It causes the narrowing of the energy gap between $0p$ and $1s$ orbits. This is necessary for reproducing the observation that $^{11}$Li has large mixture of the $(1s)^2$ component besides the usual closed-shell component $(0p)^2$. We need to explain why positive-parity neutron orbit can be lowered than usual. Two kinds of Pauli-blocking effect have been proposed for this explanation. One is the effect of blocking of pairing excitation of $^9$Li core which was originally proposed in Ref. [12] for explaining the parity inversion of the $^{11}$Be ground state. The other is the effect of blocking of neutron-proton tensor excitation of $^9$Li core [13] which coexists with the blocking effect of the two-neutron pairing excitation. Both kinds of Pauli blocking cause blocking for $0p$ orbit but no blocking for $1s$ orbit, which results in the narrowing of the energy gap between $0p$ and $1s$ orbits. Ref. [13] reports that these two kinds of Pauli-blocking effects contribute to the narrowing of the $0p - 1s$ energy gap with roughly the same amount of magnitude and the calculation well reproduces the experimental data which include the binding energy, matter radius, Coulomb breakup cross sections, and the $(1s)^2$ probability of the wave function [14]. It is to be noted that the approximate degeneracy of $0p$ and $1s$ orbits is favorable for the spatial correlation of dineuteron. We thus see that there are close relations among the three characteristics of $^{11}$Li, neutron halo, magicity breaking, and dineuteron.

## 4. Relation of mean-field dynamics and clustering dynamics

The investigatation of the clustering correlation on the basis of the mean field is of course a fundamentally important way to study the relation between mean-field dynamics and clustering dynamics. Here, however, we would like to point out the fact that there exists a deep relation between mean-field nature and clustering nature of nuclei. This fact has been known for a long time but its fundamental importance has been emphasized only recently [15, 16, 17]. It is the duality of the ground-state wave function of the non-heavy nucleus that the wave function can be expressed in two ways, one in the mean-field-like type and the other in the cluster-model type. A typical example is the doubly-closed-shell wave function of $^{16}$O. We have the following relation;

$$\det|(0s)^4(0p)^{12}| \;=\; c_L\mathcal{A}\left[R_{4,L}(r_{\mathrm{C}-\alpha},3\nu)[Y_L(\widehat{r}_{\mathrm{C}-\alpha})\phi_L(^{12}\mathrm{C})]_{J=0}\phi(\alpha)\right]g(\mathbf{X}_G,16\nu), \qquad (2)$$

where $L$ is arbitrary among $L = 0$, 2, and 4, $R_{N,L}(r, \gamma)$ stands for the radial harmonic oscillator function with width parameter $\gamma$ and $N = 2n + L$ the number of oscillator quanta, $g(\mathbf{r}, \gamma) \propto \exp(-\gamma r^2)$, and $\nu$ is the single-nucleon oscillator parameter. This duality is known as the Bayman-Bohr theorem [18].

The observed quantity which shows clearly the duality of the ground state wave function is the monopole transition strength between the ground state and excited clustering states in light nuclei. The observed transition strengths are all large and are of the order of single-nucleon strength [19]. In the description by the mean-field-type model, cluster states have (superpositions of) many-particle many-hole configuration(s), and thus it is very difficult to explain why almost all the observed monopole strengths are of single-nucleon strength. However, if we notice the duality of the ground state wave function, this observed fact that monopole transition strengths are all large and are of the order of single-nucleon strength is quite easy to explain.

We explain the above statement in the case of the monopole transition between the Hoyle state and the ground state of $^{12}$C. The observed strength is $M(E0; \text{g.s.} \leftrightarrow \text{Hoyle}) = 5.4\ \text{fm}^2$. This is of the order of single-nucleon strength since the single-nucleon strength is estimated, by using schematic single-particle wave function of constant demsity, as $M(E0) \sim \langle u_f | r^2 | u_i \rangle \sim (3/5)R^2 = 5.4\ \text{fm}^2$, where the nuclear radius $R = 3$ fm is used. The theoretical calculation of $M(E0)$ value is performed as follows by using the duality of the ground state wave function which is represented by its dominant component given by the SU(3) shell model wave function, $|(0s)^4(0p)^8, (\lambda, \mu) = (04) J = 0\rangle$. The duality is expressed as

$$|(0s)^4(0p)^8, (04) J = 0\rangle = N_0 \mathcal{A}\left\{R_{4,0}(\xi_1, (8/3)\nu) R_{4,0}(\xi_2, 2\nu) \phi(\alpha_1)\phi(\alpha_2)\phi(\alpha_3)\right\} g(\mathbf{X}_G, 12\nu), \quad (3)$$

where $\xi_k$ are Jacobi coordinates of three $\alpha$ clusters, $\xi_1 = \mathbf{X}_1 - (\mathbf{X}_2 + \mathbf{X}_3)/2$, $\xi_2 = \mathbf{X}_2 - \mathbf{X}_3$, with $\mathbf{X}_j$ standing for the center-of-mass coordinate of $j$-th $\alpha$ cluster. The Hoyle state is described by the $3\alpha$ THSR [5] wave function

$$\Phi(3\alpha\text{THSRw.f.}) = n_B \mathcal{A}\{\exp[-\frac{2}{B^2}(\mathbf{x}_1^2 + \mathbf{x}_2^2 + \mathbf{x}_3^2)]\ \phi(\alpha_1)\phi(\alpha_2)\phi(\alpha_3)\}, \quad \mathbf{x}_k = \mathbf{X}_k - \mathbf{X}_G. \quad (4)$$

From the forms of the wave functions of the ground and Hoyle states, we see that the monopole transition between these states is nothing but the monopole transition between two relative wave functions

$$R_{4,0}(\xi_1, (8/3)\nu) R_{4,0}(\xi_2, 2\nu) \quad \longleftrightarrow \quad \exp[-\frac{2}{B^2}(\mathbf{x}_1^2 + \mathbf{x}_2^2 + \mathbf{x}_3^2)]. \quad (5)$$

This fact means that only two degrees of freedom of inter-$\alpha$ relative motion is relevant to the monopole transition between the ground and Hoyle states, while in the description by the mean-field-type model we have to treat much larger number of degrees of freedom of nucleonic motion. The relevance of only two degrees of freedom is the reason why the monopole transition between the ground and Hoyle states has the strength of the order of single-nucleon strength. This argument can be seen by the analytical expression of the monopole matrix element $M(E0; \text{g.s.} \leftrightarrow \text{Hoyle})$ given as [17]

$$M(E0; \text{g.s.} \leftrightarrow \text{Hoyle}) = \sqrt{\frac{7}{6}} \sqrt{\frac{\langle F_4 \rangle}{\langle F_5 \rangle}} \sigma_5 \langle R_{40}(r, \nu) | r^2 | R_{60}(r, \nu) \rangle, \quad (6)$$

where

$$\langle F_n \rangle = \langle Q_n | \mathcal{A}\{Q_n\} \rangle, \quad Q_n = F_n(\xi_1, \xi_2)\phi(\alpha_1)\phi(\alpha_2)\phi(\alpha_3),$$
$$F_n(\xi_1, \xi_2) = \frac{1}{4\pi} \sum_{n_1+n_2=n} \sqrt{\frac{(2n_1+1)!!(2n_2+1)!!}{2n_1!!2n_2!!}} R_{2n_1,0}(\xi_1, (8/3)\nu) R_{2n_2,0}(\xi_2, 2\nu),$$
$$|\text{Hoyle}(0_2^+)\rangle = \sum_{n=5}^{\infty} \sigma_n (e_n \mathcal{A}\{Q_n\}), \quad \|e_n \mathcal{A}\{Q_n\}\| = 1. \quad (7)$$

It is to be noted here that, when we rewrite the $E0$ transition operator $O(E0;^{12}\mathrm{C}) = (1/2)\sum_{i=1}^{12}(\mathbf{r}_i - \mathbf{X}_G)^2$ as $O(E0;\alpha_1) + O(E0;\alpha_2) + O(E0;\alpha_3) + (1/2)((8/3)\xi_1^2 + 2\xi_2^2))$, there is no contribution from $\alpha$-cluster degrees of freedom, $O(E0;\alpha_1) \sim O(E0;\alpha_3)$. The quantity $\langle F_n \rangle$ represents the effect of the antisymmetrization, but in the above analytical formula, it appears in the form of ratio, $\langle F_4 \rangle / \langle F_5 \rangle$, whose magnitude is close to unity. Thus the effect of antisymmetrization has only little influence on the $M(E0)$ value. The quantity $\sigma_5$ is the amplitude of the $2\hbar\omega$ - jump component contained in the Hoyle state and its magnitude is around 0.25.

## 5. Summary

We have discussed various types of many body correlations from dilute to dense nuclear systems. The existence of various types of many body correlations shows the richness of the nuclear dynamics and system.

I close my talk by wishing Peter happy retirement and many more healthy and successful years.

## References

[1] Horiuchi H, 1974 *Prog. Theor.Phys.* **51** 1266; 1975 **53** 447.
[2] Fukushima Y and Kamimura M, 1978 *Supple. of J. Phys. Soc. Japan* **44** 225; Kamimura M, 1981 *Nucl. Phys.* A **351** 456.
[3] Uegaki E, Okabe S, Abe Y, and Tanaka H, 1977 *Prog. Theor. Phys.* **57** 1262; Uegaki E, Abe Y, Okabe S, and Tanaka H, 1978 *Prog. Theor. Phys.* **59** 1031; 1979 **62** 1621.
[4] Funaki Y, Tohsaki A, Horiuchi H, Schuck P, and Röpke G, 2003 *Phys. Rev.* C **67** 051306(R).
[5] Tohsaki A, Horiuchi H, Schuck P, and Röpke G, 2001 *Phys. Rev. Lett.* **87** 192501.
[6] Furuta T and Ono A, 2006 *Phys. Rev.* C **74** 014612.
[7] Hansen P G and Jonson B, 1987 *Europhys. Lett.* **4** 409.
[8] Bertsch G F and Esbensen H, 1991 *Ann. Phys.* **209** 327: Esbensen H and Bertsch G F, 1992 *Nucl. Phys.* A **542** 310.
[9] Aoyama S, Mukai S, Katō K, and Ikeda K, 1995 *Prog. Theor. Phys.* **93** 99; Myo T, Aoyama S, Katō K, and Ikeda K, 2002 *Prog. Theor. Phys.* **108** 133.
[10] Nakamura T et al., 2006 *Phys. Rev. Lett.* **96** 252502.
[11] Hagino K, Sagawa H, Carbonell J, and Schuck P, 2007 *Phys. Rev. Lett.* **99** 022506; Hagino K, Sagawa H, and Schuck P, 2010 *J. Phys.* G **37**, 064040.
[12] Sagawa H, Brown B. A., and Esbensen H, 1993 *Phys. Lett.* B **309** 1.
[13] Ikeda K, Myo T, Kato K, and Toki H, 2010 *Lecture Notes in Physics* **818** 165.
[14] Simon H, et al., 1999 *Phys. Rev. Lett.* **83** 496.
[15] Horiuchi H, 2009 *Int. Journ. Mod. Phys.* A **24** 2205.
[16] Horiuchi H, 2010 *Lecture Notes in Physics* **818** 57.
[17] Yamada T, Funaki Y, Horiuchi H, Ikeda K, and Tohsaki A. 2008 *Prog. Theor.Phys.* **120** 1139.
[18] Bayman B. F. and Bohr A., 1958/59 *Nucl. Phys.* **9** 596.
[19] Kawabata T, et al., 2007 *Phys. Lett.* B **646** 7.

Journal of Physics: Conference Series **321** (2011) 012043 doi:10.1088/1742-6596/321/1/012043

# Universality of short-range correlations in light nuclei

**W. Horiuchi**[1,2] **H. Feldmeier**[1], **T. Neff**[1], **and Y. Suzuki**[3,4]

[1]GSI Helmholtzzentrum für Schwerionenforschung GmbH, Planckstraße 1, 64291 Darmstadt, Germany

[3]RIKEN Nishina Center, RIKEN, Wako 351-0198, Japan

[4]Department of Physics, Niigata University, Niigata 950-2181, Japan

E-mail: `whoriuchi@riken.jp (W. Horiuchi)`

**Abstract.** In this contribution, we investigate the structure of short-range correlations in many-body states. We obtain the highly correlated many-body states with an explicitly correlated basis which enables us to get a precise solution of a many-body Schrödinger equation for a realistic interaction. We show two-body density distributions calculated from three- and four-body states to investigate the short-range correlations between nucleon pairs. At distances below 1 fm a universal behavior is found which does not depend on the many-body states. The universality is also seen in the high momentum components of the two-body momentum distributions.

## 1. Introduction

In nuclei, the nuclear interaction induces strong short-range correlations among the nucleons. Realistic nucleon-nucleon interactions, which reproduce the nucleon-nucleon scattering phase-shifts and deuteron properties, contain short-range repulsive and tensor components. Due to the short-range repulsion, nucleon pairs will not be found at distances below 0.5 fm. This is reflected by a high momentum component in the momentum distribution. The tensor correlations induce further momenta above the Fermi momentum. Though these correlations can only be measured indirectly, some physical observables may reflect the high momentum component. These days short-range correlations attract increasing interest. Recent measurements which try to extract information on the short-range correlations have been carried out at JLab [1]. These correlations also provide important information on the saturation property in nuclear matter.

We investigate the structure of short-range correlations in many-body states. The highly correlated many-body states are represented with an explicitly correlated basis which enables us to get a precise solution of a many-body Schrödinger equation for a realistic interaction [2, 3]. We show two-body density distributions calculated from three- and four-body states to investigate the short-range correlations between nucleon pairs. At distances below 1 fm a universal behavior is found which does not depend on the many-body states. Two-body momentum distributions reflecting the high momentum transfers of the short-range repulsive and tensor correlations are also discussed.

[2] Present address: RIKEN Nishina Center, RIKEN, Wako 351-0198, Japan, supported by the Special Postdoctoral Researchers Program of RIKEN.

Published under licence by IOP Publishing Ltd

## 2. Numerical Calculations

We assume that an $A$-nucleon state can be expanded in terms of a combination of basis states, each of which is a product of space, spin and isospin parts,

$$|\Phi; JM\rangle = \sum_{i=1}^{K} C_i \mathcal{A}\Big\{ \Big[ \Big|\psi_i^{(\text{space})}\psi_i^{(\text{spin})}\Big\rangle\Big]_{JM} \Big|\psi_i^{(\text{isospin})}\Big\rangle \Big\}. \tag{1}$$

Here $\mathcal{A}$ is the antisymmetrizer and the square bracket $[\cdots]$ stands for the angular momentum coupling. The spin and isospin parts are expanded using the basis of successive coupling, e.g.,

$$\Big|\psi_i^{(\text{spin})}\Big\rangle = |[\cdots[[[\tfrac{1}{2}\tfrac{1}{2}]_{S_{12}}\tfrac{1}{2}]_{S_{123}}]\cdots]_{S_iM_S}\rangle, \tag{2}$$

where the set of intermediate spins $(S_{12}, S_{123}, \ldots)$ takes all possible values compatible with the total spin $S_i$ of the $i$th basis. The isospin mixing is ignored in this contribution, so that the total isospin $T_i$ is kept fixed to $T$. The orbital part $\psi_i^{(\text{space})}$ is expressed in terms of the explicitly correlated Gaussian basis [2, 3]. The variational parameters in the basis are determined by a stochastic variational method [2, 4] and the basis dimension is increased until good convergence is reached. We investigate the wave functions of $^3$H, $^3$He and $^4$He using the Argonne V8$'$ interaction [5]. We also calculate the first excited state of $^4$He that exhibits dilute 3+1 cluster structures [6].

The antisymmetrized many-body state $|\Phi; JM\rangle$ contains all information about the nuclear system. Its $A$-body density is a function of $A$ position or momentum vectors and $4 * A$ spin-isospin possibilities and hence can not be visualized easily. Therefore we integrate and sum over $A-2$ single-particle degrees of freedom and are left with the two-body density. This represents an average over all particle pairs in the many-body state. In addition we integrate over the center of mass position of the pair and obtain the two-body densities for the four spin-isospin channels which are possible for a nucleon pair. The two-body density for the relative motion is given by

$$\rho^{\text{rel}}_{SM_S,TM_T}(\boldsymbol{r}) = \frac{1}{2J+1}\sum_M \langle\Phi; JM|\sum_{i<j}^{A} \hat{P}_{ij}^{SM_S}\hat{P}_{ij}^{TM_T}\delta^3(\hat{\boldsymbol{r}}_i - \hat{\boldsymbol{r}}_j - \boldsymbol{r})\,|\Phi; JM\rangle\,. \tag{3}$$

This is the probability density to find a nucleon pair at the relative position $\boldsymbol{r} = \boldsymbol{r}_1 - \boldsymbol{r}_2$ in the spin $S, M_S$ and isospin $T, M_T$ channel. The operators $\hat{P}_{ij}^{SM_S}$ and $\hat{P}_{ij}^{TM_T}$ project on spin and isospin of the pair, respectively. The corresponding distribution of the relative momentum $\boldsymbol{k} = (\boldsymbol{k}_1 - \boldsymbol{k}_2)/2$ of the particle pair with total spin $S, M_S$ and isospin $T, M_T$ is defined as

$$n^{\text{rel}}_{SM_S,TM_T}(\boldsymbol{k}) = \frac{1}{2J+1}\sum_M \langle\Phi; JM|\sum_{i<j}^{A} \hat{P}_{ij}^{SM_S}\hat{P}_{ij}^{TM_T}\delta^3\big(\tfrac{1}{2}(\hat{\boldsymbol{k}}_i - \hat{\boldsymbol{k}}_j) - \boldsymbol{k}\big)\,|\Phi; JM\rangle\,. \tag{4}$$

## 3. Results

Here we focus on the two-body density distributions with the $S$=1, $T$=0 channel. In Fig. 1 the spatial two-body densities $\rho^{\text{rel}}_{11,00}(\boldsymbol{r})$ of four different states are displayed. The striking observation is that at short distances they look very similar independently of the many-body state. That means that the correlation felt by a particle pair in the $S$=1, $T$=0 channel is the same at short distances independently of the remaining particles in the system. In regions where the potential is attractive, $\boldsymbol{r} \approx (0, 0, \pm 1$ fm$)$, the densities are large and in regions where the interaction is repulsive or close to zero the probability of finding the particle pair is small. At

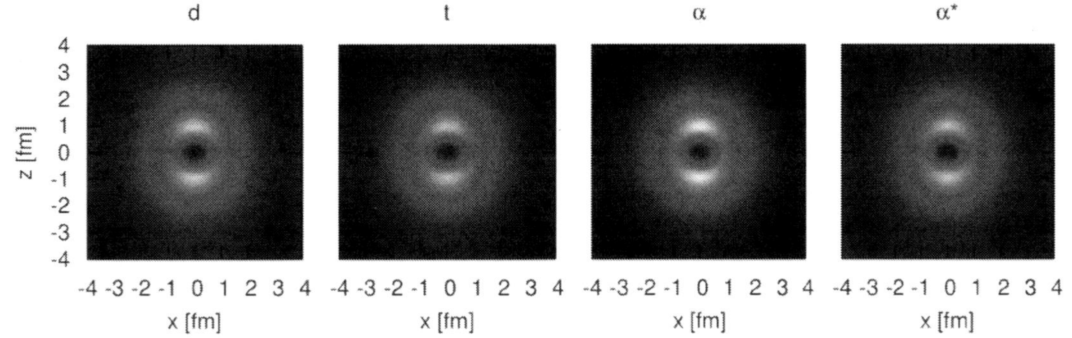


**Figure 1.** (Color online) From left to right: Two-body densities in coordinate space for a pair of nucleons with $S$=1, $M_S$=1 and $T$=0 in the ground states of $^2$H, $^3$H and $^4$He and the 20.21 MeV excited state of $^4$He denoted by d, t, $\alpha$ and $\alpha^*$, respectively. The densities have rotational symmetry around the $z$-axes and range from black = 0 to bright (yellow) = maximum. Maxima assume values of 0.008 fm$^{-3}$ for d, 0.015 fm$^{-3}$ for t, 0.035 fm$^{-3}$ for $\alpha$ and 0.015 fm$^{-3}$ for $\alpha^*$.

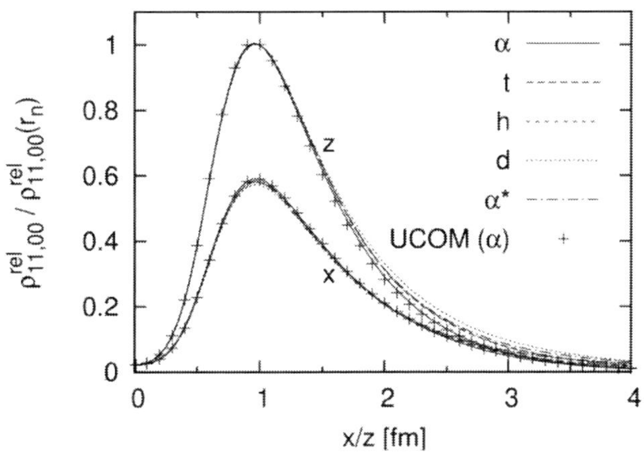


**Figure 2.** Cuts of $\rho^{\rm rel}_{11,00}(\boldsymbol{r})/\rho^{\rm rel}_{11,00}(\boldsymbol{r}_n)$ at $\boldsymbol{r}_n$=(0,0,1fm) for deuteron (d), triton (t), $^3$He (h), $^4$He ($\alpha$) and the first excited state of $^4$He ($\alpha^*$). The result with the unitary correlation operator method (UCOM) for $^4$He is also displayed.

short distances below 0.5 fm the AV8′ potential is so strongly repulsive that the pair densities in all many-body states are pushed down towards zero. One should bear in mind that in a simple shell model many-body state these correlations can not be represented and the two-body densities have actually their maximum at relative distance $\boldsymbol{r} = 0$.

Fig. 2 shows cuts of the normalized two-body density $\rho^{\rm rel}_{11,00}(\boldsymbol{r})/\rho^{\rm rel}_{11,00}(\boldsymbol{r}_n)$ along the $z$-direction and the $x$-direction. We normalize the quantities at $\boldsymbol{r}_n = (0, 0, 1{\rm fm})$, where the densities are close to their maximum value. It is surprising to see in Fig. 2 that for small distances all five densities practically coincide along the $z$-axis. The same holds true when going along the $x$-axis, although the normalization was done on the $z$-axis. This means that not only the central correlations but also the angular dependence of the tensor correlations are almost identical at short distances. The short-range central and tensor correlations exhibit universal behavior at short distances below about 1 fm. They do not depend on the nuclear many-body states for which they have been calculated.

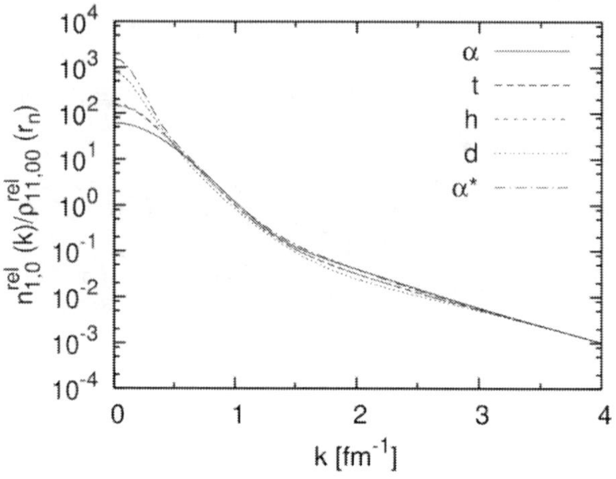


**Figure 3.** Two-body densities as a function of relative momentum $k$ for the $S=1$, $T=0$ channel multiplied by the same normalization factors for the coordinate space. Ground state densities of $^2$H, $^3$H, $^3$H, and $^4$He are denoted by d, t, h, and $\alpha$, respectively. The excited state of $^4$He is labeled with $\alpha^*$.

This universality confirms the basic assumption of the unitary correlation operator method (UCOM) [7]. In the UCOM short-range correlations are described explicitly with central and tensor correlation operators. In Fig. 2, we also display the two-body density of $^4$He calculated with the UCOM. The uncorrelated wave function is assumed to be a $(0s)^4$ harmonic oscillator shell model wave function. The UCOM result is in very good agreement with the accurate calculation obtained by our correlated basis approach.

Fig. 3 shows the normalized two-body density in the momentum space. The same normalization factors as for the two-body densities in coordinate space are used. As seen in the figure, we also find the universality at high momenta beyond 3 fm$^{-1}$ for all many-body states.

## 4. Summary

We have investigated the structure of short-range correlations in many-body states with the explicitly correlated basis using a realistic nucleon-nucleon interaction. We have shown two-body density distributions for the $S=1$, $T=0$ channel calculated from three- and four-body states to investigate the short-range correlations between nucleon pairs. At distances below 1 fm a universal behavior is found which does not depend on the many-body states. Two-body momentum distributions also exhibit the universal behavior at high momenta beyond 3 fm$^{-1}$. A study of the other spin-isospin channels are interesting and will be reported soon.

## References

[1] Subedi R *et al.* 2008 *Science* **320** 1476
[2] Varga K and Suzuki Y 1995 *Phys. Rev. C* **52** 2885
[3] Suzuki Y, Horiuchi W, Orabi M, and Arai K 2008 *Few Body Syst.* **42** 33
[4] Suzuki Y and Varga K 1998 *"Stochastic Variational Approach to Quantum-Mechanical Few-Body Problems"*, Lecture Notes in Physics, Vol. 54, Springer, Berlin Heidelberg
[5] Wiringa R B, Stoks B G J and Schiavilla R 1995 *Phys. Rev. C* **51** 38
[6] Horiuchi W and Suzuki Y 2008 *Phys. Rev. C* **78** 034305
[7] Roth R, Neff T, and Feldmeier H 2010 *Prog. Part. Nucl. Phys.* **65** 50

Journal of Physics: Conference Series **321** (2011) 012044 doi:10.1088/1742-6596/321/1/012044

# Microscopic Cluster Models: application to the structure of the $^{16}$B nucleus

**M. Dufour**(1) **and P. Descouvemont**(2)
(1)IPHC Bat27, IN2P3-CNRS/Université de Strasbourg BP28,
F-67037 Strasbourg Cedex 2 - France.
(2)Physique Nucléaire Théorique et Physique Mathématique, C.P. 229,
Université Libre de Bruxelles (ULB), B 1050 Brussels, Belgium.

E-mail: marianne.dufour@iphc.cnrs.fr

**Abstract.** General aspects of microscopic cluster models based on the combination of the Generator-Coordinate-Method and of the $R$-matrix method are presented. The adequacy of such methods to describe the physics of exotic light nuclei is illustrated with the unbound $^{16}$B nucleus.

## 1. Introduction

Microscopic Cluster Models (MCM) based on the combination of the Generator-Coordinate-Method (GCM)[1] and of the Microscopic R-matrix method (MRM) [2] are very efficient tools to investigate the nuclear many-body problem.

The principle is to treat the wave function of the $A$-nucleon system at the cluster approximation. In other words, the $A$ nucleons are assumed to be divided in clusters, described by shell-model wave functions. The cluster approximation is at the origin of the Resonating Group Method (RGM) proposed by Wheeler [3]. A significant breakthrough in microscopic cluster theories has been achieved by the introduction of the GCM [1] which is equivalent to the RGM. The GCM leads to simpler and more systematic calculations. The RGM relative functions are expanded over Gaussian functions in the GCM. The Gaussian behaviour is corrected through the MRM for scattering states [2].

The MCM framework presents numerous advantages. The Pauli principle is exactly treated through the antisymmetrization of all nucleons. The center of mass is exactly separated and the quantum numbers associated with the colliding nuclei are restored. Furthemore, the GCM wave functions of the bound and scattering states present a correct asymptotic behaviour.

Microscopic cluster models present a wide range of applications. In nuclear structure, numerous calculations attest their adequacy to describe the physics of exotic nuclei, and in particular of halo nuclei. Molecular states, which are known to be strongly deformed, and present a marked cluster structure are successfully described by such investigations [4]. They are also well suited to describe nuclear reactions of astrophysical interest where antisymmetrization effects are expected to be important.

The present paper presents a short overview of the MCM theoretical framework (Section 1). For the sake of simplicity, we limit the expressions to a two-cluster model with identical oscillator parameters. More extensive presentations can be found in Refs. [2, 5]. The recent study of the $^{16}$B resonances above the $^{15}$B+n threshold is choosen as a typical exemple of nuclear structure calculations well adapted to the microcopic cluster approach [6]. Indeed, the $^{16}$B being an unbound nucleus, only resonances are expected. In such conditions, wave-functions required a rigourous treatment of their asymptotic part that the MCM is able to performed in contrast to the traditional shell-model (SM) approaches (Section 2).

## 2. Short overview of the theoretical framework

In microscopic cluster models, the Hamiltonian of the $A$-nucleon system is usually written as

$$H = \sum_{i=1}^{A} T_i - T_{c.m.} + \sum_{i>j=1}^{A} (V_{ij}^{C} + V_{ij}^{SO} + V_{ij}^{Coul}), \tag{1}$$

where $T_i$ is the kinetic energy, $T_{c.m.}$ is the kinetic energy of the total center of mass, $V_{ij}^{C}$ is the central nucleon-nucleon part, $V_{ij}^{SO}$ is an effective spin-orbit interaction and $V_{ij}^{Coul}$ is the Coulomb force which is exactly treated. The central part is generally taken as a combination of Gaussian form factors. Typical interactions are the Volkov V2 and Minnesota interactions [7, 8].

In the following, we consider a typical process between a nucleus (1) and a nucleus (2) with $A_1$ and $A_2$ nucleons, respectively. The notation (1,2) refers to the unified nucleus with $A = A_1 + A_2$ nucleons. In the present framework, the wave functions of the unified nucleus are expressed as sums of multichannel functions. More precisely, in partial spin and parity $(J, \pi)$ vawe, they can be written in RGM notation as

$$\Psi^{JM\pi} = \sum_{c\ell I} \mathcal{A} \; g_{c\ell I}^{J\pi}(\rho) \; \left[ [\phi_{c,1}^{I_1} \otimes \phi_{c,2}^{I_2}]^{I} \; \otimes \; Y_\ell(\Omega_\rho) \right]^{JM}, \tag{2}$$

where $\mathcal{A}$ is the $A$-nucleon antisymmetrizor, $c$ labels the various channels, $\ell$ is the angular momentum between (1) and (2), $I_1$ and $I_2$ are the spin of (1) and (2) respectively, $I$ is channel spin, and $g_{c\ell I}^{J\pi}(\rho)$ are the relative functions depending on the relative coordinate $\rho$.

In the GCM, the $g_{c\ell I}^{J\pi}(\rho)$ functions are expanded over Gaussian functions $\Gamma_\ell(\rho, R)$

$$g_{c\ell I}^{J\pi}(\rho) = \int f_{c\ell I}^{J\pi}(R) \, \Gamma_\ell(\rho, R) \, dR, \tag{3}$$

where $R$ is the generator coordinate defined as the distance between the two clusters and $f_{c\ell I}^{J\pi}(R)$ are the generator functions obtained through the MRM (see Ref.[2] for more details). In practice, Eq. (3) is replaced by a finite sum over the generator coordinates $R$. It results a Gaussian behavior of the wave function which is corrected through the MRM.

The principle of the MRM is to divide the space into two regions delimited by a surface defined by $\rho = a$ where $a$ is the channel radius. In the internal region, the wave functions are given by the GCM. In the external region, radial functions are known to be the Coulomb functions. Matching conditions at $\rho = a$ provide the collision matrix and the cross section. Let us notice that the results do not depend on $a$. This method ensures the correct treatement of the asymptotic properties of the wave functions and leads to an unified description of bound states and resonances [2].

## 3. Application to the structure of the $^{16}$B

Experiments involving the $^{16}$B nucleus are difficult to perform. Its unbound nature has been confirmed by Kryger *et al.* [9]. Besides it exists mainly two experimental investigations of the low-lying structure. The first one performed by Kalpachieva *et al.* [10] shows the existence of a very narrow structure some 40 keV above the $^{15}$B+n threshold. Another peak is identified at

2.32 MeV. A second experiment performed more recently by Lecouey *et al.* [11] confirms the existence of the narrow structure at some 85 keV above the $^{15}$B+n threshold.

On the theoretical side, shell-model calculations have been performed [11]. However the unbound nature of the $^{16}$B nucleus raises the question of the accuracy of such an approach to describe states where the asymptotic behavior of the wave function cannot be neglected. MCM approach seems then better adapted.

In such a context, we have investigated the structure of $^{16}$B nucleus with an Extended Two-Cluster Model which includes many $^{15}$B+n channels [6]. The $^{16}$B wave functions are defined as a sum of $^{15}$B+n channel functions. $^{15}$B internal wave functions are obtained with 3 protons in the $p$ shells, the $s$ shell being filled for the proton part and 2 neutrons in the $sd$ shell, the $s$ and $p$ shells being filled for the neutron part. This description involves 1320 different Slater determinants leading to 87 $^{15}$B+n different channels corresponding to the $^{15}$B ground state and 86 $^{15}$B excited states. All the calculations are performed with the Volkov interaction.

Our calculations reproduce the $^{15}$B $3/2^-$, $5/2^-$ and $7/2^-$ ground-state band in fair agreement with experiment. It is well known that a good description of $^{15}$B is important in order to get a good description of the $^{16}$B nucleus.

$^{16}$B spectra are shown in Fig. 1. Experimental values of Kalpachieva *et al.* [10] and Lecouey *et al.* [11] are displayed together with the present GCM calculations. SM results [11] are also reproduced in order to compare both theoretical approaches. In the present work, the $^{16}$B resonance analysis is performed in terms of eigenphase shifts which are obtained from the diagonalization of the matrix of collision. The number of eigenphases is equal to the number of open channels at a given energy. To illustrate this point, we show typical curves obtained for the $0^-$ and $1^-$ states in Fig. 2. These curves are expected to provide information on the resonant structure for a given $J^\pi$ (see more details in Ref.[6]).

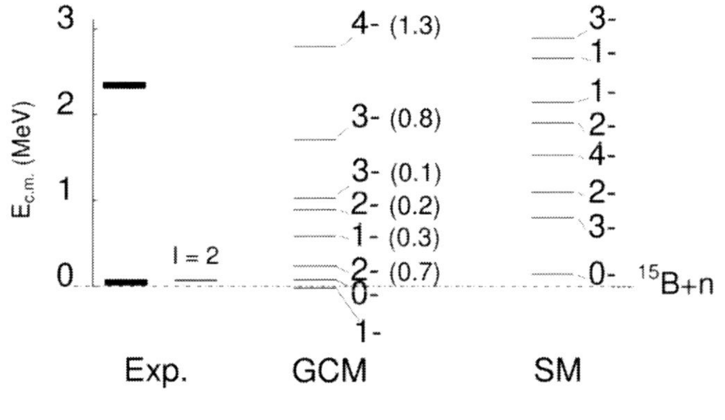


**Figure 1.** $^{16}$B spectra. Experimental values are taken from references [10, 11]. The widths in parenthesis are given in MeV.

Several resonances are obtained at low energy with the GCM. SM calculations also lead to a similar result but the ordering of the states is different. The existence of a very narrow resonance with a $\ell = 2$ dominant component above the $^{15}$B+n threshold is confirmed by the GCM. The spin is assigned to $0_1^-$. The corresponding value of the width is $\Gamma = 1.26 \times 10^{-2}$ keV in agreement with experiment.

All other resonances predicted by the GCM near the threshold present dominant $\ell = 0$ components. Let us notice that the $1_1^-$ state is even sligtly bound ($E_{c.m.} = -0.01$ MeV). Taking

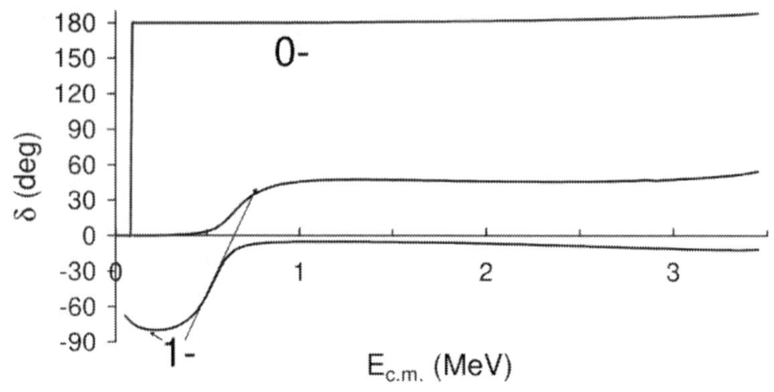


**Figure 2.** $^{16}$B eigenphaseshift analysis of the $0^-$ and $1^-$ states as a function of the energy.

into account the accuracy of our calculations, one can just predict the existence of a $1_1^-$ state near the threshold which could be the ground state. All these results are not in agreement with the SM ones. In particular, the $1^-$ states are predicted at higher energy with the SM. The main origin of these differences is probably the fact that the SM is not adapted to describe broad resonances because of the non accurate asymptotic behavior of the wave functions. New experiments would be necessary to clarify the situation.

## 4. Conclusion

The MCM framework presents many advantages to solve the nuclear many body problem. It is particularly well adapted to the study the structure of light nuclei. In such context, we aim to apply this framework to the physics of states presenting a condensate structure.

## References

[1] Horiuchi H 1977 *Prog. Theor. Phys. Suppl.* **62** 90
[2] Descouvemont P and Baye D 2010 *Rep. Prog. Phys.* **73** 036301
[3] Wheeler J A 1937 *Phys. Rev.* **52** 1083
[4] Dufour M, Descouvemont P and Nowacki F 2010 *Nucl. Phys. A* **836** 242
[5] Descouvemont P and Dufour M 2011 *Clusters in Nuclei, Vol.2* (Springer)
[6] Dufour M and Descouvemont P 2011 *Physics Letters B* **696** 237
[7] Volkov A B 1965 *Nucl. Phys.* **74** 33
[8] Thompson D R, LeMere M and Tang Y C 1977 *Nucl. Phys. A* **286** 53
[9] Kryger R A, Azhari A, Brown J, Caggiano J, Hellström M, Kelley J H, Sherrill B M, Steiner M and Thoennessen M 1996 *Phys. Rev. C* **53** 1971
[10] Kalpakchieva R, Bohlen H, von Oertzen W, Gebauer B, von Lucke-Petsch M, Massey T, Ostrowski A, Stolla T, Wilpert M and Wilpert T 2000 *Eur. Phys. J. A* **7**(4) 451
[11] Lecouey J L, Orr N, Marqués F, Achouri N, Angélique J C, Brown B, Carstoiu F, Catford W, Clarke N, Freer M, Fulton B, Grévy S, Hanappe F, Jones K, Labiche M, Lemmon R, Ninane A, Sauvan E, Spohr K and Stuttgé L 2009 *Phys. Lett. B* **672** 6

Journal of Physics: Conference Series **321** (2011) 012045 doi:10.1088/1742-6596/321/1/012045

# Microscopic calculations of elastic scattering between light nuclei based on a realistic nuclear interaction

**Jérémy Dohet-Eraly**[1], **Jean-Marc Sparenberg and Daniel Baye**

Physique Nucléaire et Physique Quantique, CP229, Université Libre de Bruxelles (ULB), B-1050 Brussels Belgium.

E-mail: jdoheter@ulb.ac.be, jmspar@ulb.ac.be and dbaye@ulb.ac.be

**Abstract.** The elastic phase shifts for the $\alpha + \alpha$ and $\alpha + {}^3$He collisions are calculated in a cluster approach by the Generator Coordinate Method coupled with the Microscopic $R$-matrix Method. Two interactions are derived from the realistic Argonne potentials AV8' and AV18 with the Unitary Correlation Operator Method. With a specific adjustment of correlations on the $\alpha + \alpha$ collision, the phase shifts for the $\alpha + \alpha$ and $\alpha + {}^3$He collisions agree rather well with experimental data.

## 1. Introduction

The aim of this work is to give a microscopic description of elastic collisions based on a realistic interaction while keeping rather simple basis wave functions. We choose to limit our approach to light colliding nuclei and we illustrate it on the $\alpha + \alpha$ and $\alpha + {}^3$He collisions.

For a long time the elastic scattering between light nuclei has been studied in the framework of the Generator Coordinate Method (GCM) [1]. In this model, the system made up of the colliding nuclei is described by a superposition of antisymmetrized states of two clusters separated by a variable distance. Each cluster is described by a Slater determinant of harmonic-oscillator states. After elimination of a Gaussian center-of-mass (c.m.) factor, the wave function of the system becomes invariant under translation (if the oscillator parameters of the two clusters are equal) and can be projected on angular momentum and parity [2]. While it has the required symmetries, the wave function does not have the correct asymptotic behavior of a scattering wave function. This problem is solved by the Microscopic $R$-matrix Method (MRM) [3, 4].

With the GCM, a large number of elastic collisions are described with a good precision [5, 6]. However, such results are obtained by using phenomenological interactions and fitting one or two parameters for each collision or each partial wave.

To obtain a more physical description of collisions and to approach an ab initio calculation, the phenomenological interaction has to be replaced by a realistic interaction reproducing nucleon-nucleon phase shifts and properties of the deuteron, such as the Argonne [7, 8] interactions. However, the study of elastic scattering with a realistic interaction is much more difficult and only a few collisions have been studied up to now [9, 10, 11].

The GCM basis is not suitable for realistic interactions. Thus, we adapt the wave function by applying a unitary operator or equivalently we adapt the interaction by applying the

[1] F.R.S.-FNRS Research Fellow.

Journal of Physics: Conference Series **321** (2011) 012045 doi:10.1088/1742-6596/321/1/012045

corresponding unitary transform, following the Unitary Correlation Operator Method (UCOM) [12, 13]. This method was used with success in nuclear structure calculations based on a realistic interaction [14] and very recently for reactions [11].

In this paper, we apply this correlation method to elastic scattering. To be specific, we study microscopically with the GCM coupled with the MRM, the $\alpha + {}^3$He and $\alpha + \alpha$ collisions with correlated interactions coming from the Argonne potentials AV8' [7] and AV18 [8].

## 2. Generator Coordinate Method

The microscopic approach is based on the Schrödinger equation for $A$ nucleons

$$H\Psi(1, 2, \ldots, A) = E\Psi(1, 2, \ldots, A), \tag{1}$$

with

$$H = \sum_{i=1}^{A} t_i - T_{cm} + \sum_{i<j}^{A} \left[ v_{ij}^N + v_{ij}^C \right], \tag{2}$$

where $t_i$ is the kinetic energy of nucleon $i$, $T_{cm}$ is the c.m. kinetic energy, $v_{ij}^N$ is the nuclear interaction (the AV8' or AV18 Argonne potential) and $v_{ij}^C$ is the Coulomb potential.

The nuclei are described by Slater determinants of $0s$ states of harmonic oscillator with oscillator parameter $b$, noted $\phi_{S_1}$ and $\phi_{S_2}$. $S_1$ and $S_2$ designate the spins of nuclei 1 and 2.

The collision wave function $\Psi$ is expanded into the GCM basis [1]. The GCM basis functions are projected Slater determinants given by

$$\psi_{lS}^{JM}(R) = \mathcal{A} \left[ [\phi_{S_1} \otimes \phi_{S_2}]^S \otimes Y_l(\Omega_\rho) \right]_M^J \Gamma_l(\rho, R), \tag{3}$$

where $J$ is the total angular momentum, $M$ is its projection, $l$ is the orbital angular momentum of the relative motion of nuclei and $S$ is the total spin. $\mathcal{A}$ is the antisymmetrization projector. $\boldsymbol{\rho} = (\rho, \Omega_\rho)$ is the relative coordinate between the nuclei. $R$ is called the generator coordinate. The $\Gamma_l$ function is given by

$$\Gamma_l(\rho, R) = \left( \frac{\mu}{\pi b^2} \right)^{3/4} e^{-\mu(\rho^2 + R)^2/2b^2} i_l \left( \frac{\mu \rho R}{b^2} \right), \tag{4}$$

where $i_l$ is a spherical Hankel function and $\mu$ is the reduced mass.

The phase shifts calculations can be performed systematically and fast with the GCM which makes this method very efficient. However, the GCM basis is not able to describe properly the colliding nuclei for realistic interactions. Indeed, a Gaussian wave function is not suitable for the repulsive core and the tensor component of the realistic nucleon-nucleon potential has no effect on the energy with this type of cluster. These problems are tackled with the UCOM. Moreover, the GCM wave function has not the correct asymptotic behavior of a scattering function. It is corrected by the MRM.

## 3. Unitary Correlation Operator Method

The aim of the UCOM [12, 13] is to insert the correlation with a unitary operator $C$ in the uncorrelated GCM states $\Psi$

$$\widetilde{\Psi} = C\Psi. \tag{5}$$

The tilde symbol will be systematically used to represent a correlated object. $C$ is the product of a radial unitary correlator $C_r$ [12] and a tensor unitary correlator $C_\Omega$ [13]

$$C = C_\Omega C_r. \tag{6}$$

The radial correlator modifies the wave function to take the repulsive core of the nuclear potential into account and the tensor correlator adds a $D$-wave component to the clusters.

In an equivalent way, the UCOM can be used to create an effective realistic interaction since

$$\langle\widetilde{\Psi}|H|\widetilde{\Psi}\rangle = \langle\Psi|C^\dagger HC|\Psi\rangle. \quad (7)$$

The correlated Hamiltonian $C^\dagger HC$ has the same eigenvalues as $H$ and is phase equivalent to $H$. Moreover, $C^\dagger HC$ holds the symmetries of reflection, rotation and time reversal as $H$, because $C$ holds them too. $C^\dagger HC$ has the advantage to have a softer repulsive core and to be more suitable for the GCM basis. But $C^\dagger HC$ has the drawback to contain three- and more-body terms even if $H$ is limited to two-body terms

$$C^\dagger HC = \widetilde{H}^{[1]} + \widetilde{H}^{[2]} + \widetilde{H}^{[3]} + \ldots \quad (8)$$

To keep the calculations relatively easy, we neglect the three- and more-body terms.

## 4. Microscopic $R$-matrix Method

The principle of the MRM [3, 4] is to divide the configuration space into two regions at the channel radius $a$. In the internal region ($\rho < a$), the wave function is expanded as a sum of $N$ correlated GCM basis functions

$$\widetilde{\Psi}_{\text{int}} = \sum_n f_n^{JlS}\widetilde{\psi}_{lS}^{JM}(R_n), \quad (9)$$

where $f_n^{JlS}$ are generator parameters and $R_n$ are selected discrete values of the generator coordinate $R$.

In the external region ($\rho > a$), the antisymmetrization is neglected and the interaction between the two clusters is considered purely Coulombic

$$\widetilde{\Psi}_{\text{ext}} = \phi_{\text{c.m.}}\left[\left[\widetilde{\phi}_{S_1}\otimes\widetilde{\phi}_{S_2}\right]^S \otimes Y_l(\Omega_\rho)\right]^J_M \left[\cos\delta_{lJ}F_l(\eta, k\rho) + \sin\delta_{lJ}G_l(\eta, k\rho)\right], \quad (10)$$

where $\delta_{lJ}$ is the $lJ$-wave phase shift, $\eta$ is the Sommerfeld parameter, $k$ is the wavenumber and $F_l$ and $G_l$ are the regular and irregular Coulomb functions. The phase shifts are calculated by matching the internal and external pieces of the wave function at the channel radius.

## 5. Elastic phase shifts

The radial and tensor correlators are adjusted on the $\alpha + \alpha$ phase shifts. The calculated phase shifts are compared in Fig. 1 with experimental data [15].

The adjustment is made for AV8'. So, we detail the results for this case. The results for AV18 are only slightly different. The calculated $0^+$ phase shifts reproduce the experimental data, including the energy of the resonance at 92 keV with a precision of about 1 keV. The $2^+$ phase shifts are still in good agreement with the experimental data while the $4^+$ phase shifts are smaller than experiment. The too high $4^+$ resonance energy is probably explained by the excessive simplicity of our wave function. The space of basis functions is too limited to reproduce the rotational constant of the $\alpha + \alpha$ rotational band.

By keeping the same correlators, thus without specific adjustment, we calculate the $\alpha + {}^3$He phase shifts. These are compared in Fig. 2 with experimental data [16]. The calculated phase shifts reproduce the general behavior of the experimental data. The precision is very good for the $1/2^-$ and $3/2^-$ waves and slightly less good for the $1/2^+$ wave.

For this nuclear system, the comparison with experiment can be extended to the ${}^7$Be bound states. For the AV8' potential, the bound-state energies are $E_{3/2^-} = -1.348$ MeV

Journal of Physics: Conference Series **321** (2011) 012045 doi:10.1088/1742-6596/321/1/012045

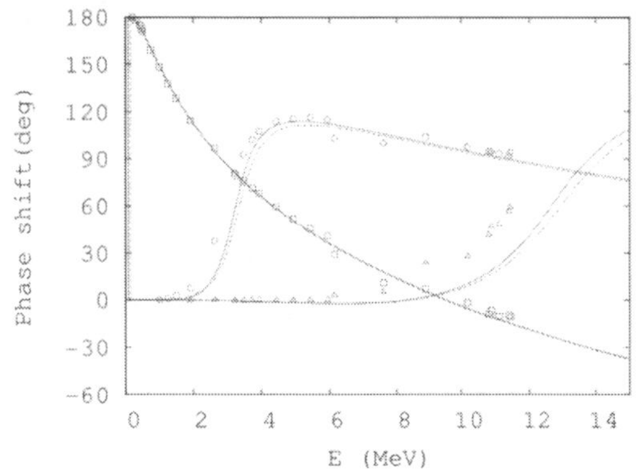


**Figure 1.** Elastic phase shifts for the $\alpha + \alpha$ collision. The phase shifts are calculated for the interactions derived from AV8'(——) and AV18(- - - -). Experimental data [$0^+$(□), $2^+$(○) and $4^+$(△)] from [15].

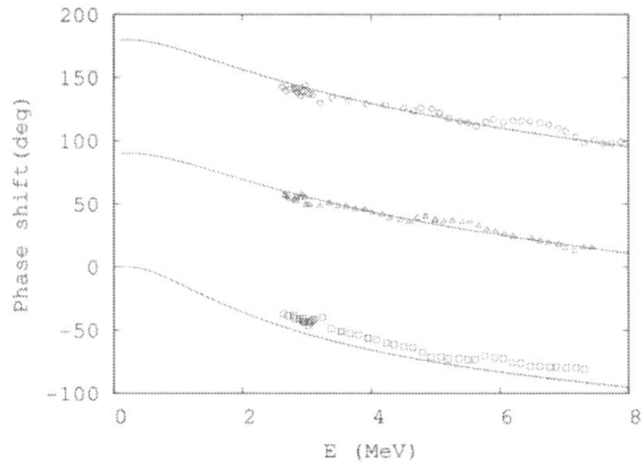


**Figure 2.** Elastic phase shifts for the $\alpha + {}^3$He collision. The phase shifts are calculated for the interactions derived from AV8'(——) and AV18(- - - -). Experimental data [$1/2^+$(□), $1/2^-$(○) and $3/2^-$(△)] from [16]. The $3/2^-$ phase shifts are shifted by $-90^\circ$ for clarity.

and $E_{1/2^-} = -0.764$ MeV with respect to the threshold energy. They are slightly higher than the experimental values $E_{3/2^-} = -1.587$ MeV and $E_{1/2^-} = -1.156$ MeV. These results display an improvement of the GCM model by using a realistic interaction. The Minnesota potential with its parameters adjusted to the $\alpha + {}^3$He phase shifts leads to strongly overbound states [6].

There exists no *ab initio* study of the $\alpha + {}^3$He collision. However, a calculation of the elastic phase shifts based on a UCOM potential derived from the AV18 potential has recently been performed [11]. In spite of a much larger basis of wave functions, the phase shifts do not agree significantly better with experiment than ours and the bound-state energies are only slightly improved.

## References


[1] Horiuchi H 1977 *Prog. Theor. Phys. Suppl.* **62** 90
[2] Brink D 1966 *Proc. Int. School Enrico Fermi 36, 1965 (Varenna)* (New York: Academic Press) p 247
[3] Baye D and Heenen P-H 1974 *Nucl. Phys.* A **233** 304
[4] Baye D, Heenen P-H and Libert-Heinemann M 1977 *Nucl. Phys.* A **291** 230
[5] Tang Y C 1981 *Lecture Notes in Physics: Topics in Nuclear Physics II* vol. 145 (Berlin: Springer) p 571
[6] Kamouni R and Baye D 2007 *Nucl. Phys.* A **791** 68
[7] Pudliner B S, Pandharipande V R, Carlson J and Pieper S C and Wiringa R B 1997 *Phys. Rev.* C **56** 1720
[8] Wiringa R B, Stoks V G J and Schiavilla R 1995 *Phys. Rev.* C **51** 38
[9] Nollett K M, Pieper S C, Wiringa R B, Carlson J and Hale G M 2007 *Phys. Rev. Lett.* **99** 022502
[10] Quaglioni S and Navrátil P 2008 *Phys. Rev. Lett.* **101** 092501
[11] Neff T 2011 *Phys. Rev. Lett.* **106** 042502
[12] Feldmeier H, Neff T, Roth R and Schnack J 1998 *Nucl. Phys.* A **632** 61
[13] Neff T and Feldmeier H 2003 *Nucl. Phys.* A **713** 311
[14] Roth R, Neff T, Hergert H and Feldmeier H 2004 *Nucl. Phys.* A **745** 3
[15] Afzal S A, Ahmad A A Z and Ali S 1969 *Rev. Mod. Phys.* **41** 247
[16] Spiger R J and Tombrello T A 1967 *Phys. Rev.* **163** 964

Journal of Physics: Conference Series **321** (2011) 012046 doi:10.1088/1742-6596/321/1/012046

# Clusters in $^{18,20}$O and $^{22}$Ne nuclei using the quantum mechanical fragmentation theory

**Manie Bansal, Raj Kumar and Raj K. Gupta**
Physics Department, Panjab University, Chandigarh-160014, INDIA.
E-mail: bansalmanni@gmail.com

**Abstract.** We examine for the first time the possibility of $^{14}$C clustering, in the resonant excited states of $^{18,20}$O and $^{22}$Ne nuclei, on the basis of the quantum mechanical fragmentation theory (QMFT). The important point is to include appropriate pairing strength $\delta$ in the temperature-dependent liquid drop energy.

## 1. Introduction

The discovery of $\alpha$-decay of heavy-nuclei prompted the idea that clusters of nucleons (here, two protons and two neutrons) might be preformed prior to emission, and that all nuclei could be considered to be composed of $\alpha$-particle clusters. In fact some of the earliest structure-models of nuclei extended such a principle. In order for the clusters to be fully formed, the proximity of cluster state to the decay threshold energy is crucial, which has become encapsulated in what is known as the Ikeda diagram. The "exotic" $\alpha$-clustering shapes in light $\alpha$-like nuclei at higher excitation energies have been nicely presented [1] in the well-known "Ikeda-diagram", which has been recently extended [2] to include also the $^{16}$O clustering in neutron-rich light nuclei. The $\alpha$ and $\alpha$-nucleus clustering in both the ground and excited states of $N = Z$, $\alpha$-nuclei have also been predicted by the recent relativistic mean-field calculations [3]. More recently, the Ikeda-diagram is proposed [4] to include also the $^{14}$C clusterization of light neutron-rich nuclei. In this contribution, we examine for the first time the possibility of $^{14}$C clusters, in the resonant excited states of $^{18,20}$O and $^{22}$Ne nuclei, on the basis of the quantum mechanical fragmentation theory (QMFT) of one of us (RKG) and collaborators [5, 6]. The important point is the inclusion of appropriate temperature-dependent pairing strength $\delta(T)$ in the liquid drop energy.

## 2. The quantum mechanical fragmentation theory (QMFT)

The QMFT is worked out in terms of the collective coordinates of mass and charge asymmetries $\eta = (A_1 - A_2)/(A_1 + A_2)$ and $\eta_Z = (Z_1 - Z_2)/(Z_1 + Z_2)$, the relative separation $R$, the multipole deformations $\beta_{\lambda i}$ ($\lambda$=2,3,4,...; $i$=1,2) and orientations $\theta_i$ of two nuclei, with the fragmentation potential $V_R(\eta, T)$, appearing in Schrödinger wave equation, calculated as

$$V(\eta, R, \ell, T) = -\sum_{i=1}^{2} B_i(A_i, Z_i, T) + V_c(R, Z_i, \beta_{\lambda i}, \theta_i, T) + V_p(R, A_i, \beta_{\lambda i}, \theta_i, T) + V_\ell(R, A_i, \beta_{\lambda i}, \theta_i, T). \quad (1)$$

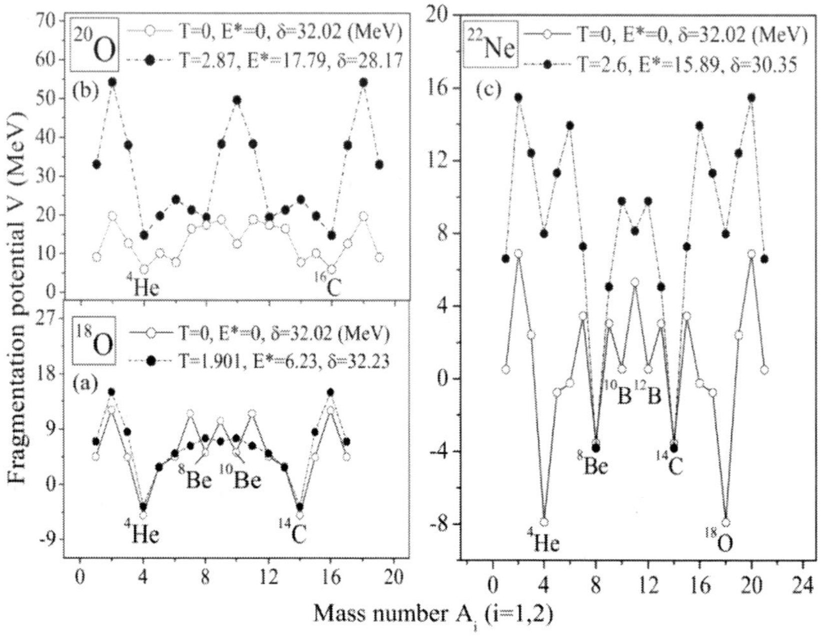


**Figure 1.** Fragmentation potentials for $^{18,20}$O and $^{22}$Ne nuclei at $T$=0 and at $T$-values of their corresponding excited state energies, for spherical nuclei.

Here $V_c$, $V_p$ and $V_\ell$ are the Coulomb, proximity and angular-momentum-dependent potentials. $B_i$ (i=1,2) are the binding energies of two nuclei, which at $T$=0 (ground state) are the experimental data [7], and, if the data were not available, or, else, $T$-dependence on $B$'s is to be included, the theoretical values [8] could be used. Thus, Eq. (1) for $T \neq 0$, and at a fixed $R$, is written as

$$\begin{aligned} V_R(\eta,T) &= \sum_{i=1}^{2} V_{LDM}(A_i,Z_i,T) + \sum_{i=1}^{2} \delta U \exp\left(-\frac{T^2}{T_0^2}\right) + V_c(R,Z_i,\beta_{\lambda i},\theta_i,T) \\ &\quad + V_p(R,A_i,\beta_{\lambda i},\theta_i,T) + V_\ell(R,A_i,\beta_{\lambda i},\theta_i,T), \end{aligned} \tag{2}$$

where $V_{LDM}(A_i,Z_i,T)$ is the liquid drop part of the binding energy and $\delta U$, the shell corrections. Note that the experimental binding energies contain both the macroscopic (liquid drop) and the microscopic (shell correction) parts.

The probability of cluster preformation in ground-state ($\nu$=0, $\nu$ refers to excited states),

$$P_0(A_2) = |\psi_R^{(0)}(A_2)|^2 \sqrt{B_{\eta\eta}(A_2)}\frac{2}{A}, \tag{3}$$

is obtained by solving the following stationary Schrödinger equation in $\eta$, at a fixed $R = R_a$,

$$\Big\{-\frac{\hbar^2}{2\sqrt{B_{\eta\eta}}}\frac{\partial}{\partial\eta}\frac{1}{\sqrt{B_{\eta\eta}}}\frac{\partial}{\partial\eta} + V_R(\eta,T)\Big\}\psi_R^{(\nu)}(\eta) = E_R^{(\nu)}\psi_R^{(\nu)}(\eta), \tag{4}$$

where $R_a = R_1(\alpha,T) + R_2(\alpha,T) + \Delta R(\eta,T) = R_t(\alpha,T) + \Delta R(\eta,T)$, with the radius vectors

$$R_i(\alpha_i,T) = R_{0i}(T)\Big[1 + \sum_{\lambda}\beta_{\lambda i}Y_\lambda^{(0)}(\alpha_i)\Big]. \tag{5}$$

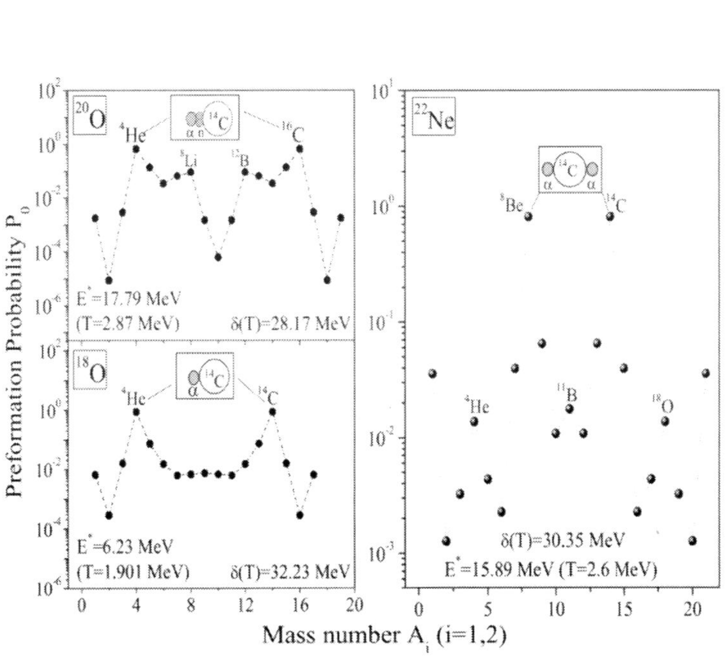


**Figure 2.** . Same as for Fig. 1, but for the preformation probability.

$R_{0i}(T)$ are the $T$-dependent nuclear radii for the equivalent spherical nuclei, given as [9],

$$R_{0i}(T) = [1.28A_i^{1/3} - 0.76 + 0.8A_i^{-1/3}](1 + 0.0007T^2), \tag{6}$$

with $T$ calculated from the excitation energy $E^*$ of resonant state, using $E^* = \frac{A}{8}T^2 - T$.

## 3. Calculations

For clustering effects in nuclei, we look for the energetically favored potential energy minima in fragmentation potential $V(\eta)$, or, for the largest preformation factor, the maxima in cluster preformation probability $P_0(A_2)$. The fragmentation potential in Fig. 1(a) shows that in the ground state ($T$=0) of $^{18}$O (with pairing constant $\delta$=32.02 MeV), the predominant cluster configuration is $^4$He+$^{14}$C, which is in addition to another weaker $^8$Be+$^{10}$Be clustering. However, at $T$=1.90 MeV ($E^*$=6.23 MeV), the above noted two cluster-configurations reduce to a single $\alpha$+$^{14}$C at a larger $\delta$=32.23 MeV, which means supporting an increased pairing at higher excitations. Similarly, for $^{20}$O at $E^*$=17.79 MeV and $\delta$=28.17 MeV in Fig. 1(b), the most probable clustering is $\alpha$+$^{16}$C, which, in view of the $^{18}$O nucleus in Fig. 1(a), could perhaps sequentially result in $\alpha$+2n+$^{14}$C clusters, in agreement with the extended Ikeda-diagram.

$^{22}$Ne nucleus in Fig. 1(c) presents a more interesting example. The predominant $^4$He+$^{18}$O, $^8$Be+$^{14}$C and $^{10}$B+$^{10}$B cluster configurations at $T$=0, become the only predominant $^8$Be+$^{14}$C cluster configuration at $E^*$=15.89 MeV ($T$=2.6 MeV) with $\delta$=30.35 MeV. The only other weaker clustering is for $^4$He+$^{18}$O. If we look at $^8$Be as two $\alpha$-particles, the most probable cluster configuration in $^{22}$Ne at $E^*$=15.89 MeV (at a smaller $\delta$=30.35 MeV), is $\alpha$+$^{14}$C+$\alpha$, which again supports the extended Ikeda-diagram to include $^{14}$C clustering. These results are better presented in Fig. 2, in terms of the preformation factor $P_0(A_i)$, showing the largest preformation yield for $\alpha$+$^{14}$C and $\alpha$+2n+$^{14}$C, respectively, in $^{18}$O and $^{20}$O, and $\alpha$+$^{14}$C+$\alpha$ in $^{22}$Ne at the observed excitation energy.

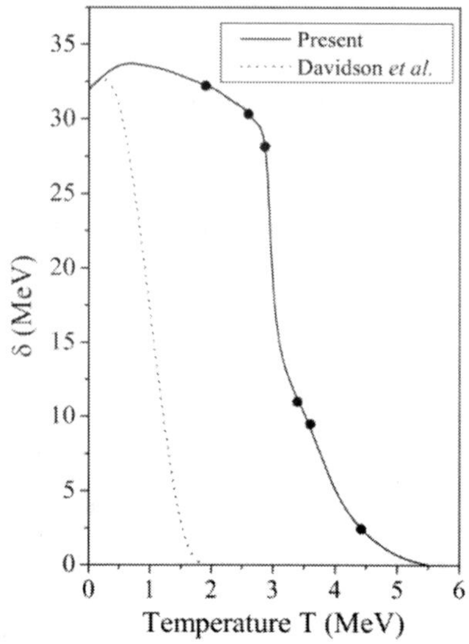


**Figure 3.** The variation of pairing constant $\delta$ with $T$ obtained from the present and earlier QMFT-based model [10] calculations, compared with that used by Davidson *et al.* [11].

Finally, Fig. 3 shows the modified $\delta(T)$ (solid line), with the lower three points corresponding to the earlier calculations on Ba nuclei [10] using a QMFT-based model, and the three higher points for the present $^{18}$O, $^{20}$O and $^{22}$Ne calculations, compared with that used by Davidson *et al.* [11] in their $T$-dependent liquid drop model (dotted line). Apparently, QMFT suggests a slower fall of pairing with temperature.

## 4. Conclusion

QMFT seems to support not only the $\alpha$-nucleus, but also the $^{14}$C clustering, provided a proper pairing constant were allowed in the temperature-dependent liquid drop binding energy.

## 5. Acknowledgments

The financial support from the Department of Sc. & Tech. (DST), Govt. of India, and Council of Sc. & Industrial Research (CSIR), New Delhi, is gratefully acknowledged.

## References

[1] H. Horiuchi and K. Ikeda, Prog. Theor. Phys. (Japan) 40, 277 (1968).
[2] W. von Oertzen, Eur. Phys. J. A. **11**, 403 (2001).
[3] P. Arumugam, B.K. Sharma, S.K. Patra and R.K. Gupta, Phys. Rev. C **71**, 064308(2005).
[4] C. Beck, *et. al*, 45$^{th}$ Zakopane Nucl. Phys. Conf., Poland, Aug. 30-Sept. 5, 2010.
[5] R.K. Gupta and W. Greiner, in *Heavy elements and related new phenomena*, editors W. Greiner and R.K. Gupta, (World Scientific, Singapore 1999), Vol. I, Ch. 14, p. 536.
[6] R.K. Gupta, in *Lecture Notes in Physics 818*, "Clusters in Nuclei", editor C. Beck, (Springer Verlag, Berlin 2010), Vol. I, p. 223.
[7] G. Audi and A.H. Wapstra, Nucl. Phys. **A** 595, 4 (1995).
[8] P. Möller, J. R. Nix, W. D. Myers and W. J. Swiatecki, At. Data Nucl. Data Tables **59**, 185 (1995).
[9] G. Royer and J. Mignen, J. Phys. G: Nucl. Part. Phys. **18**, 1781 (1992).
[10] R.K. Gupta, S.K. Arun, R. Kumar and Niyti, Int. Rev. Phys. (IREPHY) **2**, 369 (2008).
[11] N.J. Davidson, S.S. Hsiao, J. Markram, H.G. Miller and Y. Tzeng, Nucl. Phys. **A** 570, 61c (1994).

Journal of Physics: Conference Series **321** (2011) 012047 doi:10.1088/1742-6596/321/1/012047

# Cluster structures of excited states in $^{14}$C

**Tadahiro Suhara**[1] **and Yoshiko Kanada-En'yo**[2]

[1]Yukawa Institute for Theoretical Physics, Kyoto University, Kyoto 606-8502, Japan
[2]Department of Physics, Kyoto University, Kyoto 606-8502, Japan

E-mail: `suhara@yukawa.kyoto-u.ac.jp`

**Abstract.** We investigate structures of excited states in $^{14}$C with a method of $\beta$-$\gamma$ constraint antisymmetrized molecular dynamics in combination with the generator coordinate method. Various excited states with the developed $3\alpha$-cluster core structures are suggested in positive- and negative-parity states. In the positive-parity states, triaxial deformed and linear-chain structures are found to construct excited bands. Interestingly, $^{10}$Be+$\alpha$ correlation is found in the cluster states above the $^{10}$Be+$\alpha$ threshold energy. By applying a generalized molecular orbital model to the linear-chain state of $^{14}$C, we confirm the $^{10}$Be+$\alpha$ correlation.

## 1. Introduction

Owing to the experimental progress of unstable nuclei, various exotic structures have been discovered in neutron-rich nuclei where excess neutrons play important roles. In light neutron-rich nuclei, cluster structure with excess neutrons is one of the hot subjects in experimental and theoretical studies, where searching for new cluster states in excited states has been performed.

In this point of view, neutron-rich C isotopes such as $^{14}$C are an interesting subject because they have excess neutrons compared with $^{12}$C, which is already well known to show various structures owing to the coexistence of shell-model and cluster features. The ground state of $^{12}$C has mainly a shell-model feature of the $p_{3/2}$-subshell closed configuration, whereas, in the excited states above the $3\alpha$ threshold energy, various configurations of $3\alpha$-cluster structures were suggested in many theoretical works [1–11]. For example, the $\alpha$ condensation of weakly interacting three $\alpha$ clusters suggested in the $0_2^+$ and $2_2^+$ states [6–8], an equilateral-triangular structure of three $\alpha$ clusters in the $3_1^-$ state, and a linear-chainlike structure with an obtuse-angle-triangular $3\alpha$ configuration in the $0_3^+$ state [9,10] were suggested. Therefore, it is expected that $^{14}$C with two excess neutrons may also exhibit rich phenomena in the excited states. In particular, from an analogy of $^{12}$C, various cluster structures may appear in $^{14}$C as well as shell-model structures.

For the excited states of $^{14}$C, there are many experimental studies that indicate the appearance of cluster states such as $^{10}$Be+$\alpha$ cluster states [12–16]. Also on the theoretical side, cluster structures of $^{14}$C were suggested [17]. In the study of $^{14}$C with $3\alpha$+$2n$ cluster models, Itagaki *et al.* predicted that an equilateral-triangular structure of the well-developed three $\alpha$ clusters surrounded by excess neutrons is formed constructing $K^\pi = 0^+$ and $K^\pi = 3^-$ rotational bands in the excited states. Another interesting problem is whether a linear-chain $3\alpha$ structure with two excess neutrons exists in the excited states of $^{14}$C. Although there were discussions on the possibility of the linear-chain structure in neutron-rich C isotopes [12,18,19], there is no clear conclusion for stability of the linear-chain structure in $^{14}$C. Thus, the excited states of $^{14}$C are attracting much interest recently, and therefore, systematic study of the ground and excited states is required.

Our aim is to investigate the ground and excited states of $^{14}$C while focusing on cluster features. In the present study, we apply a combination of the $\beta$-$\gamma$ constraint AMD and the generator coordinate method (GCM), which we call the $\beta$-$\gamma$ constraint AMD + GCM. This method has already been proved to be a powerful approach to describe various structures such as cluster and shell-model-like structures [20]. In particular, it is useful for the systematic study of cluster states in excited states owing to the superposition of basis AMD wave functions on the two-dimensional $\beta$-$\gamma$ plane. Therefore, it is suitable for the study of $^{14}$C.

## 2. Framework

In this section, we explain the $\beta$-$\gamma$ constraint AMD + GCM, briefly. For more details, the reader is referred to Ref. [20].

In the method of AMD, a basis wave function of an $A$-nucleon system $|\Phi\rangle$ is described by a Slater determinant of single-particle wave functions $|\varphi_i\rangle$ as $|\Phi\rangle = \frac{1}{\sqrt{A!}}\det\{|\varphi_1\rangle, \cdots, |\varphi_A\rangle\}$. The $i$-th single-particle wave function $|\varphi_i\rangle$ consists of the spatial part $|\phi_i\rangle$, spin part $|\chi_i\rangle$, and isospin part $|\tau_i\rangle$ as

$$|\varphi_i\rangle = |\phi_i\rangle|\chi_i\rangle|\tau_i\rangle, \tag{1}$$

$$\langle \boldsymbol{r}|\phi_i\rangle = \left(\frac{2\nu}{\pi}\right)^{\frac{3}{4}} \exp\left[-\nu\left(\boldsymbol{r} - \frac{\boldsymbol{Z}_i}{\sqrt{\nu}}\right)^2 + \frac{1}{2}\boldsymbol{Z}_i^2\right], \tag{2}$$

$$|\chi_i\rangle = \xi_{i\uparrow}|\uparrow\rangle + \xi_{i\downarrow}|\downarrow\rangle, \tag{3}$$

where $\nu$ is the width parameter and is taken to be a common value for all the single-particle Gaussian wave functions in the present work. The spin orientation is given by the parameter $\boldsymbol{\xi}_i$, while the isospin part $|\tau_i\rangle$ is fixed to be up (proton) or down (neutron). In a basis wave function $|\Phi\rangle$, $\{X\} \equiv \{\boldsymbol{Z}, \boldsymbol{\xi}\} = \{\boldsymbol{Z}_1, \boldsymbol{\xi}_1, \boldsymbol{Z}_2, \boldsymbol{\xi}_2, \cdots, \boldsymbol{Z}_A, \boldsymbol{\xi}_A\}$ are complex variational parameters and they are determined by the energy optimization.

To describe various cluster and shell-model structures which may appear in the ground and excited states of $^{14}$C, we perform the energy variation with the constraints on the quadrupole deformation parameters, $\beta$ and $\gamma$.

In the calculations of energy levels, we superpose the parity and total-angular-momentum projected AMD wave functions $\hat{P}^J_{MK}|\Phi^{\pm}(\beta,\gamma)\rangle$. Thus, the final wave function for the $J^{\pm}_n$ state is given by a linear combination of the basis wave functions as $|\Phi^{J\pm}_n\rangle = \sum_K \sum_i f_n(\beta_i,\gamma_i,K)\hat{P}^J_{MK}|\Phi^{\pm}(\beta_i,\gamma_i)\rangle$. The coefficients $f_n(\beta_i,\gamma_i,K)$ are determined using the Hill-Wheeler equation. In the limit of sufficient basis wave functions on the $\beta$-$\gamma$ plane, it corresponds to the GCM with the two-dimensional generator coordinates of the quadrupole deformation parameters, $\beta$ and $\gamma$.

## 3. Results

In this section, we explain the calculated results for the positive-parity states. For more details and results for the negative-parity states are described in Ref. [21].

We show the calculated energy levels for the positive-parity states of $^{14}$C in Fig. 1 as well as the experimental levels [13–16,22]. We also display the density distributions of the intrinsic wave functions of bands. Four columns on the left are the experimental data and six columns on the right are the theoretical results. States in the three columns from the left are the known positive-parity states. States which were discovered in the $^{10}$Be+$\alpha$ break-up reactions are displayed in the fourth column from the left. The lower three states are those with a strong population of $^{10}$Be($0^+_1$) and upper three states are $^{10}$Be($2^+_1$). We classified the theoretical states into five

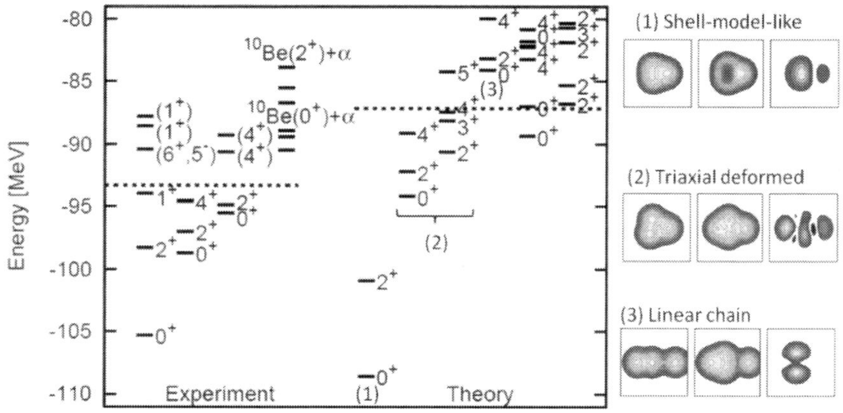


**Figure 1.** Energy levels for the positive-parity states of $^{14}$C and the density distributions of the intrinsic wave functions of bands. The dotted lines on the left and right show the experimental and theoretical $^{10}$Be+$\alpha$ threshold energies, respectively. The proton density $\tilde{\rho}_p$, neutron density $\tilde{\rho}_n$, and difference between the neutron and proton densities $\tilde{\rho}_n - \tilde{\rho}_p$ are illustrated.

groups by analyzing the intrinsic structures and $E2$ transition strengths. It is found that the ground band (1) has a shell-model-like structure, while three $\alpha$ clusters develop well in most of the excited states. The second and third column is the $K^\pi = 0^+$ and $K^\pi = 2^+$ side band (2), which has a triaxial structure. In these states, three $\alpha$ clusters have the isosceles-triangle configuration and excess neutrons occupy the $sd$-like orbitals. The fourth column is the $K^\pi = 0^+$ band (3), in which developed three $\alpha$ clusters have a linear-chainlike structure. It is interesting that the excess neutrons distribute around two of the three $\alpha$ clusters. This indicates the $^{10}$Be correlation in the linear-chain states. For the excited states of $^{14}$C, there are many experimental studies that indicate the appearance of $^{10}$Be+$\alpha$ cluster states. Our results suggest that excited states with $^{10}$Be+$\alpha$ correlation are the candidates for these states.

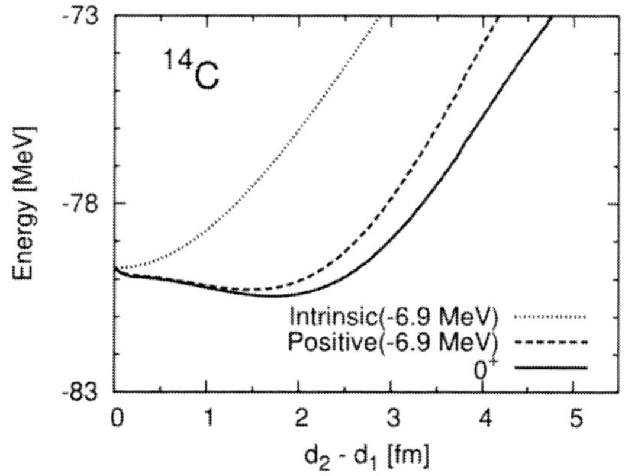


**Figure 2.** Energy curves of $^{14}$C with the linear-chain structure whose length $d_1 + d_2$ is fixed to be 5.2 fm, as a function of the difference of the $\alpha$-cluster intervals, $d_2 - d_1$.

Next, to confirm the $^{10}$Be+$\alpha$ in the linear-chain structure, we investigate the linear-chain structure in $^{14}$C with a generalized molecular orbital model which can describe asymmetric configurations of the linear-chain structure. In the previous calculation for $^{14}$C with AMD, there is a rotational band of linear-chain structure with $^{10}$Be+$\alpha$ correlation. However, there is a question whether $^{10}$Be+$\alpha$ correlation really exist. That is, in the $\beta$-$\gamma$ constraint AMD+GCM, detailed behavior of valence neutrons in the molecular orbitals which distribute whole system is not sufficiently described because a single particle wave function is approximated by a Gaussian wave packet. The calculated energy surface of $^{14}$C is shown as functions of the difference of the $\alpha$-cluster intervals $d_2 - d_1$ in Fig. 2. The intrinsic energy curve has the energy minimum at the symmetric configuration, $d_2 - d_1 = 0$ fm, while the positive-parity and the $0^+$ energy curves have energy minima at asymmetric configurations, $d_2 - d_1 > 0$ fm. At the energy minimum, $d_2 - d_1 = 1.7$ fm, the ratio of excess neutron distribution around left, middle, and right $\alpha$ clusters is l : m : r = 1.0 : 4.3 : 0.6. Valence neutrons gather around the middle $\alpha$ cluster. Since the middle $\alpha$ cluster is shifted to one side, the concentration of the amplitude at the middle $\alpha$ cluster means that valence neutron distribution leans toward the narrow side of two of three $\alpha$ clusters. That is, the asymmetric configuration of linear three $\alpha$ clusters and valence neutrons following the middle $\alpha$ cluster make $^{10}$Be+$\alpha$ correlation. More details are explained in Ref. [23].

## 4. Summary

We investigated $^{14}$C with $\beta$-$\gamma$ constraint AMD+GCM. Characteristic structures such as a triaxial structure and a linear-chain structure with the $^{10}$Be+$\alpha$ correlation appear in the excited bands. We confirmed the $^{10}$Be+$\alpha$ correlation in the linear-chain states of $^{14}$C by applying a generalized molecular orbital model to the system. As a future problem, we are estimating decay widths of $^{14}$C into $^{10}$Be+$\alpha$. We will report the detailed results in a future paper.

## Acknowledgments

The computational calculations of this work were performed by supercomputers in YITP and KEK. This work was supported by the YIPQS program in YITP. It was also supported by the Grant-in-Aid for the Global COE Program from the MEXT of Japan, and Grant-in-Aid for Scientific Research (Nos. 22·84, 18540263, and 22540275) from JSPS.

## References

[1] H. Horiuchi, Prog. Theor. Phys. **51**, 1266 (1974); Prog. Theor. Phys. **53**, 447 (1975).
[2] E. Uegaki, S. Okabe, Y. Abe, and H. Tanaka, Prog. Theor. Phys. **57**, 1262 (1977).
[3] Y. Fukushima and M. Kamimura, J. Phys. Soc. Jpn. **44**, 225 (1978).
[4] P. Descouvemont and D. Baye, Phys. Rev. C **36**, 54 (1987).
[5] Y. Kanada-En'yo, Phys. Rev. Lett. **81**, 5291 (1998).
[6] A. Tohsaki, H. Horiuchi, P. Schuck, and G. Röpke, Phys. Rev. Lett. **87**, 192501 (2001).
[7] Y. Funaki, A. Tohsaki, H. Horiuchi, P. Schuck, and G. Röpke, Phys. Rev. C **67**, 051306(R) (2003).
[8] Y. Funaki, A. Tohsaki, H. Horiuchi, P. Schuck, and G. Röpke, Eur. Phys. J. A **24**, 321 (2005).
[9] T. Neff and H. Feldmeier, Nucl. Phys. **A738**, 357 (2004).
[10] Y. Kanada-En'yo, Prog. Theor. Phys. **117**, 655 (2007).
[11] C. Kurokawa and K. Kato, Nucl. Phys. **A792**, 87 (2007).
[12] W. von Oertzen *et al.*, Eur. Phys. J. A **21**, 193 (2004).
[13] M. Milin *et al.*, Nucl. Phys. **A730**, 285 (2004).
[14] N. Soić *et al.*, Nucl. Phys. **A738**, 347 (2004).
[15] D. L. Price *et al.*, Phys. Rev. C **75**, 014305 (2007).
[16] P. J. Haigh *et al.*, Phys. Rev. C **78**, 014319 (2008).
[17] N. Itagaki, T. Otsuka, K. Ikeda, and S. Okabe, Phys. Rev. Lett. **92**, 142501 (2004).
[18] N. Itagaki, S. Okabe, K. Ikeda, and I. Tanihata, Phys. Rev. C **64**, 014301 (2001).
[19] W. von Oertzen, Z. Phys. A **357**, 355 (1997).
[20] T. Suhara and Y. Kanada-En'yo, Prog. Theor. Phys. **123**, 303 (2010).
[21] T. Suhara and Y. Kanada-En'yo, Phys. Rev. C **82**, 044301 (2010).
[22] F. Ajzenberg-Selove, Nucl. Phys. A **522**, 1 (1991).
[23] T. Suhara and Y. Kanada-En'yo, arXiv:1103.1358 [nucl-th].

Journal of Physics: Conference Series **321** (2011) 012048 doi:10.1088/1742-6596/321/1/012048

# Nuclear clustering and generalization of the Geiger-Nuttall law 100 years after its formulation

**Chong Qi, R.J. Liotta, R. Wyss**
Royal Institute of Technology, Alba Nova University Center, SE-10691 Stockholm, Sweden
E-mail: chongq@kth.se

**Abstract.** The formation of clusters in nuclei is discussed and conditions to assess clusterization are formulated. By using these conditions a generalization of the Geiger-Nuttall law is deduced. This universal decay law (UDL) is a linear relation between the half-lives of the decaying clusters (including alpha particles) and the corresponding $Q$-values. The agreement between the UDL and available experimental data is excellent. This allows us to distinguish the role played by pairing collectivity in the clustering process.

The first striking correlation between the half-lives of radioactive decay processes and the $Q$-values of the emitted particle was found in $\alpha$-decay systematics by Geiger and Nuttall [1] as,

$$\log T_{1/2} = aQ_\alpha^{-1/2} + b, \tag{1}$$

where $a$ and $b$ are constants. However, the Geiger-Nuttall law in the form of Eq. (1) has limited prediction power since the coefficients $a$ and $b$ change for the decays of each isotopic series [2, 3]. Intensive works have been done trying to generalize the Geiger-Nuttall law for a universal description of all detected $\alpha$ decay events [4, 5]. Here we present a truly universal formula valid for the radioactivity of all clusters, including $\alpha$-particles, by considering the clusterization and corresponding cluster-decay of nucleons within a nucleus [6, 7]. This allows us to search for new cluster decay modes and to carry out a simple and model-independent study of the decay properties of nuclei over the whole nuclear chart.

The relation between cluster decay and clusterization was first noticed by analyzing the decay processes in terms of the R-matrix theory. One thus obtains for the half-life of the decaying nucleus the expression [8],

$$T_{1/2} = \frac{\hbar \ln 2}{\Gamma_c} \approx \frac{\ln 2}{\nu} \left| \frac{H_l^+(\chi,\rho)}{RF_c(R)} \right|^2, \tag{2}$$

where $\nu$ is the outgoing velocity of the emitted particle. $H_l^+$ is the Coulomb-Hankel function and its arguments are standard [6]. The penetrability is proportional to $|H_l^+(\chi,\rho)|^{-2}$. Eq. (2) is valid for all clusters and for spherical as well as deformed cases [9].

The quantity $F_c(R)$ is the formation amplitude of the decaying cluster at distance $R$. The important feature that we used in the derivation of the UDL is that the decay width is independent upon the matching radius $R$. Our aim is to find few quantities that determine the half-life. Expanding in these quantities in terms of small parameters we hope to be able to find, at the lowest order of perturbation, an expression of the half-life which is as simple as the

Geiger-Nuttall law but valid in general, i.e., for all isotopic series as well as all type of clusters. With this in mind we notice that the Coulomb-Hankel function can be well approximated by an analytic formula, which for ground state to ground state transitions in even-even nuclei, i. e. for $l = 0$, reads [10],

$$H_0^+(\chi, \rho) \approx (\cot\beta)^{1/2} \exp\left[\chi(\beta - \sin\beta\cos\beta)\right], \tag{3}$$

where the cluster $Q$-value is $Q_c = \mu\nu^2/2$. One sees that $\cos^2\beta$ would be a small quantity if $Z_cZ_d$ is large, i.e., for heavy and superheavy systems [6, 7]. In this case one can expand the last term in a power series of $\cos\beta$. By defining the quantities $\chi' = Z_cZ_d\sqrt{A/Q_c}$ and $\rho' = \sqrt{AZ_cZ_d(A_d^{1/3} + A_c^{1/3})}$ where $A = A_dA_c/(A_d + A_c)$, one gets, after some simple algebra,

$$\log T_{1/2} = a\chi' + b\rho' + \log\left(\frac{\cot\beta \ln 2}{\nu R^2|F_c(R)|^2}\right) + o(3), \tag{4}$$

where $a = e^2\pi\sqrt{2m}/(\hbar\ln 10)$ and $b = -4e\sqrt{2mR_0}/(\hbar\ln 10)$ are constants ($m$ is the nucleon mass). The first two terms dominate the Coulomb penetration and $o(3)$ corresponds to the remaining small terms. But still the strong dependence of the half-life upon the formation probability in the third term of Eq. (4) has to be taken into account.

One sees from the equations above that in this theory the decay width is obtained by assuming a two-step process. In the first step the formation of the cluster and its motion in the daughter nuclear surface is established. In the second step the cluster, with the formation amplitude and corresponding wave function thus determined, is assumed to penetrate the centrifugal and Coulomb barriers. This second step is well understood since the pioneering work of Gamow [11]. Its great importance in radioactive decay studies lies in the fact that within a given cluster the penetrability process is overwhelmingly dominant. This explains the great success of macroscopic models in describing radioactive decay. However, within these effective theories the concept of "clusterization" is ignored and the decaying cluster is assumed to be already formed inside the mother nucleus. Moreover, usually the cluster is assumed to be like a little ball moving in a sort of vacuum within the mean field potential, bouncing on and reflected off the internal wall of the potential. The frequency of escape attempts is a free parameter which is formulated to reproduce experimental data.

Although successful, this semiclassical picture collides with basic quantum mechanics, since even if a cluster existed on the mother nucleus the Pauli principle would hinder any free motion of the cluster within the potential that traps the cluster inside the nucleus. What is missing in this picture is the possibility that the cluster is not "pre-formed" in the mother nucleus. In other words, one has to evaluate the probability that the cluster indeed is present on the nuclear surface. This can be done within the framework of the shell model.

It is interesting to notice that the problem of evaluating the $\alpha$ decay width properly played a fundamental role in the evolution of the shell model itself. The standard microscopic model in the 1950s was the shell model, which had just been proposed and found to perform well. But the shell model is not simply one of the many conceivable models even now: a shell-model basis is the most natural exact representation of nucleonic motion, to be used in solving the many-body nuclear Schrödinger equation. The shell model is more than a model featuring single-particle motion; in principle, with a large enough basis, it should be capable of taking into account any correlations, including correlations of inducing clustering. This was clear to the pioneers who started to use the shell model for the description of $\alpha$ decay. It aroused optimism at the beginning that very simple shell-model calculations were able to describe the low-energy spectra of not only the light nuclei but also of many heavy nuclei. Drawing on the success in spectroscopy, in the first applications of the shell model to the description of the mother nucleus of $\alpha$ decay only one configuration was used. The results were discouraging since the theoretical decay rates were smaller than the corresponding experimental values by 4-5 orders of

Journal of Physics: Conference Series **321** (2011) 012048 doi:10.1088/1742-6596/321/1/012048

magnitude [12], depending on the "reasonable" value to be chosen for the nuclear radius. Due to this failure, doubt arouse about the validity of the shell model itself [13]. With the benefit of hindsight, we now understand what the problem was. Since the matching radius $R$ in Eq. (2) has to be chosen at a distance beyond the point where the cluster was formed, i. e. beyond the range of the nuclear force and Pauli exchanges, the formation amplitude had to be or should have been evaluated at rather large distances. But that would have required shell models for the mother and daughter nuclei with large bases. With the very limited shell-model spaces used at that time, the region of prominent four-particle correlation was not reached at all, and, as a compromise, the radius $R$, was chosen to be well within the interaction region. One of the problems with those early microscopic calculations was that the residual nucleon-nucleon interaction was not known well enough. Soon after the pairing interaction had been adapted to nuclei [14, 15]. It was also applied to $\alpha$ decay [16]. It was then found with great relief that the pairing interaction, which links many shell-model configurations together, highly enhances the calculated $\alpha$-decay width. Although the calculated widths were still too small by orders of magnitude, it was clear that this discrepancy had to be attributed to the small shell-model spaces allowed by the computing facilities at that time. To their credit, the pioneers had deep insight into the role of configuration mixing. This can be best illustrated by a quotation, e.g., from Ref. [17]: ". . .the pure shell-model wave function fails badly in representing the amount of nucleon clustering in the vicinity of the tail of the matter distribution, but it may, with addition of more configuration mixture, fairly well represent the $\alpha$-clustering on the nuclear surface". Looking at the contemporary state of the art, this conclusion is far from trivial.

The fundamental role of configuration mixing was only confirmed by actual large-scale calculations [18, 19]. The physics behind the enhancement induced by configuration mixing is that, with the participation of high-lying configurations, the pairing interaction clusters the two neutrons and the two protons on the nuclear surface [20]. The two-neutron and two-proton wave-function terms add up constructively in the surface region. It was found that the mechanism that induces clustering is the same that produces the pairing collective, which is manifested in an strong increase in the form factor corresponding to the corresponding transfer cross section. This property gives rise to a giant pairing resonances, which corresponds to the most collective of the pairing states lying, as the standard (particle-hole) giant resonances, high in the spectrum.

The formation amplitude $F_c(R)$ can be extracted from the experimental half-lives data by,

$$\log|RF_c(R)| = \frac{1}{2}\log\left[\frac{\ln 2}{\nu}|H_0^+(\chi,\rho)|^2\right] - \frac{1}{2}\log T_{1/2}^{\text{Expt.}}. \tag{5}$$

We found that Eq. (4) can be written as a simple linear formula which properly takes into account the strong dependence of the formation amplitude upon the cluster as well as the mother nuclear structure to a first order of approximation. This we have archived by exploiting the property that for a given cluster $N_0 \equiv RF_c(R)/H_0^+(\chi,\rho)$ does not depend upon $R$. Proceeding as above one readily obtains the relation,

$$\log|RF_c(R)| \approx \log|R'F_c(R')| + \frac{2e\sqrt{2m}}{\hbar\ln 10}\left(\sqrt{R_0'} - \sqrt{R_0}\right)\rho', \tag{6}$$

where $R' = R_0'(A_d^{1/3} + A_c^{1/3})$ is a value of the radius that differs from $R$. Since for a given cluster any nuclear structure would be carried by the terms $RF_c(R)$ and $R'F_c(R')$ in exactly the same fashion, Eq. (6) implies that the formation amplitude is indeed linearly dependent upon $\rho'$. Therefore one can write,

$$\log T_{1/2} = a\chi' + b\rho' + c. \tag{7}$$

Journal of Physics: Conference Series **321** (2011) 012048 doi:10.1088/1742-6596/321/1/012048

This equation holds for all cluster radioactivities. We will call this relation the universal decay law (UDL). A straightforward conclusion from the UDL is that $\log T_{1/2}$ depends linearly upon $\chi'$ and $\rho'$. This to be valid should include the Geiger-Nuttall law as a special case. One sees that this is indeed the case since $\rho'$ remains constant for a given $\alpha$-decay chain and $\chi' \propto Q_c^{-1/2}$. We have chosen the parameters $a$, $b$ and $c$ by fitting experimental data of g. s. to g. s. cluster decay transitions in even-even nuclei. For details of these procedure and of the formalism itself see Refs. [6, 7]. We analyzed g.s. to g.s. radioactive decays of even-even nuclei starting with alpha-decay, for which there is a large amount of experimental data. The UDL reproduces the available experimental half-lives within a factor of about 2.2.

We found that although the UDL reproduces nicely most available experimental data, as expected, there is a case where it fails by a large factor. This corresponds to the $\alpha$ decays of nuclei with neutron numbers equal to or just below $N = 126$ [21]. The reason for this large discrepancy is that in $N \leq 126$ nuclei the $\alpha$ formation amplitudes are much smaller than the average quantity predicted by the UDL. The case that shows the most significant hindrance corresponds to the $\alpha$ decay of the nucleus $^{210}$Po. We found that the formation amplitude in $^{210}$Po is hindered with respect to the one in $^{212}$Po due to the hole character of the neutron states in the first case. This is a manifestation of the mechanism that induces clusterization, which is favored by the presence of high-lying configurations. Such configurations are more accessible in the neutron-particle case of $^{212}$Po than in the neutron-hole case of $^{210}$Po. This is a general feature in nuclei where neutrons and protons occupy different low-lying major shells.

In conclusion, in this paper we have first showed the mechanisms that induce clusterization of nucleons inside the nucleus. Using the conditions that are required to produce clusters, we have then obtained a simple formula that provides with great precision the half-lives corresponding to cluster decay. The formula is valid for all kind of clusters and for all isotopic series, as expected since we derived it from the general description of the decay half-life.

## Acknowledgments

This work has been supported by the Swedish Research Council (VR) under grant Nos. 623-2009-7340 and 2010-4723.

## References

[1] Geiger H and Nuttall J M 1911 *Philos. Mag.* **22** 613.
[2] Viola V E and Seaborg G T 1966 *J. Inorg. Nucl. Chem.* **28** 741.
[3] Buck B, Merchant A C and Perez S M 1990 *Phys. Rev. Lett.* **65** 2975.
[4] Poenaru D N, Plonski I H and Greiner W 2006 *Phys. Rev. C* **74** 014312, and references therein.
[5] Poenaru D N, Nagame Y, Gherghescu R A and Greiner W 2002 *Phys. Rev. C* **65** 054308.
[6] Qi C, Xu F R, Liotta R J and Wyss R 2009 *Phys. Rev. Lett.* **103** 072501.
[7] Qi C et al. 2009 *Phys. Rev. C* **80** 044326.
[8] Thomas R G 1954 *Prog. Theor. Phys.* **12** 253.
[9] Lovas R G, Liotta R J, Insolia A, Varga K, and Delion D S 1998 *Phys. Rep.* **294** 265.
[10] Fröman P O 1957 *Mat. Fys. Skr. Dan. Vidensk. Selsk.* **1** 3.
[11] Gamow G 1928 *Zeits. f. Physik* **52** 510.
[12] Mang H J 1960 *Phys. Rev.* **119** 1069.
[13] Wilkinson D H 1961 in Birks J B (Ed.), Proc. Rutherford Jubilee Int. Conf., Heywood and Co., London.
[14] Bohr A, Mottelson B R and Pines D 1958 *Phys. Rev.* **110** 936.
[15] Belyaev S T 1958 *Kgl. Danske Videnskab. Selskab, Mat.-Fys. Medd.* **31** 11.
[16] Mang H J and Rasmussen J O, *Kgl. Danske Videnskab. Selskab, Mat.-Fys. Skifter* **2** 3.
[17] Rasmussen J O 1965 in Siegbahn K (Ed.), Alpha-, Beta- and Gamma-Ray Spectroscopy, vol. 1 North-Holland, Amsterdam, p. 701.
[18] Tonozuka I and Arima A 1979 *Nucl. Phys. A* **323** 45.
[19] Janouch F A and Liotta R J 1979 *Phys. Lett.* **82B** 329.
[20] Janouch F A and Liotta R J 1983 *Phys. Rev. C* **27** 896.
[21] Qi C et al. 2010 *Phys. Rev. C* **81** 064319.

Journal of Physics: Conference Series **321** (2011) 012049 doi:10.1088/1742-6596/321/1/012049

# Atoms as many-body systems

**M Ya Amusia**

Racah Institute of Physics, The Hebrew University, Jerusalem, Israel;
Ioffe Physical-technical institute, RAS, St. Petersburg, Russia

E-mail: amusia@vms.huji.ac.il

**Abstract**. Contrary to common wisdom, not everything is clear and simple in the structure of many-electron atoms. Complexity in atoms is mainly a result of interelectron interaction that leads to rather unusual behaviour. Most transparently this is manifested in photo-ionization processes of many-electron atoms and some multi-atomic objects e.g. endohedrals. Particular attention will be given to the approach describing the interaction of photons with many-electron atoms in the frame of the many-body theory based on the Feynman diagrams technique. As a suitable one-electron approximation the Hartree - Fock (HF) approach will be presented. On its ground we will include the so-called electron correlation effects and discuss the frequently used Random Phase Approximation with Exchange - RPAE. Some results of recent calculations will be presented.

## 1. Introductory remarks

Historically, atoms with nuclear charge more than one were the first objects for developing quantum – mechanical many-body theory. A number of ideas and methods from atomic theory were then successfully applied in other areas of physics, particularly in nuclear physics. Here as good examples serve the shell model, Hartree-Fock (HF) method, spin-orbit interaction. The latter became now a fertilizing idea in solid-state physics. In its turn, atomic physics theory acquired Random phase approximation from electron gas theory and nuclear physics. It appeared that analogs to nuclear *Giant* dipole resonances were found in a number of atoms. Very successful was the application of the general quantum many-body theory with inclusion of Feynman diagrams. This approach permitted to understand deeply the processes of photoionization and electron scattering, paving the way to prediction of a number of different resonances, such as *interference, continuous spectrum autoionization* and *spin-doublet resonances* [1].

Specific features of complex atoms, contrary to nuclei, are the high accuracy, with which the inter-electron interaction is known, and the fact that the electron distribution is strongly inhomogeneous. Therefore, one has to start with the numerical implementation of HF and build on it many-body approaches self-consistently in order to escape the danger of double counting of the same effects. Among such theories the random-phase approximation with exchange (RPAE) [2] is the most successful. The use of this approach not only permits to describe most essential atomic processes, such as photoionization, electron scattering and vacancy decay, but to understand deeply the methods applied, which is essential in operating with this method in other domains, e.g. nuclear physics.

As objects of concrete calculations, we consider closed and semi-closed shell atoms and spherical endohedrals, i.e. fullerenes $C_{60}$ with noble gas atoms stuffed inside.

## 2. Specifics of HF

The HF or self-consistent field equation looks like a simple one particle Schrödinger-type equation, with at first glance non-essential differences. Namely, they are nonlinear and non-local, as is seen in (1), presented in atomic units, with electron mass $m$, charge $e$ and Planck constant $\hbar$ being equal to 1:

$$[-\Delta/2 - Z/r + \int d\vec{r}'\rho(\vec{r}')/|\vec{r}'-\vec{r}| - E_k]\phi_k(\vec{r}) = \sum_{i\le F}\int d\vec{r}'\phi_i^*(\vec{r}')\phi_k(\vec{r}')\phi_i(\vec{r})/|\vec{r}'-\vec{r}|. \tag{1}$$

Here Z is the nuclear charge and the summation on the r.h.s. is performed over all occupied states.

These modifications lead to profound alterations in qualitative features of the solutions of (1) [3]. The exchange leads to alteration of the asymptotic behavior, from depending upon the respective binding energy of state $k$ $\phi_{kl}(r)\big|_{r\to\infty} \sim \left[\exp(-\sqrt{2|E_k|}r)\right]/r^{l+1}$ to

$$\phi_{kl}(r)\big|_{r\to\infty} \sim \sum_o a_{kl,o}[\exp(-\sqrt{2|E_o|}r)]/r^{l+1}, \tag{2}$$

where $o$ denotes outer relative to $k$ atomic states and $E_o$ is the outermost binding energy. This alteration modifies dramatically the ionization of inner electrons $|E_k| >> |E_o|$ under the action of a strong field.

The exchange increases the number of zeroes $n_l$ in the radial wave function, leading to a violation of a well-known relation, which now reads $n_l \ge n-l-1$, where $n$ is the principal quantum number and $l$ is the angular momentum of the electron state considered. Due to non-locality, the HF equations become Gauge non-invariant in the sense that instead of the familiar relation $\vec{v}=\vec{p}$ one has $\vec{v}=i\left[\hat{H},\vec{r}\right]\ne\vec{p}$. As a result, we have two instead of one definitions of the photon-electron interaction operator.

Due to exchange, the Levinson theorem for elastic scattering phase shift at zero energy $\delta_l(E=0)$ is violated, becoming $\delta_l(E=0)=(q_l+t_l)\pi \ge q_l\pi$, where $q_l$ is the number of projectile-atom bound states (see [3] and references therein).

## 3. RPAE equations and photoionization resonances

The RPAE equation can be presented symbolically in a very simple operator form [1]:

$$\hat{D}(\omega) = d + \hat{D}(\omega)\hat{\chi}(\omega)U \text{ or } \hat{D}(\omega) = d/\left[1-\hat{\chi}(\omega)U\right], \tag{3}$$

where $\hat{D}$, $d$ are the photon absorption operators in RPAE and HF, respectively, $\hat{\chi}(\omega)$ is the electron-vacancy propagator and $U$ is the combination of direct and exchange Coulomb inter-electron interaction. Powerful broad structures that appear in the amplitude $\hat{D}(\omega)$ and in the photoionization cross-section are called Giant resonances. They are results of the fact that at some $\omega \equiv \omega_{GR}$ the following relation holds: $\left[1-\hat{\chi}(\omega_{GR})U\right]=0$.

Of interest are also cases when matrix elements of $\hat{D}(\omega)$ are close to zero, while those of $d$ are not. The corresponding structures are called interference resonances. Particularly strong is the mutual effect of subshells closely located in energy. Note that, depending upon absorbed photon frequency, the multi-electron subshells can act as resonators, mirrors and amplifiers. As a result, the photoionization cross-sections become quite complex functions of photon frequency (see [1] and references therein).

The investigation of atomic photoionization helped to reveal a number of other resonances, like Giant autoionization, Continuous spectrum photoionization and Intra doublet resonances. As far as I know, their counterparts in nuclear physics are neither found nor even predicted.

## 4. Non-dipole contribution to photoionization

The multi-electron effects in atomic photoionization are strong in the frequency region, where the dipole approximation is valid. Quadrupole corrections are suppressed by the factor $(\omega R/c)^2 << 1$, where $R$ is the atomic radius and $c$ is the speed of light. Therefore, the quadrupole contribution itself to the photoionization cross-section is almost non-observable. However, the angular distribution of photoelectrons is formed by superposition of different photoelectron partial waves. As a result, the ratio of contributions of quadrupole-to-dipole terms in the angular distribution is much bigger, namely of the order of $(\omega R/c) << 1$. The general expression for the angular photoelectron distribution due to the absorption of non-polarized light is given by the equation (see, e.g. [1])

$$\frac{d\sigma_{nl}(\omega)}{d\Omega} = \frac{\sigma_{nl}(\omega)}{4\pi}\left\{1 - \frac{\beta_{nl}(\omega)}{2}P_2(\cos\theta) + \frac{\omega}{c}\left[\gamma_{nl}(\omega)P_1(\cos\theta) + \eta_{n0}(\omega)P_3(\cos\theta)\right]\right\}, \quad (4)$$

where $nl$ are the ionized subshell quantum numbers and $P_{j=1,2,3}(\cos\theta)$ are the Legendre polynomials. The parameters in (4) are complicated functions of dipole and quadrupole matrix elements and photoelectron scattering phases. A specific feature of (4) is an asymmetry of the non-dipole terms relative to forward-backward direction of the photoelectrons ($\theta \to \pi - \theta$), which results in a preferential direction of the photoelectrons along or opposite to the photon flux momentum. It leads to macroscopic electric currents that were named "drag" currents, a not yet observed phenomenon [4].

As an example, we present the simplest expression for the angular anisotropy parameter – for s-subshells

$$\gamma_{n0}(\omega) = -\eta_{n0}(\omega) = \frac{6Q}{5D}\cos(\delta_1 - \delta_2), \quad (5)$$

where $Q, D$ are the absolute values of quadrupole and dipole matrix elements and $\delta_1, \delta_2$ are dipole and quadrupole photoelectron phases that include scattering phases and phases of the corresponding matrix elements. Resonance enhancements of $\gamma_{n0}(\omega)$ are due to maxima in Q, minima in D and *phase resonances* in the last term in (5) that happen when $\cos(\delta_1 - \delta_2) \to 1$.

## 5. Photoionization of endohedrals

Recently, the photoionization process of endohedrals attracted considerable interest. Most attention is given to studies of a $C_{60}$ fullerene molecule stuffed with a noble gas atom A, A@$C_{60}$ (see e.g. [5]). Microscopically, such a system can be treated as a sort of an atom, entirely in the frame of HF and RPAE methods for all interacting electrons of this object moving in the fields of 60 carbon nuclei and A atom nucleus. However, such an approach is too complicated to be numerically implemented. Instead, one can take into account that the $C_{60}$ shell is thin enough and its radius $R_C$ is much bigger than the atomic radius $R_A$. For low enough $\omega$ its structure is inessential and $C_{60}$ can be substituted by a spherically-symmetric homogeneous electric charge distribution, similar to another multi-electron shell quite far from atom A [5].

Having this in mind, one can separately take into account two effects that influence the photoionization of the stuffed atom. The first is the reflection of the photoelectron leaving the atom A by the static field of the $C_{60}$ shell. This can be taken into account by adding the static field of $C_{60}$ in the HF equation for the A atom. In some cases this effect can be taken into account simply by multiplying the atomic A photoabsorption amplitude by a factor $F_l(p)$, where $l$ and $p$ are the angular momentum and the modulus of the linear momentum of the outgoing electron.

The other effect is the modification of the photon beam acting upon atom A due to polarization of the $C_{60}$ electron shell under the action of this photon beam. This effect is taken into account by introducing the *polarization* factor $S(\omega)$ that is expressed via the dipole polarizability of the $C_{60}$ electron shell.

There exist two-shell endohedrals, of which an example is A@$C_{60}$@$C_{240}$ that can be treated as an atom A surrounded by two reflecting and highly polarizable shells [6].

## 6. Resonances in endohedral atoms.

The presence of fullerene shells brings in a multitude of different resonances. Particularly rich is the structure of the photoionization cross-section for two-shell endohedrals. We limit ourselves to one-shell endohedrals, for which *confinement* (exemplified by Figure 1) and *Giant endohedral* (exemplified by Figure 2) resonances are most impressive. While the former are results of interference phenomena due to photoelectron waves scattering by the fullerenes shell, the latter reflect the combined effect of photoelectron scattering by the fullerene shell and dependence of the fullerene dipole polarizability upon $\omega$ that is taken into account by the factor $S(\omega)$. Giant endohedral resonances have cross-sections that are bigger by a factor up to 10-20 than for the isolated atom A. As is seen, the corresponding cross-section reaches 1000 Mb, while for an isolated atom it is only 80 Mb.

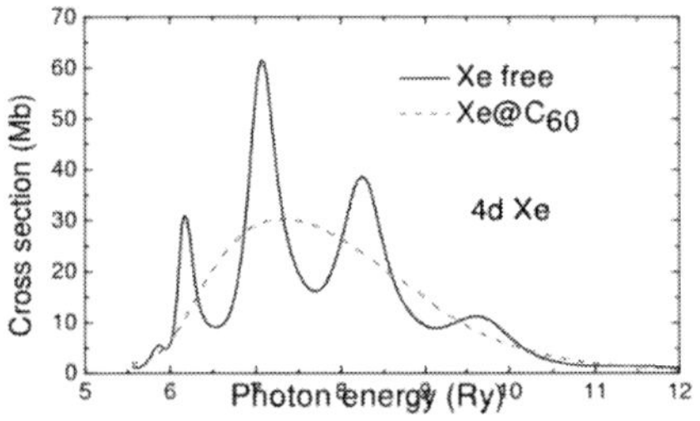


**Figure 1.** Confinement resonance of 4d-electrons for Xe@$C_{60}$, in RPAE.

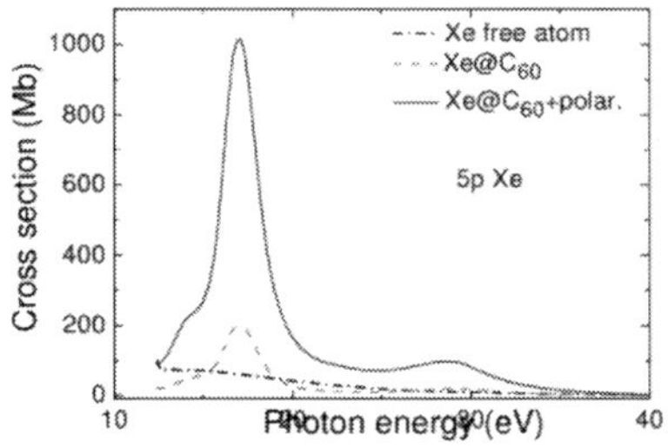


**Figure 2.** 5p- electrons Giant endohedral resonance for Xe@$C_{60}$, in RPAE

## 7. Concluding remarks

We demonstrated impressive examples of many-body effects in photoionization of atoms and endohedrals. They manifest themselves in a number of new resonances that await experimental detection. However, it is essential to note that similar effects are very important in electron scattering upon atoms and endohedrals and decay of vacancies created in the scattering processes. A number of new reaction channels open when the inter-electron interaction is taken into account. As in the considered processes of photoionization, many-electron corrections to processes, which are possible in one-electron approximation, are of great importance, too.

## References

[1] Amusia M Ya 2004 *Radiation Physics and Chemistry* **70** 237
[2] Amusia M Ya 1990 *Atomic Photoeffect* (New York-London: Plenum Press)
[3] Amusia M Ya 2009 *JETP Lett.* **90** 161
[4] Amusia M Ya, Connerade J-P 2000 *Rep. Prog. Phys.* **63**, p. 41
[5] Amusia M Ya 2007 *Journal of Electron Spectroscopy and Related Phenomena*, **159** 81
[6] Amusia M Ya, Chernysheva L V, Liverts E Z 2009 *Phys. Rev.* A **80** 032503

Journal of Physics: Conference Series **321** (2011) 012050 doi:10.1088/1742-6596/321/1/012050

# Proton-neutron mixed-symmetry states and a separable approximation for Skyrme interactions

**A. P. Severyukhin**[1], **N. N. Arsenyev**[1], **V. V. Voronov**[1], **N. Pietralla**[2] **and Nguyen Van Giai**[3]

[1]Bogoliubov Laboratory of Theoretical Physics, Joint Institute for Nuclear Research, 141980 Dubna, Moscow region, Russia
[2]Institut fr Kernphysik, Technische Universitt Darmstadt, 64289 Darmstadt, Germany
[3]Institut de Physique Nucléaire, CNRS-IN2P3, Université Paris-Sud, F-91406 Orsay Cedex, France

E-mail: sever@theor.jinr.ru

**Abstract.** Starting from an effective Skyrme interaction we present a finite rank separable approach for the QRPA calculations taking into account the effects of the phonon-phonon coupling. As an example the properties of the low-lying quadrupole states in $^{94}$Mo are considered. It is shown that the third $2^+$ state is the mixed-symmetry state.

## 1. Introduction

One of the successful tools for nuclear structure studies is the quasiparticle random phase approximation (QRPA) with the self-consistent mean-field derived from the Skyrme interaction, see a rather complete list of references on that subject in Ref. [1]. The low-energy spectrum of nuclear excitations provides the most sensitive testing ground for the nuclear structure calculations. In particular, properties of the quadrupole-collective isovector excitations of the valence shell of heavy nuclei, so-called mixed-symmetry states [2, 3], are sensitive to the proton-neutron interaction. It is interesting to study the properties of the mixed-symmetry states in the low-energy spectrum calculated with the Skyrme interaction.

The complexity of calculations taking into account the coupling between one-phonon and more complex states increases rapidly with the size of the configuration space. Making use of the finite rank separable approximation [4, 5] for the residual interaction enables one to perform the QRPA calculations in very large two-quasiparticle spaces. The approach has been generalized to take into account the coupling between the one- and two-phonon components of the wave functions [6]. Here, we use an extension of our approach by taking into account the particle-particle residual interaction [7].

One of the well-known examples is the mixed-symmetry $2^+$ state in $^{94}$Mo. The low-energy spectrum of quadrupole excitations in $^{94}$Mo is extensively studied in many experiments [3, 8, 9]. The experimental efforts have stimulated theoretical analysis based either on the interacting boson model (IBM-2) [8] or on the quasiparticle phonon model (QPM) [9, 10, 11]. In this report we describe briefly our method and present the analysis of the properties of the low-lying quadrupole states in $^{94}$Mo.

## 2. The method

The starting point of the method is the HF-BCS calculation (see for example [12]) of the ground states, where spherical symmetry is imposed on the quasiparticle wave functions. The continuous part of the single-particle spectrum is discretized by diagonalizing the HF hamiltonian in a harmonic oscillator basis. We work in the quasiparticle representation defined by the canonical Bogoliubov transformation:

$$a^+_{jm} = u_j \alpha^+_{jm} + (-1)^{j-m} v_j \alpha_{j-m} \tag{1}$$

where $jm$ denote the quantum numbers $nljm$. As effective interactions, a Skyrme interaction with tensor terms [13] in the particle-hole (p-h) channel and a density dependent zero-range force in the particle-particle (p-p) channel are used. The strength of the p-p interaction is fixed to reproduce the odd-even mass difference of nuclei in the studied region [7].

The residual interaction in the p-h channel $V^{ph}_{res}$ and in the p-p channel $V^{pp}_{res}$ can be obtained as the second derivative of the energy density functional with respect to the particle density and the pair density, accordingly. Hereafter we simplify $V^{ph}_{res}$ by approximating it by its Landau-Migdal form. For Skyrme interactions all Landau parameters with $l > 1$ are zero. We keep only the $l = 0$ terms in $V^{ph}_{res}$. In this work we study only normal parity states and one can neglect the spin-spin terms since they play a minor role [5]. The two-body Coulomb and spin-orbit residual interactions are also dropped. Therefore we can write the residual interaction in the following form:

$$V^a_{res}(\mathbf{r}_1, \mathbf{r}_2) = N_0^{-1}[F^a_0(r_1) + F'^a_0(r_1)(\tau_1 \cdot \tau_2)]\delta(\mathbf{r}_1 - \mathbf{r}_2), \tag{2}$$

where $a$ is the channel index $a = \{ph, pp\}$; $\tau_i$ are the isospin operators, and $N_0 = 2k_F m^*/\pi^2\hbar^2$ with $k_F$ and $m^*$ standing for the Fermi momentum and nucleon effective mass. The expressions for $F^{ph}_0, F'^{ph}_0$ and $F^{pp}_0, F'^{pp}_0$ can be found in Ref.[14] and in Ref.[7], respectively. The p-h matrix elements and the antisymmetrized p-p matrix elements can be written as the separable form in the angular coordinates [4, 7]. Following the paper [4] the radial integrals are calculated accurately by using the $N$-point integration Gauss formula. Thus, the residual interaction can be expressed as the sum of $N$ separable terms [4, 7].

We introduce the phonon creation operators

$$Q^+_{\lambda\mu i} = \frac{1}{2}\sum_{jj'}\left(X^{\lambda i}_{jj'} A^+(jj';\lambda\mu) - (-1)^{\lambda-\mu} Y^{\lambda i}_{jj'} A(jj';\lambda-\mu)\right), \tag{3}$$

$$A^+(jj';\lambda\mu) = \sum_{mm'} \langle jmj'm' \mid \lambda\mu\rangle \alpha^+_{jm}\alpha^+_{j'm'}, \tag{4}$$

where the index $\lambda$ denotes total angular momentum and $\mu$ is its z-projection in the laboratory frame. One assumes that the ground state is the QRPA phonon vacuum $\mid 0\rangle$. We define the one-phonon excited states as $Q^+_{\lambda\mu i} \mid 0\rangle$. Making use of the linearized equation-of-motion approach one can get the QRPA equations [12]. Solutions of the set of linear equations yield the eigen energies and the amplitudes $X, Y$ of the one-phonon excited states. Using the separable form of the residual interaction, the QRPA equations can be reduced to the secular equation and the matrix dimensions never exceed $6N \times 6N$ independently of the size of the two-quasiparticle configuration space [7].

To take into account the effects of the phonon-phonon coupling in the simplest case one can write the wave functions of excited states as

$$\Psi_\nu(JM) = \left\{\sum_i R_i(J\nu)Q^+_{JMi} + \sum_{\lambda_1 i_1 \lambda_2 i_2} P^{\lambda_1 i_1}_{\lambda_2 i_2}(J\nu) \left[Q^+_{\lambda_1\mu_1 i_1} Q^+_{\lambda_2\mu_2 i_2}\right]_{JM}\right\} \mid 0\rangle \tag{5}$$

with the normalization condition

$$\sum_i R_i^2(J\nu) + 2 \sum_{\lambda_1 i_1 \lambda_2 i_2} (P_{\lambda_2 i_2}^{\lambda_1 i_1}(J\nu))^2 = 1 \tag{6}$$

Using the variational principle one obtains the set of linear equations [6] which has the same form as the QPM equations [10], but the single-particle spectrum and the parameters of the residual interaction are calculated with the Skyrme force.

We apply our approach to study characteristics of the low-lying $2^+$ states in $^{94}$Mo. As the parameter set in the particle-hole channel, we use the central Skyrme interaction SLy5 [15] and the same zero-range tensor interaction as in Ref. [13]. The strength of the surface-peaked zero-range pairing force is taken equal to -870 MeVfm$^3$ in connection with the soft cutoff at 10 MeV above the Fermi energies as introduced in Ref. [7, 16]. This value of the pairing strength is fitted to reproduce the experimental pairing energies of $^{94}$Mo.

## 3. Low-lying quadrupole states in $^{94}$Mo

First, properties of low-lying quadrupole states in $^{94}$Mo are studied within the one-phonon approximation. There are only two $2^+$ states below 2.5 MeV, see [17]. The $2_1^+$ state is the collective state which has a typical isoscalar character. The dominant neutron and proton phonon amplitudes $X, Y$ of the $2_1^+$ state are in phase. In particular, the energy of the first proton two-quasiparticle pole $\{1g_{9/2}, 1g_{9/2}\}$ is greater than the energy of the neutron one $\{2d_{5/2}, 2d_{5/2}\}$ and, for the $2_1^+$ state, the largest contribution to the norm of the $2_1^+$ state (64%) comes from the neutron configuration $\{2d_{5/2}, 2d_{5/2}\}$. The dominant neutron and proton amplitudes of the fairly collective $2_2^+$ state are in phase opposition. As a consequence, the isovector character of the $2_2^+$ state is reflected in the noticeable size of the $B(M1; 2_2^+ \rightarrow 2_1^+)$ value. This analysis within the 1p-1h configuration space can help to identify the mixed-symmetry state, but it is only a rough estimate.

Let us now discuss the extension of the space to one- and two-phonon configurations. To construct the wave functions (5) of the low-lying $2^+$ states up to 3 MeV we use only the $2^+$ phonons, and all one- and two-phonon configurations with energies up to 8 MeV are included. The inclusion of high-energy configurations plays a minor role in the calculations [17]. The calculated $2^+$ state energies, the largest contributions of the wave function normalization (6), the transition probabilities $B(E2; 0_{g.s.}^+ \rightarrow 2_i^+)$, $B(M1; 2_i^+ \rightarrow 2_1^+)$ and the experimental data [8] are given in table 1. Note that the $B(M1)$ values have been calculated with the $g$-factors of free protons and neutrons. One can see that the first $2^+$ state is mainly the isoscalar $[2_1^+]_{QRPA}$, but the two-phonon contribution is appreciable. The second $2^+$ state has the dominant two-phonon

**Table 1.** Energy, structure of the low-lying quadrupole states and the transition probabilities in $^{94}$Mo. Experimental data are from Ref. [8].

| $\lambda_i^\pi = 2_i^+$ | Energy (MeV) | | Structure | $B(E2; 0_{g.s.}^+ \rightarrow 2_i^+)$ ($e^2fm^4$) | | $B(M1; 2_i^+ \rightarrow 2_1^+)$ ($\mu_N^2$) | |
|---|---|---|---|---|---|---|---|
| | Exp. | Theory | | Exp. | Theory | Exp. | Theory |
| $2_1^+$ | 0.871 | 0.6 | $83\%[2_1^+]_{QRPA}$ | 2030(40) | 1000 | | |
| $2_2^+$ | 1.864 | 1.6 | $60\%[2_1^+ \otimes 2_1^+]_{QRPA}$ | 32(7) | 30 | 0.06(2) | 0.01 |
| $2_3^+$ | 2.067 | 2.0 | $85\%[2_2^+]_{QRPA}$ | 230(30) | 250 | 0.48(6) | 0.84 |
| $2_4^+$ | 2.393 | 3.2 | $72\%[2_1^+ \otimes 2_2^+]_{QRPA}$ | 27(8) | 20 | 0.07(2) | 0.01 |

configuration $[2_1^+ \otimes 2_1^+]_{QRPA}$ composed of isoscalar phonons. The third $2^+$ state is dominated by the isovector $[2_2^+]_{QRPA}$ and can be associated with the mixed-symmetry state. It is worth to mention that the mixed-symmetry $2^+$ state in $^{94}$Mo was predicted for the first time in Ref. [8] based on the IBM-2 calculation. Although our approach describes the properties of the low-lying states less accurately than more phenomenological ones [9, 11], the results are still in a reasonable agreement with the experimental data [8, 9].

## 4. Summary

Starting from an effective Skyrme interaction we present a finite rank separable approach for QRPA calculations taking into account the effects of the phonon-phonon coupling. Choosing as an example the nucleus $^{94}$Mo, the properties of the low-energy spectrum of quadrupole excitations are analyzed. It is found that the third $2^+$ state is the mixed-symmetry state successfully predicted within the IBM-2. The structure of the low-lying $2^+$ states calculated in our approach are close to those that were calculated within the QPM. They are generally in a reasonable agreement with experimental date.

## Acknowledgments

We are grateful to Prof. R. V. Jolos and Prof. Ch. Stoyanov for valuable discussions. This work was supported in parts by the Heisenberg-Landau program and by the IN2P3-JINR agreement.

## References

[1] Paar N, Vretenar D, Khan E and Coló G 2007 *Rep. Prog. Phys.* **70** 691
[2] Iachello F 1984 *Phys. Rev. Lett.* **53** 1427
[3] Pietralla N, von Brentano P and Lisetskiy A F 2008 *Prog. Part. Nucl. Phys.* **60** 225
[4] Nguyen Van Giai, Stoyanov Ch and Voronov V V 1998 *Phys. Rev.* C **57** 1204
[5] Severyukhin A P, Stoyanov Ch, Voronov V V and Nguyen Van Giai 2002 *Phys. Rev.* C **66** 034304
[6] Severyukhin A P, Voronov V V and Nguyen Van Giai 2004 *Eur. Phys. J.* A **22** 397
[7] Severyukhin A P, Voronov V V and Nguyen Van Giai 2008 *Phys. Rev.* C **77** 024322
[8] Pietralla N *et al* 1999 *Phys. Rev. Lett.* **83** 1303
[9] Burda O *et al.* 2007 *Phys. Rev. Lett.* **99** 092503
[10] Soloviev V G 1992 *Theory of Atomic Nuclei: Quasiparticles and Phonons* (Bristol and Philadelphia: Institute of Physics)
[11] Lo Iudice N and Stoyanov Ch 2000 *Phys. Rev.* C **62** 047302
[12] Ring P and Schuck P 1980 *The Nuclear Many Body Problem* (Berlin: Springer)
[13] Coló G, Sagawa H, Fracasso S and Bortignon P F 2007 *Phys. Lett.* B **646** 227
[14] Nguyen Van Giai and Sagawa H 1981 *Phys. Lett.* B **106** 379
[15] Chabanat E, Bonche P, Haensel P, Meyer J and Schaeffer R 1998 *Nucl. Phys.* A **635** 231
[16] Krieger S J *et al* 1990 *Nucl. Phys.* A **517** 275
[17] Severyukhin A P, Arsenyev N N, Voronov V V, Pietralla N and Nguyen Van Giai in preparation

# Pairing and collectivity of the first $2^+$ state in light Sn isotopes

**D Tarpanov** [1,2], **Ch Stoyanov** [1], **N Lo Iudice**[3]
[1]Institute for Nuclear Research and Nuclear Energy, 1784 Sofia, Bulgaria
[2]Institute of Theoretical Physics, Warsaw University, ul. Hoza 69, PL-00681 Warsaw, Poland
[3] Dipartimento di Scienze Fisiche, Università di Napoli 'Federico II' and Istituto Nazionale di Fisica Nucleare, Monte S. Angelo, Via Cintia I-80126 Napoli, Italy

E-mail: tarpanov@inrne.bas.bg

**Abstract.** In recent experiments [1–4] the E2 strengths have been measured in the neutron deficient $^{106-112}$Sn isotopes. Using different models [5–8] a lot of theoretical effort was devoted in order to understand the effect of different aspects of the nuclear structure on the $B(E2, g.s. \to 2_1^+)$ strength in the Sn chain. In those works a large decrease of the B(E2) values in $^{102,104}$Sn in comparison to the heavier Sn isotopes $^{106-112}$Sn has been predicted. Here we concentrate on the effect of the pairing on the E2 strength in the Sn nuclei chain. We analyze the correlation one could observe between the pairing gap and the B(E2) strengths to the first $2^+$ excited state. The theoretical results on the B(E2) strengths are obtained within a QRPA model. These predictions are analyzed and compared to the experimental results.

## 1. Motivation

The $B(E2; 0^+ \to 2_1^+)$ values provide one of the most sophisticated tool to examine the shell structure of nuclei. In this work we try to convince the reader that these values are also correlated with the pairing in the nuclei. For our investigation we have chosen the Sn isotopic chain as it provides the longest chain of semi-magic nuclei, well studied both experimentally, and theoretically. What is of importance for us, is the well established, for almost the whole chain, experimental value of $B(E2; 0^+ \to 2_1^+)$. Recently, new measurements in $^{106,108}$Sn [4] have been done. They show a reduction in B(E2) strength when one goes towards the neutron deficient $^{106}$Sn. Still it is uncertain, what would one expect when going towards the $^{100}$Sn doubly magic nuclei. In this work we assume a simple model for the description of the ground state and the excited state of the nuclei, still we would like to point out the good description of the experimental data available up-to-date, and to show the prediction of such a model for the $^{102,104}$Sn. We use a Skyrme-Hartree-Fock scheme for the description of the ground state, on top of which pairing in constant gap approximation is included. The excited state is represented in Quasi-particle Random Phase Approximation. In this scheme we would like to emphasise the role of the pairing on the structure of the excited state, and thus its influence on the B(E2) strengths.

## 2. Theoretical Framework

We have used the Separable Quasi-particle Random Phase Approximation [9] in order to describe the excited states of the Sn nuclei. The single-particle spectrum has been calculated within a

Skyrme-Hartree-Fock approximation, a SIII [10] parameterization of the force has been adopted. The pairing has been taken in a constant gap approximation. The pairing strength is used as a constant for the whole isotpic chain. Still it is taken as independent parameter and it is not fitted on the experimental gap values. The continuous part of the single-particle spectrum is discretized by diagonalizing the HF hamiltonian on a harmonic oscillator basis. The single quasi-particle basis has been chosen in such a way that the Energy-weighted Sum Rule is depleted by 100% up to 30 MeV. In the model we use the residual particle-hole interaction, taken in a Landau-Migdal form, which in coordinate representation reads:

$$V_{ph} = N_o^{-1} \sum_l F_l + G_l \sigma_1 \cdot \sigma_2 + (F_l + G_l \sigma_1 \cdot \sigma_2)\, \tau_1 \cdot \tau_2 \delta(r_1 - r_2) \quad (1)$$

here $N_0 = \frac{2k_F m^*}{\pi^2 \hbar^2}$. In this expansion only the $l = 0$ terms are non zero. The Hamiltonian used for the QRPA calculations includes the average HF field, pairing interactions and the isoscalar and isovector particle-hole residual interaction in a N-rank seperable form. The exact form of the Hamiltonian could be found in [9, 11].

## 3. Results

In the Sn isotopes the protons form a closed shell subsystem. This of course affects the proton particle-hole excitations (PPHE), which could be excited only across the Z=50 gap. Because of this feature of the structure of the nuclei the PPHE are at a relatively high energy. On the other hand, the neutron particle-hole excitation (NPHE) in the semi-magic Sn nuclei could be excited in the valence shell, thus the NPHE are lower in energy. From this observation one would expect that the dominant contribution in the low-lying $2^+$ excitations would be neutron. This would be especially important in the first $2^+$ state. The proton contribution to the excited states has a strong effect on the B(E2) strengths, thus one would expect that even a small change in the structure of the first excited state would lead to a change in the observed B(E2)s. On the left-hand-side of Fig. 1 showing the neutron pairing gap behavior with the mass number, we can see that with the increasing of the mass number, the neutron pairing gap, first gets bigger, then it remains almost unchanged, and then falls again when one goes towards the doubly magic $^{132}$Sn. These changes influence the energy of the neutron particle-hole excitations, which are lower close to the doubly magic Sn isotopes. The neutron component in the first $2^+$ excited state is affected by the energy of NPHE, the lower the energy is the higher is the contribution from the neutrons on the $2_1^+$ state. The considerations above are only qualitative. To get a quantitative result we have used the QRPA model to describe the first excited state in the Sn isotopic chain. The strength of the particle-hole interaction, has been fitted in such a way that the energy of the $2^+$ state corresponds to the experimental value. The obtained results on the B(E2) strengths, corresponds well to the experimental values, as one could see from the right-hand-side of Fig. 1. Let us focus our attention on nuclei belonging to the steep slopes where the neutron pairing gap is small and the energy of the NPHE goes down. In those nuclei the contribution of the PPHEs in the structure of the first $2^+$ excited state goes down to $(3 - 5\%)$, compared to $(6 - 8\%)$ $^{108-124}$Sn. The latter has an imminent effect on the B(E2) strength, which falls rapidly, with the decreasing of the proton component of the wave function of the first excited $2^+$. This could be clearly seen on the right-hand-side of Fig. 1. One could see that a drastic reduction of the B(E2) strength is expected for the $^{102-104}$ Sn. From the figure it could also be seen that in these calculations the subshells closures are somehow smeared, and for example at the $2d_{5/2}$ closure in $^{106}Sn$ one cannot observe sharp increase in the NPHE, thus the B(E2) strength changes smoothly within the closed shell.

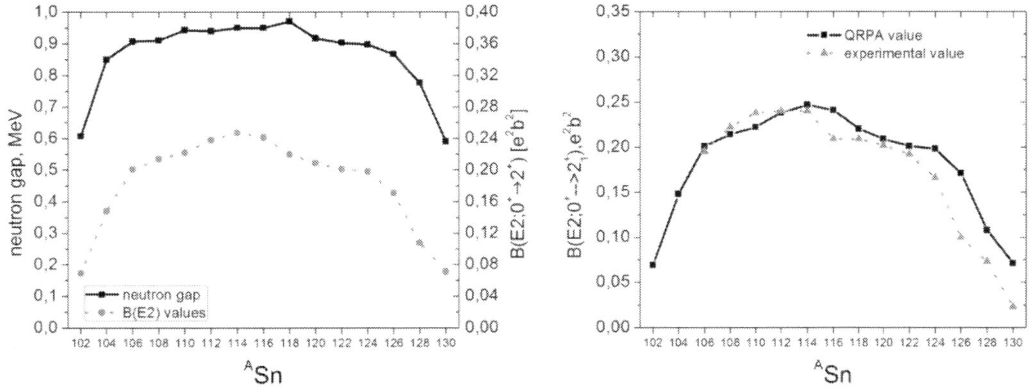


Figure 1: On the r.h.s., with black squares, is presented the evolution of the neutron gap with the mass number in the Sn isotopic chain. In red circles is presented the calculated $B(E2; 0^+ \rightarrow 2_1^+)$ strength, for the corresponding nuclei. On the l.h.s. is presented a comparison between the theoretical predictions for the B(E2) values, obtained within a QRPA model(black squares), versus the experimental observables (red circles).

## 4. Conclusion

The results from our calculations could be summarized in two points. The first result is a prediction on the B(E2) strengths, as one goes towards the neutron deficient Sn nuclei. We could expect a drastic reduction of the B(E2) strengths in the $^{102,104}$Sn isotopes as compared to the nearby nuclei $^{106,108}$Sn. The reduction we observe is only due to the increasing role of the neutron particle-hole excitation $2d^p_{5/2}2d^h_{5/2}$ which is 430keV lower in energy in $^{102,104}$Sn nuclei as compared to $^{106}Sn$. This of course increase its contribution in the structure of the first excited state by more than 8% which reduces the proton contribution in the preceding 2 nuclei by 3, 2% as compared to the last.

The second interesting result is presented on the left hand side of Fig. 1 and it supposes that the overall behavior of the B(E2) strength of the transitions to the first excited $2^+$ states is correlated with the neutron pairing gap in these nuclei. If confirmed experimentally, such a result would emphasize the crucial role of pairing also in nuclei deviating from the stability line.

## 5. Acknowledgments

This work is partly supported by CNRS (France)- BAS(Bulgaria) PICS Program No 4329

## References

[1] Banu A *et al.* 2005 *Phys. Rev.* C **72** 061305(R)
[2] Cederkall J *et al.* 2007 *Phys. Rev. Lett.* **98** 172501
[3] Vaman C *et al.* 2007 *Phys. Rev. Lett.* 162501
[4] Ekstrom A *et al.* 2008 *Phys. Rev. Lett.* **101** 012502
[5] Zelevinsky V and Volya A 2005 *Nucl. Phys.* A **725** 325c
[6] Ansari A 2005 *Phys. Lett.* B **623** 37
[7] Talmi I 1971 *Nucl. Phys.* A **172** 1
[8] Beene J R *et al.* 2004 *Nucl. Phys.* A **746** 407
[9] Giai N V, Stoyanov C and Voronov V V 1998 *Phys. Rev.* C **57** 1204
[10] Beiner M, Flocard H, Giai N V and Quentin P 1975 *Nucl. Phys.* A **238** 29
[11] Soloviev V G 1992 *Theory of atomic nuclei: Quasiparticles and phonons* (Bristol: Institute of Physics Publishing)

Journal of Physics: Conference Series **321** (2011) 012052 doi:10.1088/1742-6596/321/1/012052

# Study of the neutron skin thickness of $^{208}$Pb in mean field models

**X. Roca-Maza[1,2], M. Centelles[1], X. Viñas[1] and M. Warda[1,3]**

[1]Departament d'Estructura i Constituents de la Matèria and Institut de Ciències del Cosmos, Facultat de Física, Universitat de Barcelona, Diagonal *647*, *08028* Barcelona, Spain
[2]INFN, sezione di Milano, via Celoria 16, I-*20133* Milano, Italy
[3]Katedra Fizyki Teoretycznej, Uniwersytet Marii Curie–Skłodowskiej, ul. Radziszewskiego 10, 20-031 Lublin, Poland

E-mail: `xavier.roca.maza@mi.infn.it`

**Abstract.** We study whether the neutron skin thickness $\Delta r_{np}$ of $^{208}$Pb originates from the bulk or from the surface of the neutron and proton density distributions in mean field models. We find that the size of the bulk contribution to $\Delta r_{np}$ of $^{208}$Pb strongly depends on the slope of the nuclear symmetry energy, while the surface contribution does not. We note that most mean field models predict a neutron density for $^{208}$Pb between the halo and skin type limits. We investigate the dependence of parity-violating electron scattering at the kinematics of the PREX experiment on the shape of the nucleon densities predicted by the mean field models for $^{208}$Pb. We find an approximate formula for the parity-violating asymmetry in terms of the central radius and the surface diffuseness of the nucleon densities of $^{208}$Pb in these models.

## 1. Introduction

The neutron skin thickness of nuclei is defined as the difference between the root mean square radius of neutrons and protons,

$$\Delta r_{np} = \langle r^2 \rangle_n^{1/2} - \langle r^2 \rangle_p^{1/2} \,. \tag{1}$$

The extraction of neutron radii and neutron skins from the experiment is in general dependent on the shape of the neutron distribution used in the analysis [1, 2, 3, 4]. The data typically do not indicate unambiguously if the difference between the peripheral neutron and proton densities is caused by an extended bulk radius of the neutron density, by a modification of the width of the surface, or by some combination of both effects. In particular, the neutron skin $\Delta r_{np}$ of $^{208}$Pb is nowadays attracting significant interest in both experiment and theory since it has a close relation with the density dependence of the nuclear symmetry energy and with the equation of state of neutron-rich matter [5], which have a large impact in diverse problems of nuclear physics and astrophysics where neutron-rich matter is involved [5, 6, 7].

## 2. Method

The analysis of bulk and surface contributions to the neutron skin thickness of a nucleus requires proper definitions of these quantities based on nuclear density distributions [8]. Using the

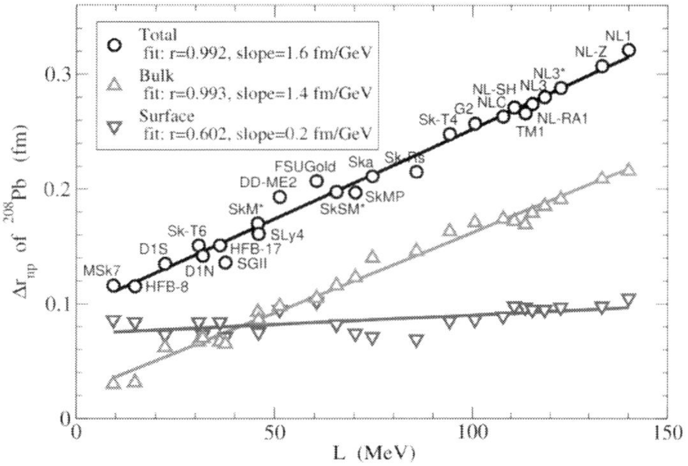


**Figure 1.** Linear correlation of $\Delta r_{np}$ of $^{208}$Pb with the density slope of the nuclear symmetry energy $L$. The dependence on $L$ of the bulk and surface contributions defined in Eqs. (2) and (3) is also displayed.

standard definitions of the equivalent sharp radius $R$ and surface width parameter $b$ of a nucleon density profile, we have shown in Refs. [8, 4] that one can write $\Delta r_{np} = \Delta r_{np}^{\text{bulk}} + \Delta r_{np}^{\text{surf}}$, where

$$\Delta r_{np}^{\text{bulk}} \equiv \sqrt{\frac{3}{5}}\,(R_n - R_p) \tag{2}$$

and

$$\Delta r_{np}^{\text{surf}} \equiv \sqrt{\frac{3}{5}}\,\frac{5}{2}\Big(\frac{b_n^2}{R_n} - \frac{b_p^2}{R_p}\Big). \tag{3}$$

We recall that $R_q$ stands for the radius of a uniform sharp distribution whose density equals the bulk value of the actual density and that has the same number of particles. Therefore, one has a natural splitting of $\Delta r_{np}$ in terms of a bulk part (2) independent of surface properties, and a part (3) of surface origin [8, 4].

## 3. Results

### *3.1. Model predictions for bulk and surface contributions*

Nonrelativistic (NRMF) and relativistic (RMF) mean field models often differ in their predictions for properties of asymmetric nuclear matter. A common example is the value predicted for the density slope $L$ of the nuclear symmetry energy at saturation density, which may show large discrepancies in the MF interactions. The $L$ parameter is defined as

$$L = 3\rho_0 \frac{\partial c_{\text{sym}}(\rho)}{\partial \rho}\bigg|_{\rho_0} \tag{4}$$

where $c_{\text{sym}}(\rho)$ is the symmetry energy and $\rho_0$ the nuclear saturation density.

In Figure 1 we show the linear correlation of $\Delta r_{np}$ of $^{208}$Pb with the parameter $L$ and demonstrate that it mainly arises from the bulk part of $\Delta r_{np}$ within a large and representative set of mean field models of very different nature. Relativistic (RMF) models: G2, NLC, NL-SH, TM1, NL-RA1, NL3, NL3*, NL-Z, NL1, DD-ME2 and FSUGold; and nonrelativistic (NRMF) models: HFB-8, MSk7, D1S, SGII, D1N, Sk-T6, HFB-17, SLy4, SkM*, SkSM*, SkMP, Ska,

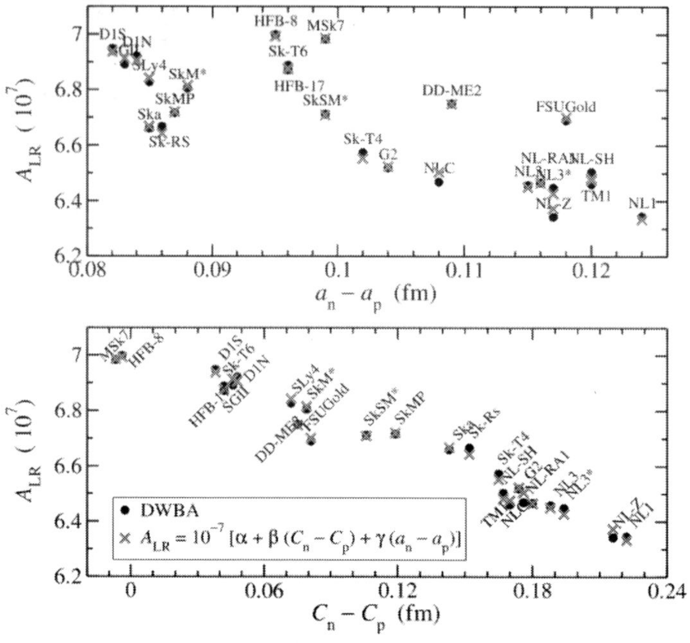

**Figure 2.** Parity violating asymmetry (DWBA) in $^{208}$Pb for 1 GeV electrons at 5° scattering angle. MF results and those obtained with the parametrized formula given in the text.

Sk-Rs and Sk-T4. The original references to the different interactions can be found in [9, 10] for the Skyrme models, [11] for the Gogny models, and [12, 13, 14, 15, 16] and Ref. [19] in [12] for the RMF models. From this figure, one sees that whereas the bulk contribution to the neutron skin thickness of $^{208}$Pb changes largely among the different mean-field interactions, the surface contribution remains confined to a narrow band of values. The shape of the neutron density is more uncertain than the proton density in $^{208}$Pb, and even if the neutron rms radius is determined (*e.g.* in PREX [17]), it can correspond to different shapes of the neutron density. As discussed in [4], from the study of the two-parameter Fermi functions fitted to the self-consistent densities of the MF models, we find that for $^{208}$Pb the central radii $C_q$ vary within the windows $C_n \approx 6.7-6.85$ fm in NRMF and $6.8-7$ fm in RMF, $C_p \approx 6.65-6.7$ fm in NRMF and $6.7-6.77$ fm in RMF, and that the diffuseness parameters $a_q$ vary within the windows $a_n \approx 0.53-0.55$ fm in NRMF and $0.55-0.59$ fm in RMF and $a_p \approx 0.43-0.47$ fm. From these results, we can conclude that most of the MF models predict a neutron distribution for $^{208}$Pb between the halo-type limit (where $C_n - C_p = 0$) and the skin-type limit (where $a_n - a_p = 0$). The halo-type is supported by models with a very soft symmetry energy. Models with a stiffer symmetry energy (larger $L$ values) have $C_n - C_p$ more and more different from zero. However, a pure skin-type is not found in any mean-field model as $a_n - a_p$ is always non-vanishing.

### 3.2. *Parity violating electron scattering at the kinematics of the PREX experiment*

Parity-violating electron scattering (PVES) probes the neutron density in a nucleus since the $Z^0$ boson couples mainly to neutrons [17]. The PREX experiment at JLab [18] aims to determine the neutron rms radius of $^{208}$Pb by PVES. We have investigated the dependence of PVES on the shape of the neutron and proton densities of $^{208}$Pb within the MF models. To this end, we compute the parity-violating asymmetry

$$A_{LR} \equiv \Big(\frac{d\sigma_+}{d\Omega} - \frac{d\sigma_-}{d\Omega}\Big)\Big/\Big(\frac{d\sigma_+}{d\Omega} + \frac{d\sigma_-}{d\Omega}\Big) \tag{5}$$

at the PREX kinematics [18]. We obtain the differentical cross sections $d\sigma_\pm/d\Omega$ for right and left-handed electrons by performing the exact phase shift analysis (DBWA) of the Dirac equation [19] for incident electrons moving in the potentials $V_\pm(r) = V_{\rm Coulomb}(r) \pm V_{\rm weak}(r)$ [4]. See Ref. [4] for the details of the calculations. We display in Fig. 2 the results for $A_{LR}$ as a function of $C_n - C_p$ and $a_n - a_p$ of the two parameter Fermi distributions fitted to the $^{208}$Pb MF densities. We have found that the results can be reasonably parametrized by the formula $10^7 A_{LR} \approx \alpha + \beta(C_n - C_p) + \gamma(a_n - a_p)$ with $\alpha = 7.33$, $\beta = -2.45$ fm$^{-1}$ and $\gamma = -3.62$ fm$^{-1}$, which is depicted by the crosses in the figure. Recently, in [20] we have analyzed systematically the correlations of $A_{LR}$ with the neutron skin thickness of $^{208}$Pb and with the slope of the symmetry energy in the nuclear MF models.

## 4. Conclusions

We have found that the known linear correlation of $\Delta r_{np}$ of $^{208}$Pb with the density derivative of the nuclear symmetry energy arises from the bulk part of $\Delta r_{np}$. This implies that an experimental determination of $R_n$ of $^{208}$Pb could be as useful for constraining $L$ as a determination of $\langle r^2\rangle_n^{1/2}$. MF models can accomodate the halo-type distribution in $^{208}$Pb if the symmetry energy is very soft but do not support a purely skin-type distribution. We find a simple parametrization of $A_{LR}$ in terms of $C_n - C_p$ and $a_n - a_p$ that would provide a new correlation between the central radius and the surface diffuseness of the distribution of neutrons in $^{208}$Pb assuming the proton density known from experiment.

## Acknowledgments

Work partially supported by grants CSD2007-00042, FIS2008-01661 and 2009SGR-1289 (Spain) and N202 231137 (Poland).

## References

[1] Trzcińska A, Jastrzębski J, Lubiński P, Hartmann F J, Schmidt R, von Egidy T and Kłos B 2001 *Phys. Rev. Lett.* **87** 082501
[2] Dong T, Chu Y, Ren Z and Wang Z 2009 *Phys. Rev. C* **79** 014317
[3] Friedman E 2009 *Hyperfine Interact.* **193** 33
[4] Centelles M, Roca-Maza X, Vinas X and Warda M 2010 *Phys. Rev. C* **82** 054314
[5] Brown B A 2000 *Phys. Rev. Lett.* **85** 5296
[6] Centelles M, Roca-Maza X, Vinas X and Warda M 2009 *Phys. Rev. Lett.* **102** 122502
[7] Warda M, Vinas X, Roca-Maza X and Centelles M 2009 *Phys. Rev. C* **80** 024316
[8] Warda M, Vinas X, Roca-Maza X and Centelles M 2010 *Phys. Rev. C* **81** 054309
[9] Xu J, Chen L W, Li B A and Ma H R 2009 *Ap. J.* **697** 1549
[10] Goriely S, Chamel N and Pearson J M 2009 *Phys. Rev. Lett.* **102** 152503
[11] Chappert F, Girod M and Hilaire S 2008 *Phys. Lett. B* **668** 420
[12] Patra S K, Centelles M, Vinas X and Estal M D 2002 *Phys. Rev. C* **65** 044304
[13] Sulaksono A, Mart T and B T J
[14] Todd-Rutel B G and Piekarewicz J 2005 *Phys. Rev. Lett.* **95** 122501
[15] Lalazissis G A, Nikšić T, Vretenar D and Ring P 2005 *Phys. Rev. C* **71** 024312
[16] Lalazissis G A, Karatzikos S, Fossion R, Arteaga D P, Afanasjev A V and Ring P 2009 *Phys. Lett. B* **671** 36
[17] Horowitz C J, Pollock S J, Souder P A and Michaels R 2001 *Phys. Rev. C* **63** 025501
[18] K. Kumar P. A. Souder, R. Michaels, and G. M. Urciuoli, spokespersons, http://hallaweb.jlab.org/parity/prex.
[19] Roca-Maza X, Centelles M, Salvat F and Vinas X 2008 *Phys. Rev. C* **78** 044332
[20] Centelles M, Roca-Maza X, Vinas X and Warda M 2011, arXiv:1103.1762.

# Microscopic - Macroscopic Mass Calculations with Wigner - Kirkwood expansion

**A. Bhagwat[1], X. Viñas[2], M. Centelles[2], R. Wyss[3] and P. Schuck[4,5]**

[1]UM-DAE Centre for Excellence in Basic Sciences, Mumbai 400 098, India
[2]Departament d'Estructura i Constituents de la Matèria and Institut de Ciències del Cosmos, Facultat de Física, Universitat de Barcelona, Diagonal 647, E-08028 Barcelona, Spain
[3]KTH (Royal Institute of Technology), Alba Nova University Center, Department of Nuclear Physics, S-10691 Stockholm, Sweden
[4]Institut de Physique Nucléaire, IN2P3-CNRS, Université Paris-Sud, F-91406 Orsay Cedex, France
[5] Laboratoire de Physique et Modélisation des Milieux Condensés, CNRS and Université Joseph Fourier, 25 Avenue des Martyrs, Boîte Postale 166, F-38042 Grenoble Cedex 9, France

E-mail: `ameeya@cbs.ac.in`

**Abstract.** The systematic study and calculation of ground state nuclear masses continues to be one of the active and important areas of research in nuclear physics. The present work is an attempt to determine the ground state masses of nuclei spanning the entire periodic table, using the Microscopic - Macroscopic approach. The semi - classical Wigner - Kirkwood (WK) $\hbar$ expansion method is used to calculate shell corrections for spherical and deformed nuclei. The expansion is achieved upto the fourth order in $\hbar$. The shell corrections, along with the pairing energies obtained by using the Lipkin - Nogami scheme, constitute the microscopic part of the nuclear masses. The macroscopic part is obtained from a liquid drop formula with six adjustable parameters. It is shown that the Microscopic - Macroscopic mass calculation thus achieved, yields reliable description of ground state masses of nuclei across the periodic table. The present status of the WK mass calculations and the possible future perspectives are discussed.

## 1. Introduction

Theoretical determination of ground state nuclear masses has been an active research area from the very early days of nuclear structure physics. These activities have become even more relevant, since it is now possible to produce and study the neutron rich and deficient nuclei spanning the entire periodic table. Besides, reliable description of the processes like the astrophysical r-process requires accurate knowledge of the masses of the highly neutron rich nuclei. Even with the recent experimental developments, these nuclei are still inaccessible, and one is compelled to rely on theoretical models for the masses.

A large number of models with varying degree of sophistication for the determination of nuclear masses are available in the literature. Generically, they are known as the 'mass formulas'. In this work, we report a mass formula, within the microscopic - macroscopic (often called 'Mic - Mac') framework. The Mic - Mac formulas are based on the Strutinsky's theorem, according to which, the binding energy of a given nucleus can be written as sum of a smooth part, and an oscillatory part. Most of the Mic - Mac mass formulas available in the literature differ from each other only in the macro part, which is adopted from the Liquid Drop formula. The micro part, almost always is computed using the well - known Strutinsky smoothing scheme. The Strutinsky smoothing scheme, amounts to convolution of single particle spectrum over a given energy interval. The energy interval, in practice, is required to be wide enough, typically a few $\hbar\omega$, to ensure that the fluctuations are washed out. This implies, that for a practical Strutinsky calculation, one needs to include the continuum as well. The problem becomes more serious, particularly for the neutron rich or deficient nuclei, where, the Fermi state may lie close to the continuum itself. The smoothing based on Wigner Kirkwood averaging, on the other hand, is free from this difficulty, since there, one does not require a knowledge of single particle spectrum. This motivates us to explore the possibility of employing the Wigner - Kirkwood expansion technique to determine the smooth part of the energy, and hence the shell corrections.

We shall develop the essential formalism in the second section. The third section contains some of the representative results. The fourth section contains the summary and future outlook.

## 2. Formalism

As stated earlier, the Strutinsky theorem plays a central role in the Mic Mac models. According to the Strutinsky theorem, binding energy $E$ of a given nucleus with $N$ neutrons and $Z$ protons, can be written as sum of a smooth part ($E_{LDM}$) and an oscillatory part ($\delta E$). The liquid drop or droplet formulas constitute $E_{LDM}$. The second term, that is, the oscillatory part comprises of shell correction, and pairing energy. The shell correction is traditionally computed by subtracting the averaged energy for given Hamiltonian, from the corresponding quantum mechanical energy of the system. The semi classical Wigner Kirkwood averaging scheme [1–8], achieves the smoothing through expansion of quantal partition function in the powers of Planck's constant, $\hbar$. In the present work, we employ the Wigner - Kirkwood expansion, upto the fourth order in $\hbar$, for a system of nucleons at zero temperature, with deformation parameters $\beta_2$, $\beta_4$ and $\gamma$. The quantum mechanical partition function for such a system is given by:

$$Z(\beta) \;=\; \mathrm{Tr}\left(\exp(-\beta\hat{H})\right). \tag{1}$$

The variable $\beta$ appearing in this equation is a variable for the purpose of Laplace inversion; $\hat{H}$ is the Hamiltonian of the system, given by:

$$\hat{H} \;=\; \frac{-\hbar^2}{2m}\nabla^2 \;+\; V(\vec{r}) \;+\; \hat{V}_{LS}(\vec{r}), \tag{2}$$

where $V(\vec{r})$ is the one-body central potential and $\hat{V}_{LS}(\vec{r})$ is the spin-orbit interaction, given by,

$$\hat{V}_{LS} \;=\; \frac{\iota\kappa\hbar^2}{2m}\left(\vec{\nabla}f \times \vec{\nabla}\right)\cdot\hat{\sigma}\,, \tag{3}$$

where $\hat{\sigma}$ is the unit Pauli matrix, $\kappa$ is the strength of spin-orbit interaction, and $f$ is the spin-orbit form factor.

The partition function can be readily expanded in powers of $\hbar$ as described in Ref. [3]. The expansion neatly separates into two parts:

$$Z^{(4)}_{WK}(\beta) \;=\; Z^{(4)}(\beta) \;+\; Z^{(4)}_{LS}(\beta)\,. \tag{4}$$

where, the two term, respectively, correspond to the mean field and the spin - orbit interaction. The explicit forms of the two terms can be found in Refs. [3,9]. The level density $g_{WK}$, particle number $N$, and energy $E$ can be calculated directly from the WK partition function by Laplace inversion. For more details, see Ref. [9]. Detailed description of choice of potential can also

Journal of Physics: Conference Series **321** (2011) 012053 doi:10.1088/1742-6596/321/1/012053

be found in the same reference. The quantum mechanical energy of the system is obtained by diagonalising the Hamiltonian in the axially symmetric deformed harmonic oscillator basis. In all, 15 oscillator shells are used for such an expansion. The pairing energy is obtained within the Lipkin - Nogami scheme [10–12].

Here, we use a simple liquid drop model, with six adjustable parameters [13]. Explicitly,

$$E_{LDM} = a_v\left[1+\frac{4k_v}{A^2}T_z\left(T_z+1\right)\right]A + a_s\left[1+\frac{4k_s}{A^2}T_z\left(T_z+1\right)\right]A^{2/3} + \frac{3Z^2e^2}{5r_0A^{1/3}} + \frac{C_4Z^2}{A}, \quad (5)$$

where the terms respectively represent: volume energy, surface energy, Coulomb energy and correction to Coulomb energy due to surface diffuseness of charge distribution. The coefficients $a_v$, $a_s$, $k_v$, $k_s$, $r_0$ and $C_4$ are free parameters; $T_z$ is the third component of isospin, and $e$ is the electronic charge. The free parameters are obtained by $\chi^2$ minimisation, with experimental binding energies taken from Audi - Wapstra evaluation [14].

## 3. Results and Discussions

In the present exploratory study, we consider a set of 367 spherical nuclei [15]. This set has been adopted from the Möller - Nix compilation [16], with a criterion that the quadrupole deformation parameter should be smaller than 0.005. The 6 liquid drop parameters described above are fitted to reproduce the experimental binding energies of these nuclei. The liquid drop parameters thus obtained, are: $a_v = -15.828$ (MeV), $a_s = 19.134$ (MeV), $k_v = -1.951$, $k_S = -2.579$, $r_0 = 1.189$ (fm) and $C_4 = 1.247$ (MeV). The *rms* deviation in the binding energies obtained here, turns out to be 634 keV. The corresponding deviation in the Möller - Nix framework is 741 keV [16, 17].

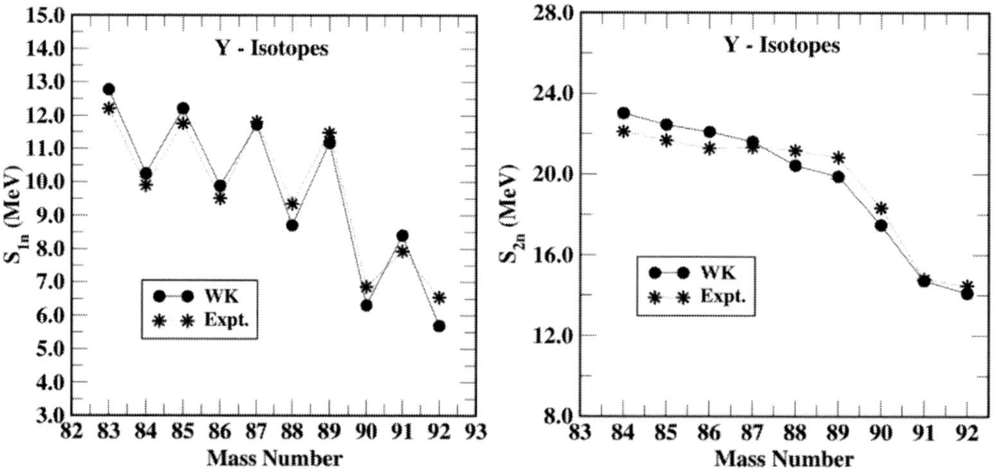


**Figure 1.** The calculated (WK) and the corresponding experimental [14] single and two neutron separation energies for Y isotopes.

As representative cases, we here present the single and two neutron separation energies for the chain of Y isotopes (considered to be spherical) in Fig. 1. It is seen that the expected odd - even staggering in the single neutron separation energies is well reproduced in the present calculations. The two neutron separation energies, too, are well reproduced, indicating that the mass formula proposed in the present work, indeed, is reliable.

Finally, we note that the same set of liquid drop parameters, modified suitably to take the deformation effects into account, reproduce the binding energies of the deformed nuclei well. As an illustration, the difference between the calculated and the corresponding experimental [14]

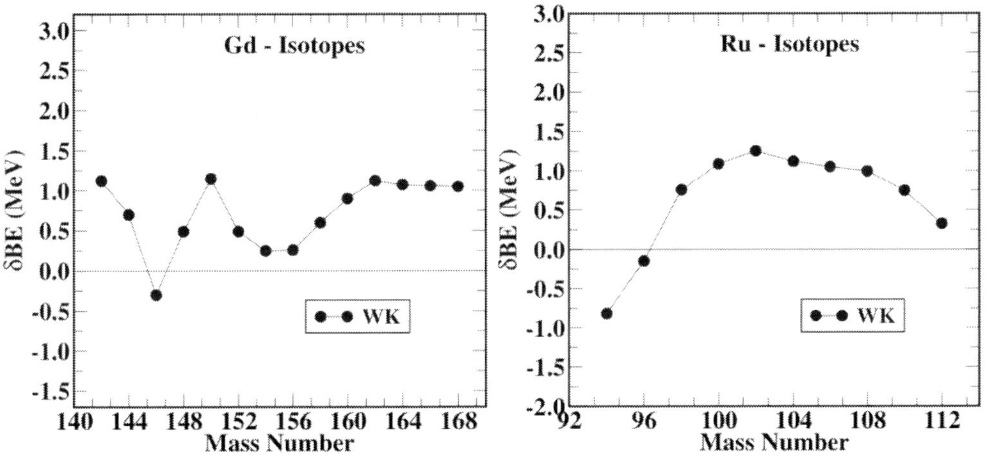


**Figure 2.** The difference between the corresponding experimental [14] binding energies for the deformed Gd and Ru isotopes.

binding energies for Gd and Ru isotopes is plotted in Fig. 2. It is seen that the calculations agree well with the experiment. The nuclei $^{102-110}$Ru turn out to have finite tri-axiality in the ground state, in agreement with the observations reported by Möller *et al.* [18].

## 4. Summary

The Wigner Kirkwood technique for obtaining averaged energy is developed upto fourth order in $\hbar$, for a system of nucleons with deformation parameters $\beta_2$, $\beta_4$ and $\gamma$. The shell corrections thus obtained, along with the pairing energies calculated by using the well known Lipkin Nogami scheme constitute the 'micro' part of the binding energy. The macro part is adopted from traditional liquid drop formula with six adjustable parameters. These parameters are determined so as to reproduce the experimental binding energy of spherical nuclei. It is found that the fit yields mean squared deviation of 634 keV, indicating that the present approach is reliable. Further investigations are in progress.

[1] E. Wigner, Phys. Rev. **40**, 749 (1932).
[2] J. G. Kirkwood, Phys. Rev. **44**, 33 (1933).
[3] B. K. Jennings, R. K. Bhaduri and M. Brack, Nucl. Phys. A **253**, 29 (1975).
[4] P. Ring and P. Schuck, *The Nuclear Many-Body Problem* (Springer-Verlag, Berlin, 1980).
[5] M. Brack, C. Guet and H. -B. Håkansson, Phys. Rep. **123**, 275 (1985).
[6] M. Brack and R. K. Bhaduri, *Semi-classical Physics* (Addison - Wesley Publishing Co., 1997).
[7] M. Centelles *et al.*, Phys. Rev. C **74**, 034332 (2006) and references cited therein.
[8] M. Centelles, P. Schuck and X. Viñas, Ann. Phys. **322**, 363 (2007).
[9] A. Bhagwat *et al.*, Phys. Rev. C **81**, 044321 (2010).
[10] H. J. Lipkin, Ann. Phys. (N.Y.) **9**, 272 (1960).
[11] Y. Nogami, Phys. Rev. **134**, B313 (1964).
[12] H. C. Pradhan, Y. Nogami and J. Law, Nucl. Phys. **A201**, 357 (1973).
[13] K. Pomorski and J. Dudek, Phys. Rev. C **67**, 044316 (2003) and references cited therein.
[14] A. H. Wapstra, G. Audi and C. Thibault, Nucl. Phys. A **729**, 129 (2003).
[15] The table of nuclei chosen can be found at http://www.ecm.ub.es/~xavier/be_WK.dat
[16] P. Möller, J. R. Nix, W. D. Myers and W. J. Swiatecki, At. Data Nucl. Data Tables **59**, 185 (1995).
[17] P. Möller, J. R. Nix and K. -L. Kratz, At. Data Nucl. Data Tables **66**, 131 (1997).
[18] P. Möller *et al.*, At. Data Nucl. Data Tables, **94**, 758 (2008).

Journal of Physics: Conference Series **321** (2011) 012054 doi:10.1088/1742-6596/321/1/012054

# Description of the ground state of axially deformed nuclei within the Relativistic Hartree-Fock-Bogoliubov model

**J.-P. Ebran**

CEA/DAM/DIF, F-91297 Arpajon, France
Institut de Physique Nucléaire, Univ. Paris-Sud, IN2P3-CNRS, F-91406 Orsay Cedex, France

**E. Khan**

Institut de Physique Nucléaire, Univ. Paris-Sud, IN2P3-CNRS, F-91406 Orsay Cedex, France

**D. Peña Arteaga**

Institut de Physique Nucléaire, Univ. Paris-Sud, IN2P3-CNRS, F-91406 Orsay Cedex, France

**D. Vretenar**

Physics Department, Faculty of Science, University of Zagreb, 10000 Zagreb, Croatia

E-mail: jean-paul.ebran@cea.fr

**Abstract.** The Relativistic Hartree-Fock-Bogoliubov model for axially deformed nuclei (RHFBz) is presented. The model involves a phenomenological Lagrangian with density-dependent meson-nucleon couplings in the particle-hole channel and the central part of the Gogny force in the particle-particle channel. The RHFBz equations are solved by expansion in the basis of a deformed harmonic oscillator. Illustrative RHFBz calculations are performed for Neon isotopes.

## 1. Introduction

Nuclear energy density functionals (EDF) represent a tool of choice for the microscopic description of both static and dynamic properties of nuclei over the whole nuclide chart. They subsum nucleonic short-range in-medium correlations, whereas static long-range correlations are taken into account by allowing a single-determinant state to break the symmetries of the nuclear Hamiltonian [1]. Many structure phenomena in both stable and exotic nuclei have successfully been described by EDFs involving the non-relativistic Gogny and Skyrme [2] effective interactions, as well as relativistic phenomenological Lagrangian densities [3]. The Relativistic Mean Field (RMF) framework [3] is an example of a covariant EDF. The corresponding phenomenological Lagrangians provide a quantitative description of a variety of ground-state data. However, it does not include explicitly the Fock term, but implicitly takes it into account through the fit of model parameters to structure data. A more involved approach, the Relativistic Hartree-Fock (RHF) theory [4], treats the exchange contributions explicitly.

The first RHF models used to predict under-bound nuclei in comparison to experimental data. This problem originated from the lack of a medium dependence in the corresponding effective nucleonic interaction [4]. To overcome this problem, an explicit nucleon-density dependence of the nucleon-meson couplings was included [5]. The resulting improvement brought the current RHF models to a similar level of accuracy as the standard RMF approach for a quantitative description of nuclear structure phenomena [6]. In particular, recent studies by W.H. Long *et. al.* [5, 6, 7, 8] and H. Liang *et. al.* [9, 10] have emphasized that, compared to the RMF approach, the explicit treatment of Fock terms can improve the description of nuclear matter and finite nuclei. Moreover, it explicitly includes the tensor contributions to the inter-nucleon interaction generated by the exchange of the $\pi$ and $\rho$ mesons. So far, the RHF framework has been limited to the description of spherical nuclei. We consider an extension of this approach to deformed, axially-symmetric nuclei: the Relativistic Hartree-Fock-Bogoliubov model with density-dependent meson-nucleon couplings (RHFBz) [11, 12]. In Sec. 2 the general formalism of the RHFBz model is briefly presented. In Sec. 3 we present and discuss applications of the RHFBz model to ground-state properties of neon isotopes. Finally, a short summary and discussion of possible future studies are displayed in Sec. 4.

## 2. Formalism of the RHFBz model

The RHFBz approach is based on a phenomenological Lagrangian density involving the relevant degrees of freedom for nuclear structure, namely nucleons and mesons:

$$\begin{aligned}
\mathcal{L} &= \bar{\psi}\Big\{ i\gamma^\mu\partial_\mu - M - g_\sigma(\rho_v)\sigma - g_\omega(\rho_v)\gamma_\mu\omega^\mu \\
&-g_\rho(\rho_v)\gamma_\mu\vec{\rho}.\vec{\tau}^\mu - \frac{f_\pi(\rho_v)}{m_\pi}\gamma_5\gamma_\mu\partial^\mu\vec{\pi}.\vec{\tau} - e\gamma_\mu A^\mu\frac{1-\tau_3}{2}\Big\}\psi \\
&+\frac{1}{2}\left(\partial_\mu\sigma\partial^\mu\sigma - m_\sigma^2\sigma^2\right) - \frac{1}{2}\left(\Omega_{\mu\nu}\Omega^{\mu\nu} - m_\omega^2\omega_\mu\omega^\mu\right) \\
&-\frac{1}{2}\left(\vec{\mathcal{R}}_{\mu\nu}\vec{\mathcal{R}}^{\mu\nu} - m_\rho^2\vec{\rho}_\mu\vec{\rho}^\mu\right) \\
&+\frac{1}{2}\left(\partial_\mu\vec{\pi}\partial^\mu\vec{\pi} - m_\pi^2\vec{\pi}^2\right) - \frac{1}{2}\left(\mathcal{F}_{\mu\nu}\mathcal{F}^{\mu\nu}\right) \; . \qquad (1)
\end{aligned}$$

Vectors in isospin space are denoted by arrows. The Dirac spinor $\psi$ denotes the nucleon with mass $M$. $m_\sigma$, $m_\omega$, $m_\rho$, and $m_\pi$ are the masses of the $\sigma$-meson, the $\omega$-meson, the $\rho$-meson and the $\pi$-meson, respectively. $g_\sigma$, $g_\omega$, $g_\rho$ and $f_\pi$ are the meson-nucleon coupling constants. $A^\mu$ stands for the electromagnetic 4-potential. $e^2/4\pi = 1/137.036$. The (density-dependent) coupling constants and meson masses are parameters, adjusted to reproduce nuclear matter properties and ground-state properties of finite nuclei. $\Omega^{\mu\nu}$, $\vec{R}^{\mu\nu}$, and $\mathcal{F}^{\mu\nu}$ are the field tensors of the vector fields $\omega$, $\rho$, and of the photon [4]. A nucleon-density dependence of the meson-nucleon couplings accounts for medium polarisation and three-body correlations [13, 14, 5].

## 3. Results and discussion

This section presents an application of the RHFBz model in the calculation of ground-state properties of neon isotopes. The RHFBz model is used with the PKO2 and PKO3 effective interactions [15, 16] in the particle-hole channel, and the central part of the Gogny D1S force [17] in the particle-particle channel. The PKO3 effective interaction is related to a covariant EDF including explicitly the pion degree of freedom, which is one actor of the tensor force. On the contrary, the PKO2 effective interaction corresponds to a covariant EDF where the pion degree of freedom is not treated explicitly.

The two-neutron separation energy $S_{2n} \equiv E_{tot}(Z,N) - E_{tot}(Z,N-2)$ of Ne isotopes, calculated with PKO2 and DD-ME2, are compared to data in Fig. 1. In general, the RHFBz results obtained with the PKO2 parameter set are closer to the experimental two-neutron separation energies. Both PKO2 and DD-ME2 predict $^{32}$Ne to be the last bound isotope.

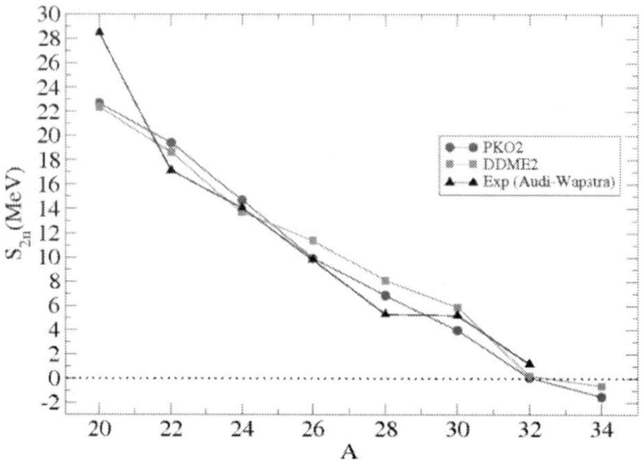


Figure 1: Two-neutron separation energy in the neon isotopic chain. The relativistic mean-field results: RHFBz with PKO2 [15, 16], and RHB with DD-ME2 [18], are compared to data (Audi-Wapstra [19]).

The evolution of the axial deformation parameter $\beta$ in the neon isotopic chain is illustrated in Fig. 2. In general, the deformation predicted by PKO3 is larger than that calculated with PKO2 and, therefore, closer to the results obtained with the DD-ME2 and Gogny D1S effective interactions. PKO3 predicts an oblate shape for $^{24}$Ne (quasi degenerate in energy with a prolate solution at $\beta = 0.3$) whereas a prolate ground-state shape for this nucleus is obtained with PKO2, DDME2, Gogny D1S (quasi degenerate in energy with an oblate solution at $\beta = -0.15$) and Skyrme SLy4 interactions. Moreover, all these interactions, except PKO3 that predicts a prolate ground-state, give no deformation for $^{26}$Ne and $^{28}$Ne. The comparison between the two PKO3 curves where the pion coupling is switched on and off shows that the prolate shape of $^{26,28}$Ne is driven by the pion.

## 4. Conclusion

The relativistic Hartree-Fock-Bogoliubov model for axially deformed nuclei (RHFBz) is based on an effective Lagrangian with density-dependent meson-nucleon couplings. In this work RHFBz calculations have been performed for neon isotopes. Results obtained with the RHF effective forces PKO2 and PKO3 have been compared to experimental $S_{2n}$. In addition, ground-state deformation has been shown in comparison with the predictions of the relativistic DD-ME2 effective interaction, as well as with the results calculated with the non-relativistic Gogny D1S and Skyrme SLy4 interactions. The effect of explicitly including the pion field has been investigated for deformation parameters. The inclusion of the tensor $\rho$-nucleon coupling will complete the model and thus enable studies of the role of tensor components of the effective inter-nucleon interaction in the evolution of shell structures in deformed nuclei.

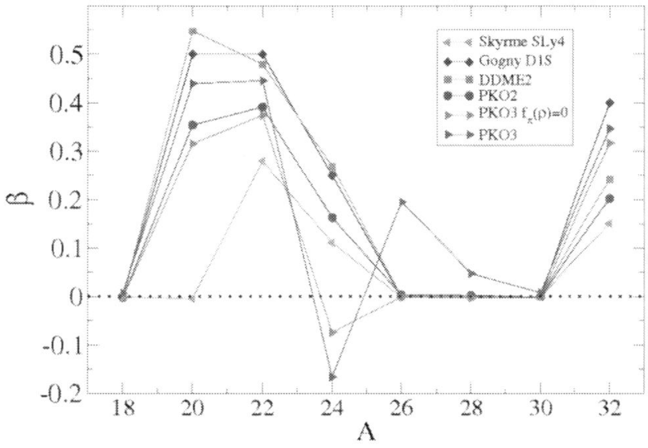


Figure 2: Axial deformation parameter $\beta$ of Ne nuclei as function of the mass number. The calculated values correspond to the PKO2 and PKO3 [15, 16], DD-ME2 [18], Gogny D1S [17], Skyrme SLy4 [20] and Skyrme SGII [21] effective interactions. PKO3 calculations with $f_\pi(\rho)$ set to 0 are also shown.

[1] Lacroix D, Duguet T and Bender M 2009 *Phys. Rev.* C **79** 044318
[2] Bender M, Heenen PH and Reinhard PG 2003 *Rev. Mod. Phys.* **75** 121
[3] Vretenar D, Afanasjev AV, Lalazissis GA and Ring P 2005 *Phys. Rep.* **409** 101
[4] Bouyssy A, Mathiot JF, Van Giai N and Marcos S 1987 *Phys. Rev.* C **36** 380
[5] Long WH, Van Giai N and Meng J 2006 *Phys. Lett.* B **640** 150
[6] Long WH, Ring P, Van Giai N and Meng J 2010 *Phys. Rev.* C **81** 024308
[7] Long WH, Sagawa H, Meng J and Van Giai N 2008 *Europhysics Letters* **82** 12001
[8] Long WH, Sagawa H, Van Giai N and Meng J 2007 *Phys. Rev.* C **76** 034314
[9] Liang H, Van Giai N and Meng J 2008 *Phys. Rev. Lett.* **101** 122502
[10] Liang H, Long WH, Meng J and Van Giai N 2010 *Eur. Phys. J.* A **44** 119-124
[11] Ebran JP, Khan E, Peña Arteaga D and Vretenar D, accepted in *Phys. Rev.* C
[12] Ebran JP 2010 PhD thesis Univ. Paris-Sud 11
[13] Typel D and Wolter HH 1999 *Nucl. Phys.* A **656** 331
[14] Nikšić T, Vretenar D, Finelli P, and Ring P 2002 *Phys. Rev.* C **66** 024306
[15] Long WH, Van Giai N and Meng J arXiv:nucl-th/0608009
[16] Long WH 2005 PhD thesis Univ. Paris-Sud 11
[17] Berger JF, Girod M and Gogny D 1991 *Comp. Phys. Comm.* **63** 365
[18] Lalazissis GA, Nikšić T, Vretenar D and Ring P 2005 *Phys. Rev.* C **71** 024312
[19] Audi G and Wapstra WH 1993 *Nucl. Phys.* A **565** 1
[20] Stoitsov MV, Dobaczewski J, Nazarewicz W, Pittel S and Dean DJ 2003 *Phys. Rev.* C **68** 054312
[21] Sagawa H, Zhou XR, Zhang XZ and Suzuki T 2004 *Phys. Rev.* C **70** 054316

Journal of Physics: Conference Series **321** (2011) 012055 doi:10.1088/1742-6596/321/1/012055

# Theoretical descriptions of isospin corrections for superallowed $\beta$ transitions

**Haozhao Liang**[1], **Nguyen Van Giai**[2], **and Jie Meng**[3,1,4]

[1]State Key Laboratory of Nuclear Physics and Technology, School of Physics, Peking University, Beijing 100871, China
[2]Institut de Physique Nucléaire, IN2P3-CNRS and Univ. Paris-Sud, 91405 Orsay, France
[3]School of Physics and Nuclear Energy Engineering, Beihang University, Beijing 100191, China
[4]Department of Physics, University of Stellenbosch, Stellenbosch, South Africa

E-mail: `hzliang@pku.edu.cn`

**Abstract.** The isospin symmetry-breaking corrections $\delta_c$ obtained by self-consistent relativistic random phase approximation (RPA) calculations are compared with those obtained by shell model calculations as well as by isospin- and angular-momentum-projected nuclear density functional theory. It is found that the present theoretical uncertainty of the isospin symmetry-breaking corrections $\delta_c$ is of the order of $\sim 0.003$, rather than $\sim 0.0002$ as indicated by the shell model calculations. Whether the unitarity of the CKM matrix is fulfilled or not is still an open question.

## 1. Introduction

The Cabibbo-Kobayashi-Maskawa (CKM) matrix [1, 2], which relates the quark eigenstates of the weak interaction with the quark mass eigenstates, is one of the key transformations in the Standard Model. Kobayashi and Maskawa were awarded the 2008 Nobel Prize in Physics "for the discovery of the origin of the broken symmetry which predicts the existence of *at least* three families of quarks in nature" [3]. If the three-generation Standard Model is complete, the rotation embodied in the CKM matrix must be unitary. Therefore, measuring the matrix elements independently and verifying its unitarity condition provide a rigorous test for the Standard Model and set limits on new physics beyond.

To date, the most precise test is that using the three top-row elements $V_{ud}$, $V_{us}$ and $V_{ub}$ [4, 5], in which the leading term $|V_{ud}|^2$ contributes $\sim 95\%$ to the unitarity sum. Thus, determining the $|V_{ud}|$ value and restraining its uncertainty is a critical issue for both experimental and theoretical researches.

Four experimental methods to determine the $|V_{ud}|$ value so far include the nuclear $0^+ \to 0^+$ superallowed $\beta$ transitions [6], neutron decay [7], pion $\beta$ decay [8] and nuclear mirror transitions [9]. In particular, the first method provides the most precise determination [4, 5]. In order to determine the $|V_{ud}|$ value with these nuclear superallowed $\beta$ transitions, apart from the experimental $ft$ values, the radiative corrections $\Delta_R^V$, $\delta'_R$, $\delta_{\rm NS}$ and the isospin symmetry-breaking corrections $\delta_c$ must also be taken into account [6]. The radiative corrections are due to the emission of real photons and the exchange of virtual photons and $Z$-bosons in semi-leptonic weak transitions [10, 11], while the isospin symmetry-breaking corrections are used to estimate

Journal of Physics: Conference Series **321** (2011) 012055 doi:10.1088/1742-6596/321/1/012055

the slight failure of the conserved vector current (CVC) hypothesis caused by the isospin $SU(2)$ symmetry-breaking in finite nuclei [11]. It is expected that the corrected $\mathcal{F}t$ values should be nucleus-independent once the radiative and isospin symmetry-breaking corrections are taken into account.

The isospin symmetry-breaking corrections $\delta_c$ characterize the slight reduction of the superallowed transition strengths $|M_F|^2$ from the ideal value $|M_0|^2$:

$$|M_F|^2 = |\langle f|\, T_{\pm}\, |i\rangle|^2 = |M_0|^2(1-\delta_c), \tag{1}$$

with $M_0 = \sqrt{2}$ for $T = 1$ states having exact isospin symmetry. Shell model calculations [11] have been generally used to evaluate the $\delta_c$ corrections for the past several decades. However, recently it was pointed out that the significant effects of radial excitations have been neglected in the treatment used in Ref. [11] and hence, the $\delta_c$ corrections as well as the $|V_{ud}|$ value have been overestimated [12]. Alternatively, the self-consistent random phase approximation (RPA) based on microscopic mean field theories is another reliable approach for evaluating the superallowed transition strength $|M_F|^2$. Such calculations have been performed for several nuclei with the non-relativistic Skyrme Hartree-Fock approach in Ref. [13]. During the last decade, great efforts have been dedicated to developing the charge-exchange (Q)RPA within the relativistic framework [14, 15, 16]. In particular, a fully self-consistent charge-exchange RPA based on the relativistic Hartree-Fock (RHF) approach [17] was established [16]. A very satisfactory description of spin-isospin resonances was obtained without any readjustment of the energy functional. Adopting these self-consistent relativistic RPA approaches, the corrections $\delta_c$ have been systematically investigated in Refs. [18, 19]. Recently, $\delta_c$ corrections have also been evaluated within isospin- and angular-momentum-projected nuclear density functional theory (DFT) [20].

Here, the isospin symmetry-breaking corrections $\delta_c$ obtained by self-consistent relativistic RPA calculations will be compared with those obtained by shell model calculations as well as by projected DFT. Together with the most up-to-date $|V_{us}|$ and $|V_{ub}|$ values, the unitarity of the CKM matrix will also be discussed.

## 2. Discussion

In Table 1, the isospin symmetry-breaking corrections $\delta_c$ for the $0^+ \rightarrow 0^+$ superallowed transitions obtained by self-consistent RHF+RPA calculations [18] with PKO1 parametrization [17] are compared with those obtained by shell model calculations (T&H) [11] and isospin- and angular-momentum-projected nuclear density functional theory [20] with the Skyrme force SV [21]. In order to clearly see the different predictions among these three approaches, the differences of the isospin symmetry-breaking corrections $\delta_c$ are shown in the left panel of Fig. 1 as a function of the charge $Z$ of the daughter nucleus, where the T&H results are taken as references, while the uncertainties of $\delta_c$ are not taken into account.

On the one hand, it is found that the RHF+RPA results are systematically smaller than those of T&H by around 0.4% in most cases. Since it was pointed out that the corrections $\delta_c$ obtained by shell model calculations may be overestimated due to the neglect of radial excitations, it is worthwhile to notice this systematic discrepancy for future investigations. Meanwhile, it should also be pointed out that, in the present relativistic RPA calculations, the effects of pairing correlations and deformation are not taken into account. In addition, the missing charge symmetry breaking (CSB) and charge independence breaking (CIB) forces are expected to enlarge the corrections $\delta_c$ by $\sim 0.1\%$ according to the calculations in Ref. [13].

On the other hand, it is found that the projected DFT results are similar to the T&H results on average, while the individual values differ from each other by about $0.1 \sim 0.2\%$ in most cases. Nevertheless, three $pf$-shell nuclei – $^{62}$Ga, $^{66}$As, and $^{74}$Rb – should be paid more attention to. In particular, the uncertainty of the experimental $ft$ value for $^{62}$Ga has become relatively small [6], thus the underestimate (or overestimate) of the correction $\delta_c$ for this

**Table 1.** Isospin symmetry-breaking corrections $\delta_c$ for the $0^+ \to 0^+$ superallowed transitions obtained by self-consistent RHF+RPA calculations [18] with PKO1 parametrization [17] are compared with those obtained by the shell model calculations (T&H) [11] and isospin- and angular-momentum-projected nuclear density functional theory [20] with the Skyrme force SV [21]. All values are expressed in %.

| | PKO1 [18] | T&H [11] | SV [20] |
|---|---|---|---|
| $^{10}C \to {}^{10}B$ | 0.082 | 0.175(18) | 0.559(56) |
| $^{14}O \to {}^{14}N$ | 0.114 | 0.330(25) | 0.303(30) |
| $^{18}Ne \to {}^{18}F$ | 0.270 | 0.565(39) | |
| $^{22}Mg \to {}^{22}Na$ | | 0.380(22) | 0.243(24) |
| $^{26}Si \to {}^{26}Al$ | 0.176 | 0.435(27) | |
| $^{30}S \to {}^{30}P$ | 0.497 | 0.855(28) | |
| $^{34}Ar \to {}^{34}Cl$ | 0.268 | 0.665(56) | 0.865(87) |
| $^{38}Ca \to {}^{38}K$ | 0.313 | 0.765(71) | |
| $^{42}Ti \to {}^{42}Sc$ | 0.384 | 0.935(78) | |
| $^{26}Al^m \to {}^{26}Mg$ | 0.139 | 0.310(18) | 0.494(49) |
| $^{34}Cl \to {}^{34}S$ | 0.234 | 0.650(46) | 0.679(68) |
| $^{38}K^m \to {}^{38}Ar$ | 0.278 | 0.655(59) | |
| $^{42}Sc \to {}^{42}Ca$ | 0.333 | 0.665(56) | 0.767(77) |
| $^{46}V \to {}^{46}Ti$ | | 0.620(63) | 0.759(76) |
| $^{50}Mn \to {}^{50}Cr$ | | 0.655(54) | 0.740(74) |
| $^{54}Co \to {}^{54}Fe$ | 0.319 | 0.770(67) | 0.671(67) |
| $^{62}Ga \to {}^{62}Zn$ | | 1.48(21) | 0.925(93) |
| $^{66}As \to {}^{66}Ge$ | 0.475 | 1.56(40) | |
| $^{70}Br \to {}^{70}Se$ | 1.140 | 1.60(25) | |
| $^{74}Rb \to {}^{74}Kr$ | 1.088 | 1.63(31) | 2.06(21) |

nucleus represents the dominant contribution to the value of chi-square per degree of freedom $\chi^2/\nu$, which is crucial to determine the confidence level of the nuclear structure-dependent corrections. Therefore, the latter corrections in $^{62}$Ga could become one of the critical cases in future theoretical investigations.

Combining the $|V_{ud}| = 0.97273(27)$ (PKO1) [18], $|V_{ud}| = 0.97425(22)$ (H&T) [6] or $|V_{ud}| = 0.97459(24)$ (SV) [20], and the most updated values of the other two top-row matrix elements $|V_{us}| = 0.2252(9)$ and $|V_{ub}| = 0.00389(44)$ [4], the sum of squared top-row elements of the CKM matrix extracted from the superallowed $\beta$ transitions is shown in the right panel of Fig. 1. These results are also compared with the most up-to-date values obtained in neutron decay [4], pion $\beta$ decay [4] and nuclear mirror transitions [9]. It is interesting to note that whether the unitarity of the CKM matrix is fulfilled or not is still an open question. Remarkably, adopting different theoretical frameworks, the theoretical uncertainty of the isospin symmetry-breaking corrections $\delta_c$ is of the order of $\sim 0.003$, rather than $\sim 0.0002$ as it was concluded from shell model calculations in Ref. [5].

## 3. Summary

The isospin symmetry-breaking corrections $\delta_c$ obtained by the self-consistent relativistic RPA calculations are compared with those obtained by shell model calculations as well as by isospin- and angular-momentum-projected nuclear density functional theory. It is pointed out that the

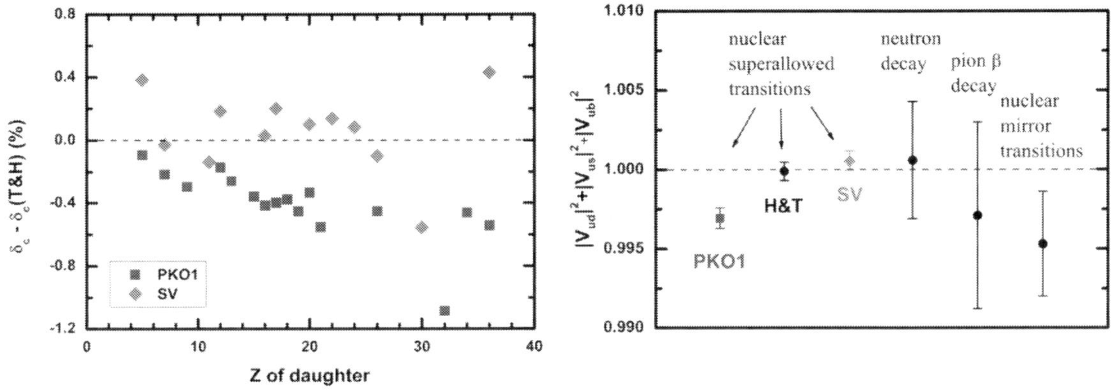


**Figure 1.** Left panel: Differences of the isospin symmetry-breaking corrections $\delta_c$ as a function of the charge $Z$ of the daughter nucleus. The T&H results [11] are taken as references, while the uncertainties of $\delta_c$ are not taken into account. Right panel: The sum of squared top-row elements of the CKM matrix, where the $|V_{ud}|$ values are obtained by RHF+RPA with PKO1 [18], shell model (H&T) [6], projected DFT with SV [20], as well as neutron decay [4], pion $\beta$ decay [4], and nuclear mirror transitions [9], respectively.

present theoretical uncertainty of the isospin symmetry-breaking corrections $\delta_c$ is of the order of magnitude $\sim 0.003$, rather than $\sim 0.0002$ as suggested by shell model calculations. Thus, whether the unitarity of the CKM matrix is fulfilled or not is still an open question.

## Acknowledgements

This work was partly supported by the Major State 973 Program 2007CB815000, the NSFC under Grant No. 10975008, and China Postdoctoral Science Foundation Grant No. 20100480149.

## References

[1] Cabibbo N 1963 *Phys. Rev. Lett.* **10** 531–533
[2] Kobayashi M and Maskawa T 1973 *Prog. Theor. Phys.* **49** 652–657
[3] The Nobel Prize in Physics 2008 URL `http://nobelprize.org/nobel_prizes/physics/laureates/2008/`
[4] Nakamura K *et al.* (Particle Data Group) 2010 *J. Phys. G* **37** 075021
[5] Towner I S and Hardy J C 2010 *Rep. Prog. Phys.* **73** 046301
[6] Hardy J C and Towner I S 2009 *Phys. Rev. C* **79** 055502
[7] Thompson D 1990 *J. Phys. G* **16** 1423–1426
[8] Počanić D *et al.* 2004 *Phys. Rev. Lett.* **93** 181803
[9] Naviliat-Cuncic O and Severijns N 2009 *Phys. Rev. Lett.* **102** 142302
[10] Marciano W J and Sirlin A 2006 *Phys. Rev. Lett.* **96** 032002
[11] Towner I S and Hardy J C 2008 *Phys. Rev. C* **77** 025501
[12] Miller G A and Schwenk A 2009 *Phys. Rev. C* **80** 064319
[13] Sagawa H, Van Giai N and Suzuki T 1996 *Phys. Rev. C* **53** 2163–2170
[14] De Conti C, Galeão A P and Krmpotić F 1998 *Phys. Lett. B* **444** 14–20
[15] Paar N, Nikšić T, Vretenar D and Ring P 2004 *Phys. Rev. C* **69** 054303
[16] Liang H, Van Giai N and Meng J 2008 *Phys. Rev. Lett.* **101** 122502
[17] Long W H, Van Giai N and Meng J 2006 *Phys. Lett. B* **640** 150–154
[18] Liang H, Van Giai N and Meng J 2009 *Phys. Rev. C* **79** 064316
[19] Liang H, Van Giai N and Meng J 2010 *J. Phys. Conf. Ser.* **205** 012028
[20] Satuła W, Dobaczewski J, Nazarewicz W and Rafalski M 2011 *Phys. Rev. Lett.* **106** 132502
[21] Beiner M, Flocard H, Van Giai N and Quentin P 1975 *Nucl. Phys. A* **238** 29–69

Journal of Physics: Conference Series **321** (2011) 012056 doi:10.1088/1742-6596/321/1/012056

# Functional approach for pairing in finite systems:

## *How to define restoration of broken symmetries in Energy Density Functional theory ?*

**G. Hupin and D. Lacroix**

Grand Accélérateur National d'Ions Lourds (GANIL), CEA/DSM-CNRS/IN2P3, Bvd Henri Becquerel, 14076 Caen, France

**M. Bender**

Université Bordeaux, Centre d'Etudes Nucléaires de Bordeaux Gradignan, UMR5797, F-33175 Gradignan, France,
CNRS/IN2P3, Centre d'Etudes Nucléaires de Bordeaux Gradignan, UMR5797, F-33175 Gradignan, France

E-mail: `hupin@ganil.fr, lacroix@ganil.fr, bender@ganil.fr`

**Abstract.** The Multi-Reference Energy Density Functional (MR-EDF) approach (also called configuration mixing or Generator Coordinate Method), that is commonly used to treat pairing in finite nuclei and project onto particle number, is re-analyzed. It is shown that, under certain conditions, the MR-EDF energy can be interpreted as a functional of the one-body density matrix of the projected state with good particle number. Based on this observation, we propose a new approach, called Symmetry-Conserving EDF (SC-EDF), where the breaking and restoration of symmetry are accounted for simultaneously. We show, that such an approach is free from pathologies recently observed in MR-EDF and can be used with a large flexibility on the density dependence of the functional.

### 1. Introduction

In the last decade, configuration mixing has become a standard tool in the Energy Density Functional approach both to restore symmetries and to provide a competitive theory for nuclear spectroscopy. With the possibility to perform systematic and precise studies, detailed analysis of MR-EDF have been made, pointing out some questionable aspects related to the way MR-EDF is constructed [1, 2, 3, 4, 5]. Below, we further investigate the MR-EDF applied to particle number restoration. The Hamiltonian case is first discussed as well as some specific features that will be useful for the EDF case.

### 2. Particle number restoration in the Hamiltonian case

Starting from a two-body Hamiltonian, written here as:

$$H = \sum_{ij} t_{ij}\, a_i^\dagger\, a_j + \frac{1}{4}\sum_{ijkl} \bar{v}_{ijkl}\, a_i^\dagger\, a_j^\dagger\, a_l\, a_k \ , \tag{1}$$

where $\bar{v}$ denote the anti-symmetric two-body interaction matrix elements, pairing can be treated approximately by minimizing the energy in the space of quasi-particle vacua. For a given quasi-particle reference state of the special Bogolyubov type, denoted by $|\Phi_0\rangle$, the energy writes

$$E_{SR}[\Phi_0] = \quad E_{SR}[\rho, \kappa, \kappa^*] = \sum_i t_{ii}\rho_{ii} + \frac{1}{2}\sum_{i,j} \bar{v}_{ijij}\rho_{ii}\rho_{jj} + \frac{1}{4}\sum_{i,j} \bar{v}_{i\bar{i}j\bar{j}}\kappa^*_{i\bar{i}}\kappa_{j\bar{j}}, \tag{2}$$

where $\rho$ and $\kappa$ denote the normal and anomalous density respectively and where the set of single-particle states are those where $\rho$ is diagonal, i.e. the so-called canonical basis of $\rho$. For non-zero anomalous density, the U(1) symmetry is explicitly broken and the particle number is only imposed in average.

In a second step, this symmetry can be restored by projecting out the component with $N$ particles $|\Psi_N\rangle = P^N|\Phi_0\rangle$ where $P^N$ denotes the particle number projection operator defined through[6]

$$P^N = \frac{1}{2\pi}\int_0^{2\pi} d\varphi \; e^{i\varphi(\hat{N}-N)}. \tag{3}$$

The expectation value of any operator $O$ that conserves particle number can then be expressed as

$$\frac{\langle\Psi_N|\,O\,|\Psi_N\rangle}{\langle\Psi_N|\Psi_N\rangle} = \int_0^{2\pi} d\varphi \, \frac{\langle\Phi_0|O|\Phi_\varphi\rangle}{\langle\Phi_0|\Phi_\varphi\rangle}\mathcal{N}_N(0,\varphi), \tag{4}$$

where the shorthand

$$\mathcal{N}_N(0,\varphi) \equiv \frac{e^{-i\varphi N}}{2\pi}\frac{\langle\Phi_0|\Phi_\varphi\rangle}{\langle\Psi_N|\Psi_N\rangle}, \tag{5}$$

has been introduced. Here $\varphi$ denotes the gauge angle, whereas $|\Phi_\varphi\rangle = e^{i\varphi\hat{N}}|\Phi_0\rangle$ refers to the state $|\Phi_0\rangle$ rotated in gauge space. For $O = H$, an expression is obtained for the projected energy:

$$E_N[\Psi_N] \equiv \int_0^{2\pi} d\varphi \, E_{SR}\left[\rho^{0\varphi}, \kappa^{0\varphi}, \kappa^{\varphi 0\star}\right]\mathcal{N}_N(0,\varphi). \tag{6}$$

The energy kernel are defined through

$$E_{SR}\left[\rho^{0\varphi}, \kappa^{0\varphi}, \kappa^{\varphi 0\star}\right] \quad \equiv \quad \frac{\langle\Phi_0|H|\Phi_\varphi\rangle}{\langle\Phi_0|\Phi_\varphi\rangle}. \tag{7}$$

An equivalent form of the projected energy $E_N$ can directly be found by taking the expectation value of $H$ with the projected state, leading to:

$$\begin{aligned} E_N[\Psi_N] &= \frac{\langle\Psi_N|H|\Psi_N\rangle}{\langle\Psi_N|\Psi_N\rangle} = \sum_{ij} t_{ij}\frac{\langle\Psi_N|a_i^\dagger a_j|\Psi_N\rangle}{\langle\Psi_N|\Psi_N\rangle} + \frac{1}{4}\sum_{ijkl}\bar{v}_{ijkl}\frac{\langle\Psi_N|a_i^\dagger a_j^\dagger a_l a_k|\Psi_N\rangle}{\langle\Psi_N|\Psi_N\rangle}. \\ &= E_N\left[\rho^N, R^N\right] = \sum_{ij} t_{ij}\,\rho^N_{ij} + \frac{1}{4}\sum_{ijkl}\bar{v}_{ijkl}R^N_{kl,ij}. \end{aligned} \tag{8}$$

Therefore, we see that the projected energy can be considered either as a functional of the transition densities (Eq. (6)) or as a functional of the one- and two-body density matrices (called hereafter OBDM and TBDM) of the projected state (Eq. (8)). The two formulations are strictly equivalent and can be connected by expressing the density matrices in terms of the gauge angles as:

$$\rho^N_{ij} = \int_0^{2\pi} d\varphi\rho^{0\varphi}_{ij}\mathcal{N}_N(0,\varphi), \qquad R^N_{kl,ij} = \int_0^{2\pi} d\varphi(\rho^{0\varphi}_{ki}\rho^{0\varphi}_{lj} - \rho^{0\varphi}_{kj}\rho^{0\varphi}_{li} + \kappa^{0\varphi}_{kl}\kappa^{\varphi 0\star}_{ij})\mathcal{N}_N(0,\varphi) \tag{9}$$

Indeed, starting from Eq. (6) and combining the $\rho\rho$ and the $\kappa\kappa$ terms (Eq. 2), one can indeed directly recognize the components of the TBDM.

### 3. Particle number restoration in the EDF case

The formulation of broken symmetry and its restoration in the nuclear EDF closely follows the Hamiltonian case. In a first step, called Single-Reference EDF (SR-EDF), an energy written as a functional of an auxiliary quasi-particle state is given. Restricting here to bilinear functional based on effective interactions, it can be written as

$$\mathcal{E}_{SR}[\Phi_0] = \quad \mathcal{E}_{SR}[\rho, \kappa, \kappa^*] = \sum_i t_{ii}\rho_{ii} + \frac{1}{2}\sum_{i,j} \overline{v}^{\rho\rho}_{ijij}\rho_{ii}\rho_{jj} + \frac{1}{4}\sum_{i,j} \overline{v}^{\kappa\kappa}_{i\bar{i}j\bar{j}}\kappa^*_{i\bar{i}}\kappa_{j\bar{j}}\,. \tag{10}$$

The EDF theory differs from the Hamiltonian case by the appearance of effective vertices $\overline{v}^{\rho\rho}$ and $\overline{v}^{\kappa\kappa}$ in the mean-field and pairing channels. In particular, these vertices generally (i) differ from each other, (ii) might not be fully antisymmetric, (iii) might be density dependent.

These differences are at the heart of problems occurring when trying to extend SR-EDF theory to the Multi-Reference EDF framework. Standard approach to perform MR-EDF consists in using equation (6) where $E_{SR}$ is replaced by its functional counterpart $\mathcal{E}_{SR}$. However, this direct mapping has led to inconsistencies: First, MR-EDF can lead to jumps and/or divergences in the energy [1, 2, 3, 4, 5]. This problem originates in the strategy used to construct MR-EDF and requires a specific regularization [3, 4, 5] that is possible only for specific density dependence of the energy. Secondly, the use of density dependent term, that has been so successful in SR-EDF is at the origin of technical and conceptual difficulties. Which density should be used in the effective kernels still needs to be clarified [7]. Finally, the direct application of Eq. (6) or any equivalent for other projection does not insure in the EDF case the proper behavior of the energy with respect to the symmetry transformation [8]. None of these problems occur in the Hamiltonian case. This is clearly shown by expression (8), noting that projected densities have all the required properties (symmetry transformation, anti-symmetrization, non-divergence and regularity...).

It is therefore quite natural to see if and under which conditions, an equation similar to (8) can be obtained starting from the MR-EDF approach. We have recently shown [9], that starting with [a] a bilinear functional, [b] with density independent interaction and [c] a slight modification of the renormalization proposed in ref. [3], the MR-EDF obtained by gauge angle integration is equivalent to a functional of the projected state densities written as:

$$\begin{aligned}\mathcal{E}_N[\Psi_N] &= \sum_i t_{ii} n_i^N + \frac{1}{2}\sum_{i,j,j\neq\bar{i}} \overline{v}^{\rho\rho}_{ijij} R^N_{ijij} + \frac{1}{4}\sum_{i\neq j, j\neq\bar{i}} \overline{v}^{\kappa\kappa}_{i\bar{i}j\bar{j}} R^N_{j\bar{j}i\bar{i}} \\ &+ \frac{1}{2}\sum_i \overline{v}^{\rho\rho}_{i\bar{i}i\bar{i}} n_i^N n_i^N + \frac{1}{2}\sum_i \overline{v}^{\kappa\kappa}_{i\bar{i}i\bar{i}} n_i^N (1 - n_i^N)\,,\end{aligned} \tag{11}$$

where $n_i^N$ denotes the occupation numbers of $|\Psi_N\rangle$. This expression clearly shows that conditions [a-c] insure a well behaved functional as far as the particle number projection is concerned. It also shows that the MR-EDF can eventually be interpreted as a functional of densities constructed from an auxiliary state $|\Psi_N\rangle$ which leads to the functional sequence:

$$\Psi_N \implies (\rho^N, R^N) \implies \mathcal{E}_N. \tag{12}$$

The direct use of Eq. (11) instead of Eq. (6) is possible and will lead to result identical to MR-EDF when the regularization is applicable. This formulation where breaking and restoration are accounted for simultaneously will be called hereafter Symmetry-Conserving EDF (SC-EDF). SC-EDF can be used with density dependent interaction, including dependence in non-integer powers of the density, and can be formulated as a consistent extension of the SR-EDF approach. Then, the density entering in the effective vertices becomes the density of the projected state. An illustration of the SC-EDF is shown in figure 1 for a functional containing non-integer powers of the density. The MR-EDF results where no regularization is possible [8] is also shown as a reference.

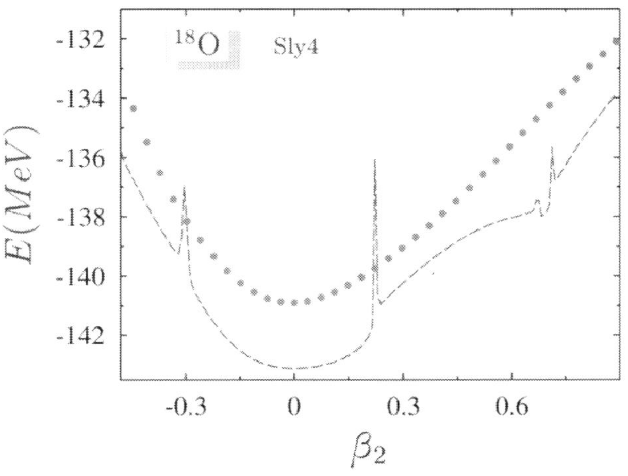


**Figure 1.** (Color online) Particle-number restored deformation energy surface of $^{18}O$ calculated using Eq. (11) when the SLy4 effective interaction is used in the particle-hole channel. The dashed line corresponds to the non-regularized MR-EDF result directly obtained by gauge angle integration using 199 points in the discretization (Eq. (6)). The filled circles correspond to the result obtained using the Symmetry-Conserved EDF based on equation (11).

## 4. Summary

The MR-EDF approach applied to particle number projection has been further analyzed here. Guided by the Hamiltonian case, we show that, under certain conditions, this approach can be formulated as a functional theory of the projected one- and two-body density matrices. This alternative formulation called Symmetry Conserving-EDF can eventually be applied to functional where the MR-EDF is ill defined. Note finally that, it has also been shown recently that the two-body density matrix of the projected state is a functional of the one-body density matrix [11, 12].

## References

[1] M. Anguiano, J. L. Egido, and L. Robledo, Nucl. Phys. A 696, 467 (2001).
[2] J. Dobaczewski, M. Stoitsov, W. Nazarewicz, and P.- G. Reinhard, Phys. Rev. C 76, 054315 (2007).
[3] D. Lacroix, T. Duguet and M. Bender, Phys. Rev. C 79, 044318 (2009).
[4] M. Bender, T. Duguet and D. Lacroix, Phys. Rev. C 79, 044319 (2009).
[5] T. Duguet, M. Bender, K. Bennaceur, D. Lacroix and T. Lesinski, Phys. Rev. C 79, 044320 (2009).
[6] P. Ring and P. Schuck, The Nuclear Many-Body Problem (Springer-Verlag, 1980).
[7] L. M. Robledo, J. Phys. G 37, 064020 (2010).
[8] T. Duguet and J. Sadoudi, J. Phys. G 37, 064009 (2010).
[9] G. Hupin, D. Lacroix and M. Bender, `arXiv:1105.4084v1`
[10] R. Balian and E. Brezin, Il Nuovo Cimento B 64, 37 (1969).
[11] D. Lacroix and G. Hupin, Phys. Rev. B 82, 144509 (2010).
[12] G. Hupin and D. Lacroix, Phys. Rev. C 83, 024317 (2011).

# BCS-BEC crossover and liquid-gas phase transition in nuclear matter

**Meng Jin**[1], **Michael Urban**[2] **and Peter Schuck**[3]

[1]Institute of Particle Physics and Physical Department, Central China Normal University, Wuhan 4300079, China

[2]Groupe de Physique Théorique, Institut de Physique Nucléaire -Centre Scientifique d'Orsay , F-91406 ORSAY

[3]Laboratoire de Physique et Modélisation des Milieux Condensés,CNRS and Université Joseph Fourier, BP 166, 38042 Grenoble Cedex, France

E-mail: `jinm@iopp.ccnu.edu.cn`

**Abstract.** The effect of nucleon-nucleon correlations in symmetric nuclear matter at finite temperature is studied beyond BCS theory. We calculate the critical temperature for a BEC superfluid of deuterons, of a BCS superfluid of nucleons, and in the crossover between these limits. The effect of the correlations on the liquid-gas phase transition is discussed. Our results show that nucleon-nucleon correlations beyond BCS play an important role for the properties of nuclear matter, especially in the low-density region.

## 1. Introduction

Pairing and nucleon-nucleon correlations are important properties of interacting nuclear systems. For example, in the weak-coupling limit, i.e., at high density, the nucleons form Cooper pairs, and below a certain critical temperature $T_c$ the system is in a superfluid phase as described by the Bardeen-Cooper-Schrieffer (BCS) theory. In the strong-coupling limit, i.e., at low density, neutrons and protons form deuteron bound states which will condense if the temperature is below the critical temperature for the corresponding Bose-Einstein condensation (BEC). It was believed that there is a smooth crossover which connecting the BCS and BEC limits. Qualitatively, especially at zero temperature, these features can be studied within the BCS (mean field) approximation. Quantitatively, however, the critical temperature obtained in this way is too high because the BCS theory does not include the existence of non-condensed pairs at finite temperature. In order to go beyond mean field, one has to consider pair correlations already above the critical temperature, as in the Nozières-Schmitt-Rink (NSR) theory [1].

It is well known that there exists a liquid-gas phase transition in nuclear matter. Within mean-field theory, we know that the BCS-BEC crossover is completely covered by the instability region of the liquid-gas phase transition. In the case of the ultracold atomic Fermi gases, because the pair correlations stabilize the gas [1] such that the system does not collapse into its solid ground state but it remains in its metastable gas state. By analogy, one expects that pair correlations will stabilize low-density nuclear matter and thus reduce the liquid-gas coexistence region. One of our subjects of investigation is to evaluate quantitatively the strength of this effect of nucleon-nucleon correlations on the liquid-gas phase transition.

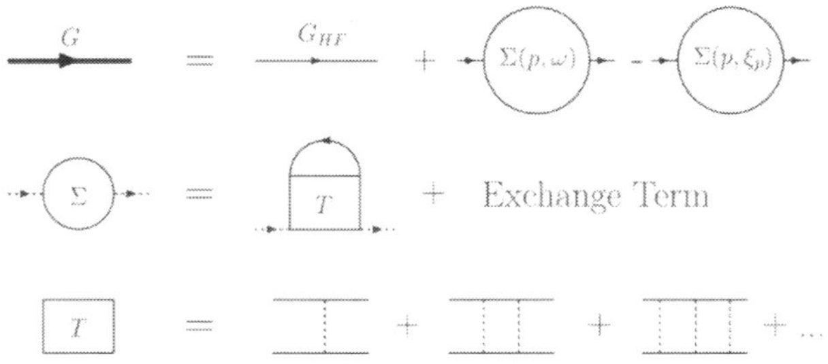

**Figure 1.** The Feynman diagrams for the Green's function (top), for the self-energy (middle), and for the $T$-matrix in ladder approximation (bottom).

## 2. Formalism

Based on the HF single particle Green's function $G_{HF}(p,\omega) = \frac{1}{\omega-\xi_p+i0}$ with the quasiparticle energy $\xi_p = \frac{p^2}{2m} - \Sigma_{HF}(p) - \mu$ and the HF self-energy $\Sigma_{HF}$, which will be calculated using the D1 Gogny force [2], we calculate the nucleon Green's function including the nucleon-nucleon correlation from the T-matrix contribution,

$$G^{-1}(p,\omega) = G^{-1}_{HF}(p,\omega) - \tilde{\Sigma}(p,\omega)\,, \tag{1}$$

Since the Gogny effective interaction on HF is supposed to account most of the correlation effect, we need to do the substraction and define the self-energy as follows,

$$\tilde{\Sigma}(p,\omega) = \Sigma(p,\omega) - Re\Sigma(p,\xi_p)\,. \tag{2}$$

The whole process can be described by the Feynman diagram in fig.1. In principle, with the Green's function (1) one can get all the properties of the system at this level of approximation. This can be done numerically with a selfconsistent Green's function method[3]. But here we make the assumption that the correlations can be treated as a small correction to the Gogny HF self-energy. This allows us to use the HF Green's function $G_{HF}$ in the calculation of the T-matrix and of the self-energy. Then, for consistency, one should also keep only the first-order term of eq.(1),

$$G(p,\omega) = G_{HF}(p,\omega) + G^2_{HF}(p,\omega)\tilde{\Sigma}(p,\omega)\,. \tag{3}$$

In order to get a simple expression for the T-matrix, we use the separable Yamaguchi potential[4], $V_\alpha(k,k') = -\lambda_\alpha v(k)v(k')$, $v(k) = \frac{1}{k^2+\beta^2}$. Then the T-matrix has the following analytical form,

$$T_\alpha(k,k',K,\omega) = \frac{V_\alpha(k,k')}{1-J_\alpha(K,\omega)}\,, \tag{4}$$

where $k$ and $k'$ are the incoming and outgoing momenta in the center of mass frame, $K$ is the total momentum, and

$$J_\alpha(K,\omega) = \int \frac{d^3k}{(2\pi)^3} V_\alpha(k,k) \frac{1 - f(\xi_{\mathbf{K}/2+\mathbf{k}}) - f(\xi_{\mathbf{K}/2-\mathbf{k}})}{\omega - \xi_{\mathbf{K}/2+\mathbf{k}} - \xi_{\mathbf{K}/2-\mathbf{k}} + i0}\,. \tag{5}$$

In terms of the T-matrix, the self-energy in eq.(2) can be written as,

$$\Sigma(p,i\omega_n) = \frac{3}{2} \sum_{\alpha={}^3S_1,\,{}^1S_0} T \sum_{n'} \int \frac{d^3p'}{(2\pi)^3} G_{HF}(p',i\omega_{n'}) T_\alpha(k,k,K,i\omega_n + i\omega_{n'})\,. \tag{6}$$

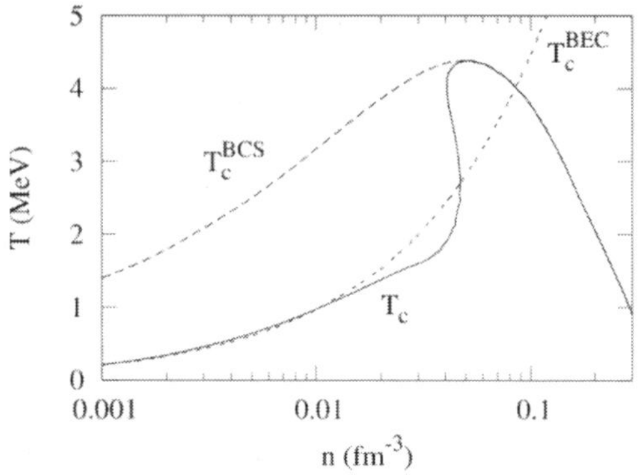


**Figure 2.** The solid line is the full calculation, while the long dashes is the BCS result. The short dashes show the critical temperature of Bose-Einstein condensation of a deuteron gas.

With the Green's function, the density of the system can be calculated with the following formula,

$$n(T,\mu) = -4T\sum_n \int \frac{d^3k}{(2\pi)^3} G(k, i\omega_n)\,. \tag{7}$$

After a lengthy derivation (see [5]), one finds the formulas for the density of the nuclear system initially given in Refs. [6]: $n = n_{HF} + n_{corr}$. Where the $n_{HF}$ is the normal HF density[7] and the correlation contribution reads

$$n_{corr} = 6\int_{K>K_{Mott}} \frac{d^3K}{(2\pi)^3} g(\omega_b(K)) - 6\int_{K>K_{Mott}} \frac{d^3K}{(2\pi)^3} g(\omega_0(K)) - 6\sum_{\alpha={}^3S_1,{}^1S_0} \int \frac{d^3K}{(2\pi)^3} \int_{\omega_0(K)}^{\infty} \frac{d\omega}{\pi} \left(\frac{d}{d\omega} g(\omega)\right) \left(\delta_\alpha - \tfrac{1}{2}\sin 2\delta_\alpha\right)\,. \tag{8}$$

In the upper formula, $g(\omega) = 1/(e^{\omega/T} - 1)$ is the Bose function, $\delta_\alpha$ is the nucleon-nucleon phase shift in the $\alpha$ channel, and $\omega_0(K) = \frac{K^2}{4m^*} - 2\mu^*$, $\omega_b(K) = E_b(K) - \omega_0(K)$ with the effective nucleon mass and effective chemical potential defined in [5].

## 3. Some results

In the above calculation, when the temperature is below some critical value, we get a divergence in the T-matrix. This pole corresponds to the formation of Cooper pairs at high density and to Bose-Einstein condensation of deuterons at low density. One can determine the critical temperature of the superfluid transition as the temperature where the T-matrix develops a pole at zero total momentum ($K = 0$) and at zero energy ($\omega = 0$). This is the well-known Thouless criterion[8] for the onset of superfluidity, coinciding with the BCS gap equation when the gap goes to zero,

$$1 - J_\alpha(K = 0, \omega = 0; T = T_c) = 0. \tag{9}$$

The critical temperature as a function of density is shown by fig.2. One can see that the phase boundary coincides with the BCS curve (long dashed line) in the high density region. This means the nucleon-nucleon correlation effect vanishes here. At very low density and temperature, the main contribution to the density comes from the deuteron bound state. Therefore the superfluid

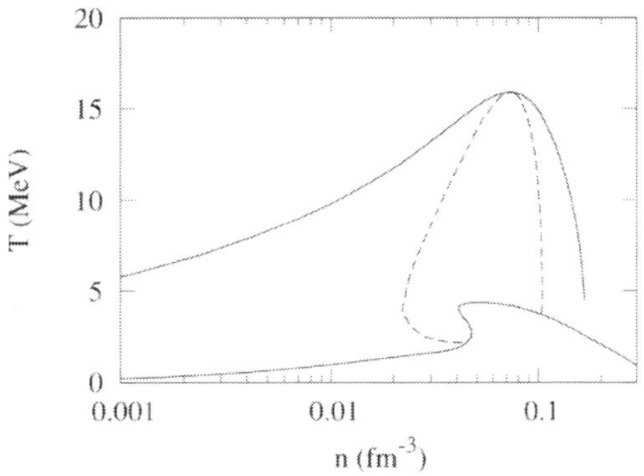


**Figure 3.** Phase diagram combining the boundary of the superfluid phase (lower curve), the liquid-gas coexistence region (upper curve), and the spinodal line (dashed curve).

critical temperature at low density coincides with the critical temperature for Bose-Einstein condensation of a deuteron gas which is shown by the short-dashed line.

Furthermore, one can get the pressure as a function of temperature and chemical potential, $P(T,\mu)$, from the number density $n(T,\mu)$ by integration over the chemical potential $\mu$[5]. With the pressure, one can determine the coexistence region of the liquid and gas phases of nuclear matter from the conditions: $P(T,n_1) = P(T,n_2)$ and $\mu(T,n_1) = \mu(T,n_2)$. At the same time, we can determine the spinodal curve from the zeros of $\partial P/\partial n$. In the region under the spinodal curve, the system cannot exist in a homogeneous phase. In the region between the phase boundary curve and the spinodal curve, the gas phase or the liquid phase can exist as a metastable state.

Combining the normal-superfluid phase transition and the liquid-gas phase transition, we get the phase structure for the nuclear matter at low temperature region shown by the fig.3. The results show that the coexistence curve crosses the superfluid curve at $n \sim n_0$, which means the homogeneous nuclear matter with pairing is stable above this density. The spinodal curve (dashed line) can be calculated until it reaches the superfluid region. Note that on the low-density side, the density region where the gas phase is metastable is strongly increased by the correlations, which indicates the correlations have a stabilizing effect. However, the BEC-BCS crossover lies still in the unstable region of the liquid-gas phase transition.

This work was supported in part by NSFC (grants 10805023 and 10975060) and by ANR (project NEXEN). M. Jin also supported by the Project-sponsored by SRF for ROCS, SEM.

## References

[1] Nozières P and Schmitt-Rink S 1985 *J. Low Temp. Phys.* **59** 195.
[2] Dechargé J and Gogny D 1980 *Phys. Rev.* C **21** 1568.
[3] Rios A, Polls A, Ramos A, and Müther H 2008 *Phys. Rev.* C **78** 044314.
[4] Yamaguchi Y 1954 *Phys. Rev.* **95** 1628.
[5] Jin M, Urban M and Schuck P 2010 *Phys. Rev.* **82** 024911.
[6] Schmidt M, Röpke G, and Schulz H 1990 Ann. Phys. (N.Y.) **202** 57; Stein H, Schnell A, Alm T, and Röpke G 1995 *Z. Phys.* A **351** 295.
[7] Ventura J, Polls A, Viñas X, Hernandez X, and Pi M 1992 *Nucl. Phys.* A **545** 247c.
[8] Thouless D J 1960 *Ann. Phys. (N.Y.)* **10** 553.

# Liquid-gas phase transition in nuclear matter in the mean-field approximation

**A Rios[1], A. Carbone[2], A. Polls[2] and I. Vidaña[3]**
[1]Department of Physics, Faculty of Engineering and Physical Sciences, University of Surrey, Guildford, Surrey GU2 7XH, United Kingdom
[2]Departament d'Estructura i Constituents de la Matèria and Institut de Ciències del Cosmos, Universitat de Barcelona, Avda. Diagonal 647, E-08028 Barcelona, Spain
[3]Centro de Física Computacional, Department of Physics, University of Coimbra, PT-3004-516 Coimbra, Portugal
E-mail: `a.rios@surrey.ac.uk`

**Abstract.** The liquid gas phase transition in nuclear systems is a unique phenomenon, at the frontier of nuclear, many-body and statistical physics. We use self-consistent mean-field calculations to quantify the properties of the transition in symmetric nuclear matter. We explore the available parameter space of critical properties by analyzing the mean-field dependence of the phase transition. The latent heat of the transition is computed and we find that it exhibits a model independent temperature dependence due to basic physical principles.

The experimental determination of the thermodynamical properties of nuclei within a heavy-ion reaction is a major challenge [1]. Significant advances in this direction will come from the next generation of radioactive beam facilities, especially due to the advances in instrumentation [2]. The access to isospin asymmetric nuclei will allow for a more detailed study of isospin diffusion and equilibration [3]. Theoretical approaches that can deal with finite temperature nuclear matter are therefore called for. A controlled and consistent approach in this direction is given by the self-consistent mean-field approximation, which is based on sound theoretical grounds (statistical mechanics together with quantum many-body theory) [4] and yields predictions that are equally valid at zero and finite temperature. This can also be casted as a finite temperature extension of density functional theory. So far, calculations for hot nuclei have been generally performed on selected spherical isotopes, mostly aimed at understanding the structural changes induced by temperature [5].

Before extensive finite nuclei calculations are implemented, however, the model dependence of the calculations should be assessed. Among the possible sources of uncertainty, the dependence associated with the underlying NN effective interaction has been scarcely investigated. Homogeneous nuclear matter within a mean-field picture provides the simplest testing ground for such an analysis, since it lies within the thermodynamical limit. It is well-known that different effective interactions predict different bulk properties for nuclear matter [6]. This statement also applies at finite temperature and, as a result, different liquid-gas critical points are found for different forces. One would also expect that a reliable determination of the critical point could be helpful in constraining the properties of the effective interaction [7].

The mean-field description of nuclear matter should not be taken as a meaningful description of the liquid-gas transition in nuclear systems, though. Clustering as well as fluctuations

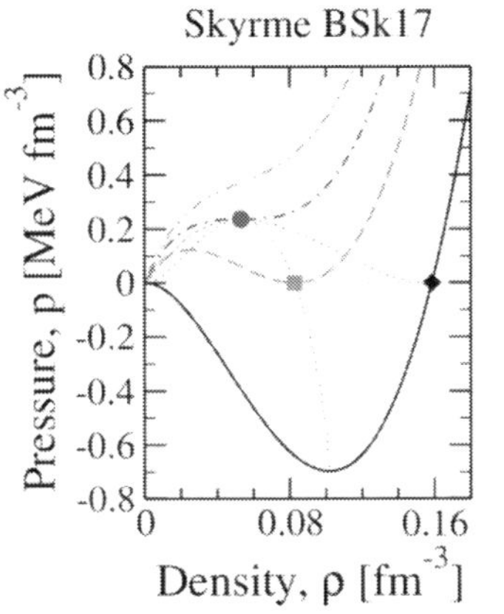


**Figure 1.** Pressure isotherms of symmetric nuclear matter for different temperatures: $T = 0$ (solid line), $T_f$ (dashed line), $T_c$ (dash-dotted line) and 18 MeV (dash-double-dotted line). The green and purple dotted lines represent the corresponding spinodal and coexistence regions.

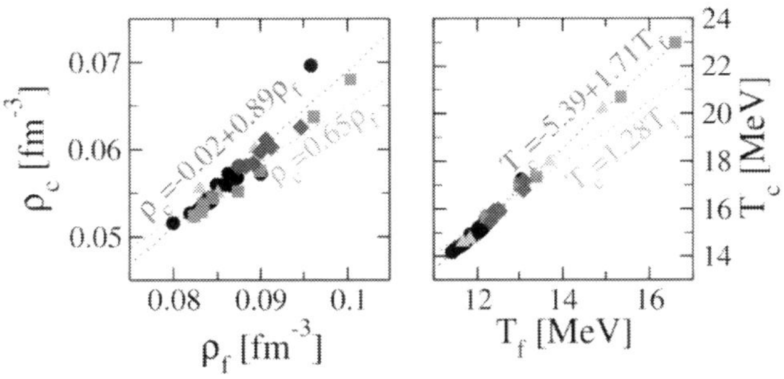


**Figure 2.** Left panel: critical density versus flashing density for different effective interactions. Right panel: critical temperature versus flashing temperature for different effective interactions.

dominate close to the critical point and none of these is taken into account within the mean-field approximation. As a matter of fact, here we do not try to provide a quantitatively accurate picture for such critical properties. Instead, we aim at investigating the effective interaction dependence of the liquid-gas transition as well as the overall behavior of its latent heat. As such, this can be taken as a first methodological contribution to test what information, if any, can be obtained from such analysis.

The density and temperature dependence of the thermodynamical properties obtained within the Hartree-Fock approximation are non-trivial. These can only be obtained numerically and they cannot be parametrized easily in any analytical form [8]. The mean-field solution describes *simultaneously* two different physical phases: a gas at low densities and a liquid at densities close to saturation [5]. This behavior is more easily visualized in the pressure isotherms of nuclear matter, as show in Fig. 1. This corresponds to results obtained with a typical Skyrme effective interaction (BSk17). Four characteristic temperatures have been chosen for illustrative purposes.

At $T = 0$ (solid lines), the pressure is a decreasing function of the density at low densities until a minimum at $\sim \rho_0/2$ (with $\rho_0$, the saturation density) is reached. This regime, for which $\frac{\partial p}{\partial \rho} < 0$, corresponds to the spinodal region, where the system is mechanically unstable. Beyond the minimum, the pressure increases with density until it changes sign at the saturation point (black diamond). As the temperature increases, the spinodal region, which is shown in green dotted lines in Fig. 1, shrinks. The isotherm at which the pressure becomes a completely positive function defines the so-called *flashing* temperature. The flashing point (red square) corresponds to the crossing of the spinodal with the zero pressure point, $p = \frac{\partial p}{\partial \rho} = 0$. For an isolated system, without an external gas to stabilize it, this would correspond to the maximum temperature at which the system could still be self-bound. For the infinite system, however, the system can exist above the flashing temperature because the gas and the liquid phase exert the same pressure on each other.

Journal of Physics: Conference Series **321** (2011) 012058 doi:10.1088/1742-6596/321/1/012058

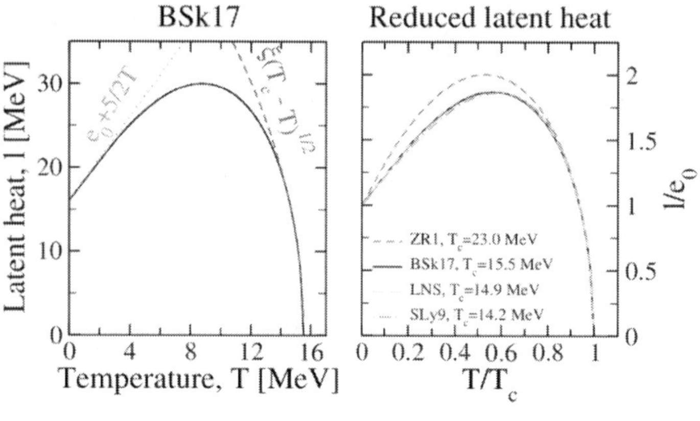


**Figure 3.** Left panel: temperature dependence of the latent heat of symmetric nuclear matter for the BSk17 interaction. The low- and high-temperature dependences correspond to the dotted and dashed lines, respectively. Right panel: reduced latent heat vs. reduced temperature for different Skyrme interactions.

According to the Maxwell criterion, the condition that the gas and liquid phases have the same pressure, together with the requirement that their chemical potentials are the same, set the coexistence points [9]. The coexistence region is shown with a purple dotted line in Fig. 1. Note that, at $T = 0$, there is a coexistence between a zero-density gas and a liquid at saturation density. Increasing the temperature results into larger gas and smaller liquid densities. Eventually, the two densities meet at the critical point (blue circle), which occurs at the critical temperature, $T_c = 15.5$ MeV for this particular Skyrme force. The critical isotherm is shown with a dashed-dotted line in Fig. 1. Above the critical temperature, the system only exists in the gas phase and the pressure isotherms become a monotonically increasing function of density. This behavior is observed in the dashed-double-dotted line, which has been computed at $T = 18$ MeV. Modern Skyrme and Gogny forces seem to favor a critical and flashing temperature in the regions $T_c \sim 14 - 17$ MeV and $T_f \sim 11 - 13$ MeV [8].

By computing the liquid-gas critical properties with a large set of Skyrme force parametrizations, one might also be able to identify behaviors which are independent of the effective interaction. An examples of such correlation is given in Fig. 2. A tight proportionality between the critical and the flashing densities (right panel) and temperatures (left panel) is found for a wide range of Skyrme and Gogny forces (blue circles). Different generations of Skyrme forces are represented by squares (old forces with no effective mass), triangles (old forces with $m^* \neq m$) and new forces (black circles). The flashing-critical correlation can be modeled by a single slope parameter or, for better quality, with a linear regression with an offset. Proportionality constants agree with those predicted by analytical models and the empirical relation $T_c/T_f = 2\rho_c/\rho_f$ is fulfilled. Finally, in the context of critical exponents, these have also been observed to be independent of the underlying mean-field [8].

The latent heat of nuclear matter is a very appealing quantity that provides a further characterization of the liquid-gas phase transition [10]. More interestingly, it can be potentially extracted from experiments, which suggests values of $l \sim 4 - 8$ MeV [1, 7, 11]. In the case of infinite nuclear matter, the latent heat per particle accounts for the amount of heat needed to take a nucleon from the liquid to the gas phase. One can describe its basic temperature dependence from basic quantum statistical mechanics arguments [10].

The Clausius-Clapeyron formula gives the latent heat per particle in terms of the product of the temperature, the difference between volume per particle of the two phases and the derivative of the pressure with respect to the temperature along the coexistence curve [9]. Alternatively, the latent heat can also be computed as the heat exchange between two phases that only involves the difference in entropies, $l = T(s_g - s_l)$ . For nuclear matter, this formula is numerically more stable than the Clausius-Clapeyron one. One can however check the thermodynamic consistency of the theory by comparing both values [10]. Values of the vaporization specific latent heat for common liquids and gases are in the region of $10 - 5000$ kJ/kg when measured at the normal boiling point. In contrast, nuclear matter has a maximum specific latent heat of the order of $\sim 30$ MeV, *i.e.* $\sim 3 \times 10^{12}$ kJ/kg, which is orders of magnitude higher and among the highest in nature. The origin of this extremely large value can be traced back to the strong force, which binds the nucleons tightly.

The latent heat, $l$, obtained with the BSk17 Skyrme parametrization is reported in the left panel of Fig. 3 (solid line). One finds that $l$ has a characteristic bump shape as a function of temperature. We have observed that, for all other Skyrme parametrizations, the qualitative behavior of $l$ is very similar [10]. The latent heat matches the value of the binding energy, $e_0$, at $T = 0$, and then rises for small temperatures. The initial rise is linear and the slope is independent of the Skyrme parametrization. As a matter of fact, the low-temperature behavior can be extracted from a careful analysis of the thermodynamics of the phase transition, yielding $l \sim e_0 + 5/2T$. This analysis, valid for any fermionic system undergoing a liquid-gas phase transition, confirms that the latent heat grows linearly with temperature and that the linear growth is controlled by a model-independent slope [10].

Further up in temperature, $l$ reaches a maximum and then drops to zero at the critical point, where the difference between the liquid and gas phases disappears. For BSk17, the latent heat maximum lies at $T = 8.7$ MeV with $l_{\text{max}} = 29.9$ MeV. In the large temperature limit, critical exponents characterize the properties of the phase transition. The latent heat can be described, in this regime, in terms of a single critical exponent [10]. As shown in Ref. [12], the critical exponent for $l$ is the same as that of the order parameter. Since in the Hartree-Fock approximation the latter is given by $\beta = 1/2$, one finds that $l \sim (T_c - T)^{1/2}$.

The $T = 0$ limit of the latent heat suggests that, if one wants to obtain interaction-independent results, $l$ might be normalized to $e_0$. A plot of the latent heat in reduced units, $l/e_0$, as a function of the reduced temperature, $T/T_c$, is presented in the right panel of Fig. 3 for a selected subset of four Skyrme forces. These represent a wide range of critical temperatures and maximum latent heats, ranging from $l_{max} = 34$ MeV for ZR1 to $l_{max} = 28$ MeV for LNS. In genera, we observe that the dependence on the mean-field is eliminated to a large extent in the dimensionless plot. For all Skyrme forces, the latent heat tends to peak within a limited region of temperatures, $T/T_c \sim 0.5 - 0.6$. Moreover, the peak is also quite narrowly distributed around the value $l_{max}/e_0 \sim 1.7 - 2$. This indicates that the latent heat is more determined by thermal correlations than by the effective force itself.

**Acknowledgments**
This work has been supported by a Marie Curie Intra European Fellowship within the $7^{th}$ Framework programme, STFC grant ST/F012012, by grant No. FIS2008-01661 (MEC, Spain) and grant No. 2009SGR1289 from Generalitat de Catalunya.

## References

[1] Pochodzalla J 1997 *Prog. Part. Nucl. Phys.* **39** 443
[2] Sfienti C *et al.* (ALADIN2000 Collaboration) 2009 *Phys. Rev. Lett.* **102** 152701
[3] Tsang M B *et al.* 2001 *Phys. Rev. Lett.* **86** 5023
[4] Vautherin D 1996 *Advances in Nuclear Physics, Vol. 22* (Springer US) chap Many-body methods at finite temperature, p 123
[5] Bonche P, Levit S and Vautherin D 1985 *Nucl. Phys. A* **436** 265
[6] Rikovska Stone J, Miller J C, Koncewicz R, Stevenson P D and Strayer M R 2003 *Phys. Rev. C* **68** 034324
[7] Natowitz J B *et al.* 2002 *Phys. Rev. Lett.* **89** 212701
[8] Rios A 2010 *Nucl. Phys. A* **845** 58
[9] Huang K 1987 *Statistical Mechanics* 2nd ed (John Wiley and Sons)
[10] Carbone A, Polls A, Rios A and Vidaña I 2011 *Phys. Rev. C* **83** 024308
[11] Bonnet E *et al.* 2009 *Phys. Rev. Lett.* **103** 072701
[12] Landau L D and Lifshitz E M 1980 *Statistical Mechanics, Part 1* 3rd ed (Butterworth-Heineman)

Journal of Physics: Conference Series **321** (2011) 012059 doi:10.1088/1742-6596/321/1/012059

# Nuclear matter properties and equation of state of neutron matter

**Satoshi Yoshida**[1] **and Hiroyuki Sagawa**[2]

[1] Science Research Center, Hosei University, 2-17-1 Fujimi, Chiyoda, Tokyo 102-8160, Japan
[2] Center for Mathematical Sciences, the University of Aizu, Aizu-Wakamatsu, Fukushima 965-8580, Japan

E-mail: s_yoshi@hosei.ac.jp

**Abstract.** Correlations among symmetric nuclear matter properties, the neutron matter EOS and the pressure of neutron matter are studied in the Skyrme Hartree-Fock (SHF) and the relativistic mean  eld (RMF) models. At present, many different parameter sets for the Skyrme interaction and RMF model have been proposed. As a result, the property of neutron matter which is one of extreme conditions depends on the parameter set. We found clear linear correlations between the energy per neutron of neutron matter and the symmetry energy and between the pressure of neutron matter and $L$ which is one of isovector nuclear matter properties at the saturation density of symmetric nuclear matter. We show that values of the energy per neutron and the pressure of neutron matter at the saturation density of symmetric nuclear matter can be obtained by accurate experimental data of the symmetry energy and $L$.

## 1. Introduction

Nuclear mean  eld models have been successful to describe properties of the ground state, collective excitations and so on. Two di  erent models are widely used. The  rst one is the Skyrme Hartree-Fock (SHF) model [1] which is one of non relativistic models. The second one is the relativistic mean  eld (RMF) model with terms of  , $\omega$ and  meson  elds in an e  ective Lagrangian [2]. The nuclear binding energy and the saturation density precisely measured are used to determine the parameter sets of the Skyrme interactions and RMF models. As a result, many di  erent parameter sets for the Skyrme interaction and RMF model have been proposed. Other nuclear matter properties such as the incompressibility, the symmetry energy are less well measured by experiments and are only used as rough guides in the determination of parameter sets. Di  erent parameter sets give fairly similar results for nuclear ground state properties. However in the extreme condition such as ground state properties of neutron rich unstable nuclei and properties of in  nite nuclear matter, results depend on parameter sets. We have discussed about correlations among isovector nuclear matter properties and the neutron skin thickness [3]. In this paper, we study relations between properties of in  nite neutron matter and nuclear matter properties of the Hamiltonian density for a large number of di  erent parameter sets of the Skyrme interaction and RMF model.

## 2. Results and discussion

The Hamiltonian density $H$ consists of the isoscalar part $H_S$ and the isovector part $H_V$ generally and in microscopic calculations $H$ is described by the following equation,

$$H(\ ) = H_S(\ ) + H_V(\ )I^2 + \ D(\ , I), \tag{1}$$

where $I$ and $D(\ , I)$ are the asymmetry parameter $I = (\ _n \quad _p)/\ $ and the term proportional to $I^4$, respectively and $_n(\ _p)$ and are densities of neutrons (protons) and $= \ _n + \ _p$, respectively. The isovector part $H_V$ can be expressed as a Taylor expansion around the saturation density $_{nm}$:

$$\frac{H_V(\ )}{} = J + \frac{L}{3}\ \frac{\quad _{nm}}{_{nm}} + \frac{K_{sym}}{18}\ \frac{\quad _{nm}}{_{nm}}^{2} + \quad . \tag{2}$$

The symmetry energy $J$ and $L$ can be obtained from following equations:

$$J = \frac{H_V(\ _{nm})}{_{nm}}, \tag{3}$$

$$L = 3\ _{nm}\frac{\partial}{\partial}\ \frac{H_V(\ )}{}\ \Big|_{= \ nm}. \tag{4}$$

On the other hand, the nuclear saturation density $_{nm}$, the binding energy per nucleon $E_0$, and the incompressibility $K$ relate to the isoscalar part of the Hamiltonian density $H_S$. $_{nm}$ and $E_0$ are de ned by following equations, respectively:

$$0 = \frac{\partial}{\partial}\ \frac{H_S(\ )}{}\ \Big|_{= \ nm}, \tag{5}$$

$$E_0 = \frac{H_S(\ _{nm})}{_{nm}}. \tag{6}$$

In our previous paper [4], we studied correlations among the isovector and isoscalar nuclear matter properties by using SHF results for 13 di erent Skyrme interactions and RMF results for seven parameter sets. In this work, we discuss those correlations by using 111 Skyrme interactions and 15 RMF parameter sets listed in Table 1. The isoscalar and isovector nuclear matter properties a ect the equation of state of nuclear matter. We study relations between

**Table 1.** Skyrme interactions and RMF parameter sets

| | | | | | | | | | | |
|---|---|---|---|---|---|---|---|---|---|---|
| Skyrme interactions | | | | | | | | | | |
| SI, | SI' | SII, | SIII, | SIII*, | SIV, | SV, | SVI, | SVII, | Skya, | Skyb, |
| SkM, | SkM*, | SkM1, | SLy4, | SLy5, | SLy6, | SLy7, | SLy10, | SLy230a, | SLy20b, | SGI, |
| SGII, | SkI1, | SkI2, | SkI3, | SkI4, | SkI5, | SkX, | SkXce, | SkXm, | SkP, | SkT, |
| SkT1, | SkT2, | SkT3, | SkT4, | SkT5, | SkT6, | SkT7, | SkT8, | SkT9, | RATP, | E, |
| Es, | T, | Rs, | Gs, | Z, | Zs, | Zs*, | SkO, | SkO', | LNS, | MSkA, |
| MSk1, | MSk2, | MSk3, | MSk4, | MSk5, | MSk5*, | MSk6, | MSk7, | MSk8, | MSk9, | v1.10, |
| v1.05, | v1.00, | v0.90, | v0.80, | v0.75, | v0.70, | SK255, | SK272, | SkSC1, | SkSC2, | SkSC3, |
| SkSC4, | SkSC4o, | SkSC5, | SkSC6, | SkSC10, | SkSC11, | SkSC14, | SkSC15, | SkRA, | SkMP, | BSk1, |
| BSk2, | BSk2', | BSk3, | BSk4, | BSk5, | BSk6, | BSk7, | BSk8, | BSk9, | BSk10, | BSk11, |
| BSk12, | BSk13, | BSk14, | BSk15, | BSk16, | BSk17, | BSk18, | BSk19, | BSk20, | BSk21, | KDE0v, |
| KDE0v1, | | | | | | | | | | |
| RMF parameter sets | | | | | | | | | | |
| NL1, | NL2, | NL3, | NLSH, | NLC, | NL−S, | NL−Z, | NL−Z2, | TW99, | TM1, | TM2, |
| DDME1, | DDME2, | PK1, | PKDD | | | | | | | |

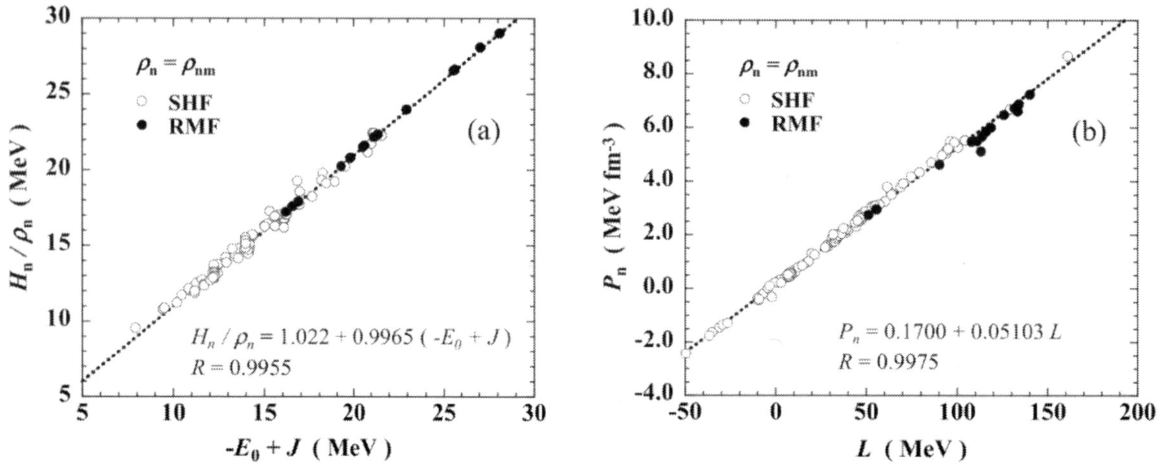


**Figure 1.** Correlations between (a) $\ E_0 + J$ and $H_n/\ \ _n$, and (b) $L$ and $P_n$ calculated with 111 Skyrme interactions (open circles) and 15 RMF parameter sets (full circles), respectively listed in Table.1. $P_n$ and $H_n/\ \ _n$ are the pressure of neutron matter and the energy of neutron matter at the neutron density equal to the saturation density of the symmetric nuclear matter. Dotted lines and equations are given by the method of the least-squares. $R$ is the correlation coe cient.

properties of in nite neutron matter and nuclear matter properties of the Hamiltonian density at the saturation density $\ _{nm}$ of symmetric nuclear matter which is determined by Eq.(5). Figure 1(a) shows the clear linear correlation between the energy per neutron $H_n(\ _{nm})/\ _{nm}$ of neutron matter at the saturation density $\ _{nm}$ and $(\ E_0 + J)$ calculated by 111 Skyrme interactions and 15 RMF parameter sets listed in Table.1. In Fig.1(a), $H_n$ is the Hamiltonian density of neutron matter and $J$ and $\ E_0$ are de ned by Eqs.(3) and (6), respectively. Because of $I = 1$ in the case of neutron matter, Eq.(1) becomes the following equation,

$$H_n(\ _n) = H_S(\ _n) + H_V(\ _n) + \ _n D(\ _n, I = 1). \tag{7}$$

We t the linear function shown by the dotted line in Fig.1(a) by the method of least-squares and obtain the following relation,

$$\frac{H_n(\ _{nm})}{\ _{nm}} = 0.9965\ (\ E_0 + J) + 1.022 \qquad (R = 0.9955), \tag{8}$$

where $R$ is the correlation coe cient. Because the value of $\ E_0$ de ned by Eq.(6) is about 16 MeV in all the parameter sets of the Skyrme interactions and RMF models listed in Table.1, the correlation between $H_n(\ _{nm})/\ _{nm}$ and $J$ becomes linear. Actually Eq.(8) is analytically obtained in the case that the neutron density $\ _n$ is equal to the saturation density $\ _{nm}$ of symmetric nuclear matter. By using Eq.(7), the energy per neutron $H_n(\ _{nm})/\ _{nm}$ of neutron matter at the saturation density $\ _{nm}$ is given by the following equation,

$$\frac{H_n(\ _{nm})}{\ _{nm}} = (\ E_0 + J) + D(\ _{nm}, I = 1). \tag{9}$$

The coe cient of $(\ E_0 + J)$ in Eq.(8) which is 0.9965 is very close to one which is that in Eq.(9). And because the second term in Eq.(8) corresponds to the second term in Eq.(9), the value of the term $D$ at $I = 1$ and $\ _{nm}$ can be obtained by the following relation,

$$D(\ _{nm}, I = 1) = 1.022. \tag{10}$$

Next, the pressure of neutron matter is as important as its energy per neutron in the discussion of neutron matter EOS. Figure 1(b) shows the clear linear correlation between the pressure $P_n(\ _{nm})$ of neutron matter at the saturation density $_{nm}$ and $L$ de ned by Eq.(4) calculated by 111 Skyrme interactions and 15 RMF parameter sets listed in Table.1. We t the linear function shown by the dotted line in Fig.1(b) by the method of least-squares and obtain the following relation,

$$P_n(\ _{nm}) = 0.05103\ L + 0.1700 \qquad (R = 0.9975), \tag{11}$$

where $R$ is the correlation coe cient. Actually Eq.(11) is also analytically obtained by using Eq.(7). In the case of neutron matter, the rst derivative $\frac{\partial}{\partial\ _n}\frac{H_n}{_n}$ of the energy per neutron of neutron matter at the saturation density $_{nm}$ is given by the following equation,

$$\left.\frac{\partial}{\partial\ _n}\frac{H_n(\ _n)}{_n}\right|_{_n=\ _{nm}} = \frac{1}{3\ _{nm}}L + \left.\frac{\partial}{\partial\ _n}D(\ _n, I)\right|_{_n=\ _{nm}}. \tag{12}$$

Because the pressure $P_n$ of neutron matter is de ned by the following equation as a function of neutron density $_n$,

$$P_n(\ _n) = \ _n^2\frac{\partial}{\partial\ _n}\frac{H_n(\ _n)}{_n}, \tag{13}$$

we can nd that the pressure $P_n(\ _{nm})$ of neutron matter at the saturation density $_{nm}$ is proportional to $L$ as shown in the following equation,

$$P_n(\ _{nm}) = \frac{1}{3}\ _{nm}L + \ _{nm}^2\left.\frac{\partial}{\partial\ _n}D(\ _n, I=1)\right|_{_n=\ _{nm}}. \tag{14}$$

The coe cient of $L$ in Eq.(11) which is 0.05103 is very close to the value of $_{nm}/3$ which is that of $L$ in Eq.(14) because the value of empirical saturation density $_{nm}$ of symmetric nuclear matter is around 0.16 fm $^{3}$. Because the second term in Eq.(11) corresponds to the second term in Eq.(14), the value of the rst derivative of the term $D$ at $I = 1$ and $_{nm}$ can be obtained by the following relation,

$$\ _{nm}^2\left.\frac{\partial}{\partial\ _n}D(\ _n, I=1)\right|_{_n=\ _{nm}} = 0.1700. \tag{15}$$

The equation of state and the pressure of neutron matter depend on parameter sets of the Skyrme interaction and RMF model we use. However if we can obtain accurate experimental data of $J$ and $L$, we will be able to nd values of the energy per neutron and the pressure of neutron matter at the saturation density of symmetric nuclear matter by using Eqs.(8) and (11).

## 3. Summary

We found clear linear correlations between the energy per neutron of neutron matter and the symmetry energy and between the pressure of neutron matter and $L$ at the saturation density of symmetric nuclear matter with 111 Skyrme interactions and 15 RMF parameter sets and were able to obtain linear functions tted by the method of least-squares. And we pointed out that values of the energy per neutron and the pressure of neutron matter at the saturation density of symmetric nuclear matter can be obtained by accurate experimental data of $J$ and $L$ with those linear functions.

## References

[1] Vautherin D and Brink M 1972 *Phys. Rev.* **C5** 626
[2] Reinhard P.G. and Rufa M, Maruhn J, Greiner W, and Friedrich J 1986 *Z. Phys.* A **32** 13
[3] Yoshida S and Sagawa H, 2004 *Phys.Rev.* **C69** 024318
[4] Yoshida S and Sagawa H, 2006 *Phys.Rev.* **C73** 044320

# AUTHOR INDEX

# AUTHOR INDEX